The New York Times

BOOK OF

Science

The New York Times

BOOK OF

Science

MORE THAN 150 YEARS
OF GROUNDBREAKING
SCIENTIFIC COVERAGE

Edited by

DAVID CORCORAN

Foreword by

BRIAN GREENE

STERLING
New York

STERLING
New York

An Imprint of Sterling Publishing
1166 Avenue of the Americas
New York, NY 10036

ISBN 978-1-4027-9321-9

Distributed in Canada by Sterling Publishing
c/o Canadian Manda Group, 664 Annette Street
Toronto, Ontario, Canada M6S2C8
Distributed in the United Kingdom by GMC Distribution Services
Castle Place, 166 High Street, Lewes, East Sussex, England BN7 1XU
Distributed in Australia by Capricorn Link (Australia) Pty. Ltd.
P.O. Box 704, Windsor, NSW 2756, Australia

For information about custom editions, special sales, and premium and corporate purchases,
please contact Sterling Special Sales at 800-805-5489 or specialsales@sterlingpublishing.com.

Manufactured in the United States of America

2 4 6 8 10 9 7 5 3 1

www.sterlingpublishing.com

For Barbara Strauch (1951–2015),
who as science editor carried *The New York Times'*
science and health report into the digital age with
boundless energy, creativity, and humor.

CONTENTS

CHAPTER 3

Biology:
The Mechanisms of Life

CHAPTER 4

Earth Science:
Plates, Poles and Oceans

CHAPTER 5

The Environment: Challenges to Life

The Natural World

The Changing Climate

CHAPTER 6

Exploration: New Worlds, Down Here and Out There

Our Own Planet

CHAPTER 7

Life on Earth:
Biology, Paleontology, Zoology

CHAPTER 13

Technology:
Invention and Revolution

The Science of Invention

The Computer Revolution

FOREWORD

Some forty years ago, during a long subway ride to a junior high school math-team competition (the geek version of an away game), our coach pierced the din of rattling metal wheels to ask a handful of us why we loved mathematics. Even in the most conducive of circumstances, "mathletes" are not the most introspective bunch. While we'd jump all over a challenge like finding triangular numbers that are also perfect squares, or calculating the probability that two people in the subway car shared the same birthday, articulating a personal reflection on, well, anything, generated a more reserved response. We let the thunderous clatter of the express train fill the silence. No one said a word.

But it did get me thinking. Why *did* I love math? The thrilling moment when the solution to a problem snaps into focus, the beauty of constructing an elegant proof, the power of understanding a hidden pattern—all of these surely resonated with my budding seventh-grade mathematical mind. Still, I remember thinking that the most compelling answer was simply this: math works. A well-posed problem has a definite answer. Regardless of the approach you take, the calculative scheme you invoke, the oblique angle you follow, barring any mistakes you will get the answer. *The* answer. The rock-bottom certainty of problem solving was, for me, a welcome anchor in a post-1960s' world that seemed awash in uncertainty.

Yet, it is that very certainty in a rolling sea of uncertainty that makes the science writer's job so challenging.

When mathematics is applied as a scientific tool to explore the real world—not to solve artificial problems encountered on exams or competitions—precision is possible only because researchers encircle themselves with thick walls of assumptions that keep undue complexity at bay. When we calculate the orbital motion of the earth, we assume the earth is a solid ball moving solely under the influence of the sun's gravity. When we want to get closer to truth, we take account of the influence of the moon and other planets, and even the earth's complex internal structure. It's a mode of operation recapitulated across the sciences: progress takes place in the ever-shifting overlap between simplification and relevance. Science is the art of knowing what to ignore.

The science writer thus has to continually strike a delicate balance between the precision of scientific results—certainty—and the morass of qualifications upon which such results rest—uncertainty. As a scientist who also writes, I've experienced this from both sides. In countless interviews with science journalists, I've repeated "Yes, but . . ." emphasizing one crucial caveat necessary to make my description of a scientific advance accurate but rendering the responsible journalist's story murky. In my own writing, I've countlessly wrestled with finding descriptions that capture the excitement of scientific advances while remaining faithful to the precise and ever-present limits that accompany each advance.

In essence, we scientists are just as protective of what we know as we are of what we don't. Our ignorance is a precious commodity. It not only defines the boundary of understanding but provides the terra incognita that beckons exploration and, on rare occasion, plays host to a remarkable new insight.

Scientists and science writers have not always done a great job of communicating this to the public. Breathless articles are surely exciting but over time they suggest that science is unstable, buffeted this way and that by a steady stream of revolutions that, one would naturally think, continually rewrite the textbooks. But the fact is that science is remarkably stable. New insights typically don't obliterate existing understanding but, instead, extend its reach a few additional steps into the realms of darkness. This is an essential quality of the whole scientific enterprise that I find is often misunderstood.

In fact, the continuity of science plays an even more vital role, something that can be difficult for a reader to tease out of even the best journalism. One of the greatest scientific achievements of the twentieth century was Albert Einstein's completion, in 1915, of the General Theory of Relativity, a new and more powerful approach to understanding the force of gravity. In 1919, Einstein's theory was confirmed through astronomical observations of distant stars during a solar eclipse. The story was widely covered, with two *New York Times* articles (both reprinted in this volume) being those I've most often seen referenced. Understandably, the articles give only modest attention to Einstein's radically new view of gravity, framed in terms of warps and curves in space and time—it often takes years of hindsight, even for scientists, to find the right language for communicating the most abstract of ideas to those without technical training. And, correctly, the articles emphasize that in the everyday manifestations of gravity we all experience, from the arc of a tossed ball to the trajectory of a dropped cup, Einstein's and Newton's theories hardly differ in their predictions.

But what the reader is unlikely to discern from the articles is that however revolutionary Einstein's discovery, Newton's approach to gravity was one of the essential guiding lights leading Einstein to his theory of relativity. Between 1912 and 1915, as he groped his way through a terrain of complex mathematics, Einstein diligently required that any new equation he developed be reduced to Newtonian gravity when applied to ordinary situations, like the motion of the moon, where Newton's ideas had already proved impressively accurate. Indeed, in applying this very requirement in 1913, Einstein committed a technical error that set back the discovery of General Relativity two years. And so, far from throwing Newtonian gravity overboard, Einstein tightly grasped the Newtonian lifeline and rode it to an unfamiliar but spectacularly beautiful shore.

The point is that there is uncertainty at the frontier of knowledge—which is what makes science exciting—but there is a core of scientific insight that you can count on. Einstein's discovery of the General Theory of Relativity does not mean your effort in high-school physics to understand Newtonian gravity was a waste of time. Unlike that ultra skinny tie in your closet, deep scientific understanding doesn't go out of fashion.

On occasion, many of us who write on science have fallen into the trap of letting the singular excitement of a breakthrough overshadow the fundamental continuity of scientific progress. I understand well the push-pull of announcing that we've crossed into virgin territory with the reality that such territory typically comprises a nearby suburb that tightly borders all that we've so far understood. The best science journalism, evidenced by so many articles in this wonderful collection, walks that border without even making it apparent that there is a line to walk.

And toeing this line is vital. In an era that will inexorably rely ever more on the insights of science and the products of technology, it is increasingly urgent for the public to have not only a familiarity with scientific results but also a sense of scientific process. Progress in science extends the reach of certainty into precisely articulated realms of uncertainty. Much like my youthful subway epiphany, the general public needs to know, and know deeply, that science works.

—*Brian Greene*

INTRODUCTION

Fittingly, the first mention of science in *The New York Times* came in its very first issue—September 18, 1851. On page 2 (out of four), what was then called *The New-York Daily Times* reported the death of the Reverend Sylvester Graham, the famed nutritionist who invented the graham cracker—and whose *Lectures on the Science of Human Life* contained "a systematic, and in some degree, a scientific exposition of the author's peculiar views."

What struck the obituary writer as "peculiar" about Graham's views is lost to history. In its early, candlelit, hand-typeset decades, *The Times* was far from becoming the journalistic powerhouse it is today, and its coverage of science, as of the news in general, could itself be quite peculiar. The paper's founder, a charismatic, ambitious, and somewhat quixotic Republican politician named Henry Jarvis Raymond, meant it to be "the best and the cheapest family news-paper in the United States," substituting "cool and intelligent judgment, for passion" (and, not so incidentally, undercutting its competitors by selling for just a penny a copy). Among Raymond's best hires was John Swinton, an editorial writer who made sure *The Times* outdid the competition in science coverage, going so far as to write three to four columns a *day* on major scientific conferences. It was Swinton who commissioned a pioneering, sympathetic and still highly readable book review (reprinted on pages 278–86) about Charles Darwin's *On the Origin of Species* in 1860, when the theory of evolution was often attacked and derided, to the extent it was understood at all. But Swinton left the paper after Raymond's death in 1869, and science news "languished for half a century," as Meyer Berger wrote in his swashbuckling, compulsively enjoyable centennial biography of the paper, *The Story of The New York Times 1851–1951*.

All of that changed with the arrival of Adolph Ochs, the young Tennessee publisher who bought the failing *Times* in 1896 with $75,000 in mostly borrowed money (the equivalent of about $2 million today) and whose descendants, the Sulzberger family, still run the paper. Ochs stands as one of journalism's most heroic visionaries. The author of the slogan "All the News That's Fit to Print" and of the proud mission statement "to give the news impartially, without fear or favor," he was a Roman candle of ideas (most of them good) and a wizard at surrounding himself with talented people who wanted to work as hard as he did.

Perhaps the most talented was Carr Van Anda, whom Ochs brought on as managing editor in 1904. Van Anda was not only a hard-driving journalist: he was a mathematician who at least twice found errors in the equations of the young Albert Einstein. Ochs and Van Anda shared an "eager curiosity for news about the unknown in the sciences and about the remote unexplored corners of the world," Berger wrote, adding:

> Without Ochs' willingness to pay almost any sum for exclusive rights to stories on modern exploration, on the advancement of science, . . . Van Anda could never have made *The Times* a leader in that kind of journalism, but Ochs gave Van Anda his head. Between them they won for *The Times* a leadership in the field that was never overtaken.

One of the paper's first Pulitzer Prizes went to Alva Johnston, "for distinguished reporting of science news"—in particular, his coverage of the 1922 meeting of the American Association for the Advancement of Science, which produced the memorable headline "Scientists Witness Smash-Up of Atoms." Johnston was a general-assignment reporter whom Van Anda plucked from the newsroom to cover the meeting, but in those days *The Times* often turned to specialists, even commissioning news articles by scientists and explorers. Waldemar Kaempffert, one of the earliest bylines in this collection, was an engineer before Ochs brought him on as an editorial writer in 1927. And William L. Laurence was hired in 1930 as the first newspaper reporter assigned exclusively to cover science. (Laurence was later nicknamed "Atomic Bill" for his assignment by the War Department in the 1940s to serve as official historian of the Manhattan Project, the crash effort to develop nuclear bombs. He could write about the project for *The Times* on the condition that he disclose nothing before the war's end—a deal it is hard to imagine a *Times* reporter making today.)

Over the decades, as *The Times* has enhanced its leadership in science journalism, the balance between generalists and specialists has shifted. Since the 1940s, with a few notable exceptions, most science reporters have been journalists first: women and men who may or may not have advanced degrees but who are imbued with the kind of passionate curiosity that drove Adolph

Ochs and Carr Van Anda. These reporters know how to find things out, how to distinguish between real news and public-relations puffery, how to cultivate expert sources who can help them grasp the significance of new discoveries and the postgraduate-level science that underlies them. And how they can write!—often on punishing deadlines that leave no margin for artful revision. John Noble Wilford, who as a young man set his sights on political writing but found himself captivated by the 1960s space race, later collected two Pulitzer Prizes for science reporting. Here he is covering a rather routine astronomical conference celebrating the tenth anniversary of the Hubble Space Telescope:

> Hubble's pictures of faraway galaxies and brooding clouds of stellar nurseries
> have impressed astronomers and ordinary people alike. One of the more
> recent pictures shows dazzling fireworks in the constellation Aquila. Rings
> of glowing hot gas and showering streamers of cooler gas are visible around
> the central stellar remnant. It is an image of what the Sun will look like in its
> death throes some six billion years from now.

And here's Natalie Angier, who won a Pulitzer for beat reporting just ten months after arriving at *The Times:*

> With its miserly metabolism and tranquil temperament, its capacity to
> forgo food and drink for months at a time, its redwood burl of a body shield,
> so well engineered it can withstand the impact of a stampeding wildebeest,
> the turtle is one of the longest-lived creatures Earth has known.

Small wonder that more than a century after Ochs's arrival, science writing is still a *Times* mainstay. Science Times, the paper's Tuesday science supplement, was born in 1978. (An evocative twenty-fifth-anniversary account of the blessed event is reprinted on pages 438–40.) One the very few freestanding science sections left on the diminished landscape of American newspapers, it remains one of the paper's most popular features.

The New York Times Book of Science collects 125—best? no, let's say most representative—articles from more than a century and a half of science reporting. Some are indisputably great: no collection of this kind could overlook John Wilford's heart-stopping lead story on the *Apollo 11* moon landing of July 20, 1969. (The landing also occasioned the boldest headline to that point in the *Times*'s history—MEN WALK ON MOON—and perhaps its only front-page poem, also reprinted here.) Wilford is well represented, as are Angier and such past and present giants as Kaempffert, Laurence, Willia Broad, Walter Sullivan, Malcolm W. Browne, Lawrence K. Altman, Nicholas Wade, Gina Kolata, and Dennis Overbye.

But if journalism is indeed "the first rough draft of history," it's important to include some stories that didn't quite get it right, or missed the mark altogether. The most irresistible, about Einstein's 1919 confirmation of his General Theory of Relativity, carries the weirdly poetic headline "Lights All Askew in the Heavens," a skein of subheads including "Men of Science More or Less Agog," "A Book for 12 Wise Men," and "No More in All the World Could Comprehend It." The article's lead sentence candidly admits defeat: "Efforts made to put in words intelligible to the non-scientific public the Einstein theory of light proved by the eclipse expedition so far have not been very successful." But at least *The Times* knew it was on to *something* important, and less than a month later it recouped in fine style by paying a visit to the great man himself and letting him explain relativity in his own words.

Nor can a collection of this length remotely do justice to the broad sweep of scientific endeavor chronicled by *The New York Times* over the past 164 years. There is no chapter on chemistry, for example; the most interesting stories we found on that elemental discipline seemed to fit more comfortably in the chapters on physics and technology. Some towering scientists and accomplishments will not be found here. This book is less survey course than nonfiction narrative, a newspaper's story—in its own words—of the evolution of science journalism over an immensely consequential period for both science and journalism. Fortunately for readers seeking more detail, the three previous books in this series—on physics and astronomy, mathematics, and medicine—have that in abundance.

—*David Corcoran*

CHAPTER 1

Archaeology: Rediscovering Civilizations

Tut-Ankh-Amen's Inner Tomb
Is Opened, Revealing Undreamed-Of
Splendors, Still Untouched
After 3,400 Years

This has been, perhaps, the most extraordinary day in the whole history of Egyptian excavation. Whatever any one may have guessed or imagined of the secret of Tut-ankh-Amen's tomb, they surely cannot have dreamed the truth as now revealed.

The entrance today was made into the sealed chamber of the tomb of Tut-ankh-Amen, and yet another door opened beyond that. No eyes have seen the King, but to practical certainty we know that he lies there close at hand in all his original state, undisturbed.

Moreover, in addition to the great store of treasures which the tomb has already yielded, today has brought to light a new wealth of objects of artistic, historical, and even intrinsic value which is bewildering.

It is such a hoard as the most sanguine excavator can hardly have pictured, even in visions in his sleep, and puts Lord Carnarvon's and Mr. Carter's discovery in a class by itself and above all previous finds.

Official Opening Sunday

Though the official opening of the sealed mortuary chamber of the tomb has been fixed for Sunday, it was obviously impossible to postpone until then the actual work of breaking in the entrance. This was a job involving some hours of work, because it had to be done with the greatest care, so as to keep intact as many of the seals as possible, and also to avoid injury to any of the objects on the other side which might be caused by the falling of material dislodged.

All this could not be done on Sunday while the official guests were kept waiting in the singularly unpleasant atmosphere of the tomb, so an agreement was made with the Egyptian authorities by which the actual breaking through of the wall should be done in their presence today.

Consequently, Howard Carter was very busy inside the tomb all morning with Professor Breasted and Dr. Alan Gardiner, whose assistance has been invaluable from the beginning of the work of examining seals and deciphering and copying inscriptions of all kinds. They had finished by noon, and the tomb was closed till after luncheon, at which Lord Carnarvon, Mr. Carter and Lady Evelyn Herbert entertained those invited to be present today.

Official Story of Inner Tomb

It was after 1 o'clock when the official party entered the tomb, and the operation was begun which was to result in such astounding discoveries of which I am able to give the following authoritative description:

Today between the hours of 1 and 3 in the afternoon the culminating moment in the discovery of Tut-ankh-Amen's tomb took place when Lord Carnarvon and Howard Carter opened the inner sealed doorway in the presence of Lady Evelyn Herbert, Abdel Hamid Soliman Pasha, Under Secretary of Public Works; Pierre Lacau, Director General of the Antiquities Department; Sir William Garstin, Sir Charles Cust, Mr. Lythgoe, the curator of the Metropolitan Museum of Art of New York; Mr. Winlock, Director of the Egyptian expedition of the Metropolitan Museum, together with other representatives of the Government.

The process of opening this doorway, bearing the royal insignia and guarded by the protective statues of the King, had taken several hours of careful manipulation under the intense heat. It finally ended in a wonderful revelation, for before the spectators was the resplendent mausoleum of the King, a spacious and beautifully decorated chamber completely occupied by an immense shrine covered with gold inlaid with brilliant blue faience. This beautiful wooden construction towers nearly to the ceiling and fills the great sepulchral hall within a short span of its four walls. Its sides are adorned with magnificent religious texts and fearful symbols of the dead and it is capped with a superb cornice and a tyrus molding like the propylaeum of a temple, in fact, indeed, the sacred monument.

Another Shrine Within

On the eastern end of this shrine are two immense folding doors, closed and bolted. Within it is yet another shrine, closed and sealed bearing the cipher of the Royal Necropolis. On this inner shrine hangs the funerary pall, studded with gold, and by the evidence of the papyrus of Rameses IV, there must be a series of these shrines within, covering the remains of the King lying in the sarcophagus.

Around the outer canopy, or shrine, stand great protective emblems of a mystic type finely carved and covered with gilt, and upon the floor lie seven oars for the King's use in the waters of the other world.

In the further end of the eastern wall of this sepulchral hall is yet another doorway, open and never closed. It leads to another chamber, the store chamber of the sepulcher. There at the end stands an elaborately and magnificently carved and gilded shrine of indescribable beauty. It is surmounted by tiers of uraei and its sides are protected by open-armed goddesses of the finest workmanship, their pitiful faces turned over their shoulders toward the invader. This is no less than the receptacle for the four canopic jars which should contain the viscera (brain, heart, &c.) of the King.

Immediately at the entrance to this chamber stands the jackal god Anubis, in black and gold, upon his shrine, which again rests upon a portable sled, strange and resplendent. Behind this again is the head of the bull, emblem of the underworld.

Stacked on the south side of the chamber in great numbers are black boxes and shrines of all shapes, all closed and sealed, save one with open doors in which are golden effigies of the King standing upon black leopards. Similarly at the end of the chamber are more of these cases, including miniature coffins, sealed, but no doubt containing funerary statuettes of the monarch; servants for the dead in the coming world. On the south side of the deity Anubis is a tier of wonderful ivory and wooden boxes of every shape and design, studded with gold and inlaid with faience, and beside them yet another chariot.

This sight is stupendous and its magnificence indescribable, and as the time was fast creeping on, and dusk was falling, the tomb was closed for further action and contemplation.

The foregoing narrative is necessarily hasty and may be subject to correction in details as a result of future investigation. The truth is that all those who were privileged to share in today's unparalleled experiences were and still are so bewildered that it is not easy for any of them to give a consecutive narrative. All, however, agree in describing as overwhelming the impression produced by the discovery of the great shrine, or canopy, splendid in its blue and gold and almost filling the entire space of the new chamber.

Observing Gives More Details

Another informant gave me the following further particulars:

"As soon as it was possible to see through the opening which was being made by Mr. Carter and Mr. Callender it became evident that some large obstacle blocked the way inside. It looked like a screen of gold inlaid with blue, in the decoration of which I noticed the well-known so-called buckle of Isis.

"In fact, it was the great shrine, or canopy, or tabernacle, or whatever you call it, made of wood, carved and gilded, and almost filling up the entire interior of the new chamber. It reached nearly to the ceiling and the space between it and the walls at the sides may have been eighteen inches. That is quite enough to permit the passage of the old Egyptian workmen and others scantily clad and slimmer than we of today, but it was narrow for us to squeeze through in our clothes.

"On entering one turned to the right, which would be to the north, and then along the east side, the passage being still as narrow as at first. I noticed that the interior of the walls of the chamber were decorated, but the painting has much deteriorated and looked to me of inferior quality. The door into the shrine, or tabernacle, is in the eastern side and had heavy bronze hinges and was opened with some difficulty. When opened it only showed another wooden wall or screen being the exterior of a second inner tabernacle, a box inside a box.

"The interior faces of the wall of the outer tabernacle are all carved and decorated with religious texts, and so far the outer face of the inner tabernacle is similarly gilded and decorated, and I can make no attempt to describe the feelings of awe, wonder and mystery with which the spectacle inspired one.

"On the door of the inner tabernacle the original seals seem yet unbroken. The robbers do not appear to have penetrated it. We made no attempt to open it today, simply because it was impossible. Apparently the inner door will not be opened until the outer wall is removed. It must have been shut and sealed and the outer wall then erected around it. It will, therefore, I conjecture, be necessary to pull down the outer screening wall before the inner shrine is opened. This will be quite an undertaking in the narrow space.

"Among the individual objects I noticed besides the oars, or paddles, for the use of the deceased in the hereafter, were some alabaster vases, seemingly of the finest quality, and a piece of some sort of jewelry lying huddled on the floor, where one may suppose they had been thrown by robbers."

From the foregoing it will be evident that we have really arrived at the sepulcher of an old Egyptian King unviolated by robbers and undisturbed through 3,000 years. In the official narrative given above, reference is made to the papyrus of Ramses IV. Egyptologists will remember that this papyrus gives a sketch of the ground plan of the tomb, which shows the resting place of the sarcophagus to be inside a series of a concentric boxes, or tabernacles, painted to imitate wood, precisely such as was found today.

Of Great Historic Interest

The historical interest of the discovery is, therefore, enormous.

What also may be taken as reasonably certain is that the construction of these successive tabernacles, which successfully baffled the ancient robbers, makes the immediate opening of them impossible. The actual sight of King Tutankh-Amen where he lies will have to be postponed to some time indefinite. The period of his long and lonely watch is not yet ended.

Each of those who entered today is enthusiastic, if rather incoherent, in admiration of the new hoard of articles lying in the further chamber. This is merely a room hewn out of the rock, never having had a door built in it. So it has always been open to access from the chamber containing the tabernacles. The view, therefore, of its contents is unobstructed.

What especially struck all who saw it was the shrine spoken of above with the four guardian goddesses, all with their faces turned to the entrance as if pleading not to be disturbed. The workmanship of this is most beautiful. The greatest admiration is also expressed for some alabaster vases, said to be much lovelier even than the lovely ones formerly discovered. They are apparently white, not having become yellow by age. The statuettes and so forth seen are undoubtedly of the greatest beauty and value, while the number of unopened boxes with their seals unbroken give promise of an unprecedented harvest of precious things.

Great Problem Faces Discoveries

With the contents of the annex to the outer chamber still awaiting attention added to this new amazing store of wonders the mere embarrassment of riches confronts Lord Carnarvon and Mr. Carter with a serious problem, even though no attempt is made to touch for awhile the tabernacle of the King. The immensity of the whole thing makes one gasp.

The actual ceremony of breaking through the sealed door started at 1:45 p.m. Luncheon was over a little after I and the party, led by Lord Carnarvon and Sir William Garstin, made their way from the staff dining room in a cache to the tomb, into the well of which all descended.

A short interval elapsed for the arrival of M. Lacau, with Abdel Hamid Soliman Pasha, Under Secretary for Public Works.

After various introductions Mr. Callender removed the grille and descended the passage in order to unlock the steel gate. All present then proceeded to take off their coats, for not only was the opening process likely to be lengthy, but the atmosphere was certain to be sultry, to say the least of it.

There was a slight hitch owing to the failure of the electric current. A few moments were full of tense suspense, and even those watching from the parapet could sense the suppressed excitement which possessed each of those standing below at the top of the steps, on the lookout for the signal when they were to descend to experience the moment for which they had waited three months, when, before their eyes, the crumbling wall would reveal the mystery that had lain behind it for 3,000 years.

At last Mr. Callender sent up word that the light was on, and Mr. Carter and then Lord Carnarvon with Lady Evelyn Herbert close at hand, followed by Sir William Garstin, Abdel Hamid Soliman Pasha, M. Lacau, Mr. Engelbach, Professor Breasted, Dr. Gardiner, Mr. Lythgoe and others, descended into the antechamber. There were twenty in all, to whom must be added the laborers who carried down the huge trays for the reception of seals, &c.

Carter Eulogizes Carnarvon

Before the actual work started, Mr. Carter made a little speech in which he stated that all that had been done and anything that the wall might reveal was entirely due to Lord Carnarvon. He thanked every one for coming to the ceremony and expressed his gratitude to the Metropolitan Museum for the great assistance it had given, and also the Egyptian Government. They had still most important work to do, he said, and much might be damaged if improperly handled. He hoped that they would be allowed to carry it to completion in peace, for after all they were all working for the sake of science.

Lord Carnarvon followed with a few words, likewise expressing thanks to those present, to the Egyptian Government, and particularly to Mr. Lythgoe and the Trustees of the Metropolitan Museum for their ready assistance and sympathy and the loan which they made of their experts, who were probably the best experts in the world, for carrying out a very important and delicate work and dealing in a proper and in a scientific manner with the treasures of this tomb.

Lord Carnarvon concluded by saying that it was entirely due to Mr. Carter that they made this scientific discovery, for Mr. Carter, by his unquenchable faith and indomitable perseverance, had "bucked him up" in the face of the many previous disappointments they had experienced.

Mr. Carter then turned to the sealed wall and began breaking it with a chisel and mallet. In a short time he had remove d a large piece; which revealed the wooden lintel of the door. By about 3 o'clock sufficient had been removed to enable Mr. Carter to enter, and shortly afterward a large portion of the wall came away, revealing to the dazzled and spellbound gaze of the spectators the wonderful spectacle described above.

The Queen of the Belgians and Prince Leopold, traveling incognito as the Countess de Rethy and Count de Rethy, and accompanied by Professor Cappart, arrived by special train this morning, having come through direct from Alexandria without changing, and with but a few minutes' halt at Cairo. She was met at the station by Abdel Aziz Bey Yeha, Governor of the province, with the Sub-Governor and other officials, and Colonel J. K. Watson. The Queen is staying at the Winter Palace Hotel. This is the Queen's second visit, the first being in 1911, when she came with the King and stayed about two months, while he went to the Sudan on a shooting expedition.

—February 16, 1923

Doors to the Ancient World Unlocked by Archaeologists

By R. L. DUFFUS

A nother link in the long chain of archaeological discovery was forged the other day when excavators dug into a house at No. 4 Street of Abundance at Pompeii and found relics which had lain hidden during more than two and a half centuries of systematic exploring of the long-buried city. Silver plates, gold bracelets, rings, necklaces, spoons, a mirror, a silver jug, a perfume container, a "fresco of surpassing beauty" and a varicolored statue of Apollo were among the finds which, with others previously made, tell us more about Pompeii than some of the Pompeiians themselves may have known.

But Pompeii is but one important chapter in a narrative full of drama and rich with meaning—the story of modern archaeology. The Rosetta Stone, which gave the key to the language of ancient Egypt; the carvings at Behistun, which performed the same service for the languages of ancient Mesopotamia; the excavations which uncovered Troy, Mycenae and the Cretan Knossos, revealing many mysteries of the Mediterranean civilization which flourished when Athens was a barbaric village; the finding of the long-lost tomb of Tut-ankh-Amen, with its incredible store of artistic riches—these are high points in a narrative as thrilling as any in modern history.

In the following article are sketched these outstanding triumphs of the great diggers and decipherers whose work has given us a perspective on our own civilization by revealing to us the great civilizations of the past.

Pompeii

On Aug. 24. AD 79, the Roman world of the emperors was near the height of its gilded glory. Vespasian was on the throne, Christianity was an obscure sect whose adherents were mainly among the submerged masses; the gods were worshiped in pillared pagan temples, and the Roman exquisites, yawning a little, declared themselves the ultimate products of a civilization which promised to endure forever. Then a mountain called Vesuvius, above what is now called the Bay of Naples, suddenly exploded, and two cities, Pompeii and Herculaneum, disappeared under a thick covering of ashes and lava. Pliny the Elder, naturalist and politician, venturing too near, was suffocated.

The Roman world lived out its span of life and perished, leaving a few arches, some broken walls and some battered statues as evidence of what it had been. Pompeii and Herculaneum, underneath the ashes and lava, were ultimately forgotten. Yet the disaster which seemed to have destroyed them actually preserved them, while the rest of the great empire went to ruin. By digging down through a few feet of earth modern archaeologists have learned more about the everyday life of the Romans of the empire than all the Latin books could tell them.

It was not, however, an archaeologist who first broke into Pompeii in modern times. Italian workingmen, digging an aqueduct across the site at the end of the sixteenth century, were amazed to find themselves tunneling through a burled city. Curiosity was aroused and some relics were removed, but it was not until the middle of the eighteenth century that anything like a systematic attempt to lay bare the ancient town was made.

The City That Was

Yet this discovery made by Domenico Fontana's pick-and-shovel gang ranks among the most significant of archaeological triumphs. If it did not carry the diggers back into the abysses of time, as some later excavations have done, it revealed most intimately and touchingly the details of life and death nearly nineteen centuries ago. Not only were there works of art but election placards, naughty descriptions scrawled by schoolboys, signs warning passers-by to beware of dogs which have not barked or bitten for fifty-six generations, skeletons lying as their owners died, and holes in the lava retaining the forms and even the features of men and women who had long ago breathed their last.

Such was and is Pompeii. To enter its cleared streets, to break in upon another of its hurled houses, is like intruding unseen upon the turbulent existence of a city which for nearly two millenniums has no longer existed.

The Rosetta Stone

Pompeii required one kind of digging, and a rather simple kind. The Rosetta Stone, found by one of Napoleon's officers near the town of that name, at the mouth of the western branch of the Nile, in 1799, called for digging of another and far more intricate sort. On this slab of black basalt, forty-five inches long and twenty-eight and one-half inches wide, there are three inscriptions, one in Greek, two in the ancient Egyptian. Of the latter the upper was in hieroglyphics, the language of the Egyptian monuments, long forgotten and until that time undecipherable. The only clue lay in the fact that the Greek could be read. By this means it was found that the stone bore a decree of Ptolemy V, about 196 BC.

Two men, Dr. Young, an English physician, and a Frenchman, Jean-François Champollion, almost literally gave their lives to the translation of the hieroglyphics. Young, after four years of arduous labor, succeeded in translating only ninety characters. The difficulties were enormous, for he was dealing not only with a dead and unknown language but with what amounted to a secret cipher. The problem was much the same as though one were required to read a Chinese newspaper with no previous knowledge of Chinese and with only a copy of an English newspaper of the same date to check it by. Young had to do a great deal of guessing. Sometimes he guessed totally wrong and had to discard the results of many months of work.

The Work of Champollion

Champollion, working at the Rosetta Stone independently, found the task one of extraordinary complexity. The picture-writing of the Egyptians had passed far beyond the primitive stage—almost as far as Chinese Itself. There was no way of telling, at first, whether a given sign stood for a letter, a syllable or a word. But Champollion lived with the Rosetta stone and thought and dreamed of it until he unraveled its mystery and produced the first modern hieroglyphic dictionary. But he wore himself out with his work in his study and in the field and died at the age of 42.

The Rosetta Stone, unearthed by chance and translated by an exercise of ingenuity which makes the most complicated modern puzzle seem a pastime for the feeble-minded, brought the history of ancient Egypt back to life. It let modern man into the secrets of the long-dead people of the Nile as the uncovering of Pompeii revealed to him those of the Romans of the first century of our era.

The Carvings at Behistun

The story of the Behistun carvings, which opened still another door to the ancient world, combined in about equal proportion the elements of high adventure and almost incredibly patient and ingenious scholarship. Behistun is in the province of Kermanshah, on the western frontier of Persia. Its essential feature is a great cliff, 1,700 feet high. On this cliff, 300 feet above the ground, were carved many centuries ago the figures of men and inscriptions in three languages. This was all that Henry Rawlinson knew when he first saw the spot in 1835. Rawlinson was then 25 years old. He had been an interpreter for the East India Company and was at that moment an officer in the Persian Army. He had a passion for linguistics.

Rawlinson determined to copy and translate the inscriptions. To do this he had to scale the cliff, which might have baffled even an alpinist. He did scale it repeatedly, hanging on by toes and fingers balancing on a ladder set on a ledge so narrow that the slightest false movement would have brought his studies to an instant conclusion, and meeting with more than one hair's breadth escape. Finally he had a faithful copy of the inscriptions. All he had to do now was to translate them.

The layman will probably never understand how he did translate them, for though he soon knew that they were the same message couched in the cuneiform writing of three ancient peoples, the Persians, the Babylonians and the Medes, neither he nor anyone else knew anything about these languages. But, aided by his knowledge of existing Persian dialects and by the fact that the inscriptions were found to contain the names of two already known kings, Rawlinson solved the puzzle.

His translation was published in 1846, eleven years after he had begun his task. The Behistun rocks, it turned out, contained the boastful record of the conquests of King Darius, engraved at the order of that arrogant monarch 500 years BC. Their inaccessibility and the dryness of the desert air had preserved them for more than twenty-four centuries.

Thus the vanity of a king and the ingenuity of Rawlinson and other modern scholars made it possible to read the cuneiform inscriptions, on stone and on clay cylinders and tablets, which allow us to gaze into the ancient civilizations of Mesopotamia.

Troy and Crete

No man ever took a more roundabout way to archaeological fame than did Heinrich Schliemann, son of a German pastor in Mecklenburg-Schwerin. Schliemann was in turn a grocer's boy, a globetrotter and a military contractor. He visited California in the early days of its history as a state and became an American citizen. With a gift for finance he accumulated a small fortune, then, because of a boyhood dream based on the reading of Homer, he determined to spend it in digging up the almost mythical city of Troy. This was in the early 1870s.

Schliemann found his way to a spot called the Hill of Hissarlik, not far from the Dardanelles. There, despite the skepticism of classical scholars, he dug. As it happened he had discovered not only the site of Homer's Troy but of five other cities and towns. He dug clear through Homer's city, past the walls around which Achilles had dragged Hector. He and those who completed his work found first, underneath the comparatively modern Trojan town, the relics of three successive villages, then of a community whose inhabitants used bronze and silver and which was destroyed by fire, and finally of a remotely ancient Neolithic settlement. Schliemann was hasty and enthusiastic, without the modern archaeologist's scientific technique. But he proved that Troy had existed and he enabled scholars to tell pretty clearly how the Trojans had lived.

But Troy was evidently not the starting point of the civilization of Homeric times. Schliemann turned to the island of Crete, toward which the trails seemed to lead. Crete lies 180 miles south of the port of Athens. In ancient times it seems to have been a stepping-stone between the Grecian and the Egyptian civilizations. Its people had a highly refined culture, when the ancestors of Pericles and Phidias were uncouth, unwashed barbarians.

Schliemann dug at Mycenae, ancient Grecian city lying on the overland route between the Gulf of Argos and the Gulf of Corinth. There he found tombs and treasures of prehistoric times, showing an unmistakable Cretan influence. Then he turned to Crete itself and stuck his spade into the ground at Knossos. But he died, and an Englishman, Sir Arthur Evans, took up the work where he left off. And Evans found the legendary palace of King Minos, who had the fabled Minotaur for a pet.

Something of the spirit of this ancient civilization of the time of Minos may be discerned in the Cretan exhibits in the Metropolitan Museum in New York City.

No one can look at them without knowing that the ancient Cretans loved color, that they had a fine and vigorous art, and that they were, despite a grim religion which demanded human sacrifices, a joyous race.

Tomb of Tut-ankh-Amen

From the days of the Rosetta Stone Egypt has been the paradise of antiquarians. It has witnessed discoveries which fairly take the breath away with their significance and beauty. There has been none, however, more splendid in itself nor more dramatic in its circumstances than the finding of the tomb of Tut-ankh-Amen by Lord Carnarvon and Howard Carter. These two men, according to their own reports, moved 70,000 tons of sand and gravel and sifted much of it with their own hands before they found what they were looking for. Even then they did not find quite what they sought—they found something far more wonderful.

Carter came upon the entrance to the tomb after fruitless weeks of digging. In the whole area of the Valley of the Kings, by instinct, science and luck combined, he hit the right spot, though it was but a few feet wide.

Though the discovery was made eight years ago its glories are still fresh in men's minds. Those glories were almost a shock to the modern consciousness, for they showed that 3,300 years ago the Egyptians had reached a cultural point which was in its way very near perfection. They had for the moment broken away from the chains of an ancient and inhibiting religion and for the first time gazed upon the beauties of nature with wide-open eyes.

Simultaneously their skill in the arts and crafts had reached a point where they could express in stone, gold, wood and colors what they saw and felt.

Thus the archaeologist the world over is enriching history in a manner not dreamed of a few decades ago. Where he digs wisely and scientifically we no longer have to depend on the written records of biased or incompletely informed historians. We see and handle the objects which were part of the daily lives of peoples now long vanished or absorbed by other races. We realize the richness of the past and perhaps abate something of the arrogance which once arose from the belief that modern civilization was the high point in human endeavor. In the arts, at least, we know now that the almost inconceivably remote past can successfully compete with us.

By penetrating the earth or reading the inscriptions on the rocks the archaeologist is adding to history a new dimension.

—December 28, 1930

Scientist Stumbles Upon Method to Fix Age of Earth's Materials

Experiments with nuclear energy have accidentally uncovered a method of determining with relative accuracy the age of earth's materials by measuring their radiocarbon content, an associate of the Institute of Nuclear Studies, University of Chicago, reported yesterday.

Dr. Willard F. Libby, 40-year-old chemistry professor, who stumbled on the technique two years ago when studying cosmic ray action on the atmosphere, said it might enable scientists to fix dates of "living materials" up to 25,000 years old. He developed a special Geiger counter in connection with the "dating" technique.

Colleagues, anthropologists and archaeologists have greeted the discovery as a tremendous advance. It now will be possible, it was said, to date the early pre-Columbian civilizations in the Western Hemisphere as well as to explore Earth's chemical composition in the Pleistocene Age (600,000 to 10,000 BC).

Dr. Libby emphasized, however, that the testing of materials with his Geiger counter would not be effective on anything beyond 25,000 years of age. He said he hoped to create a calendar to set the ages for such material as wood, charcoal, well-crystallized shell and teeth.

He reported his discovery at an opening session of the twenty-ninth International Congress of Americanists at the American Museum of Natural History before 300 scientists from thirty-five countries. Collaborating in the experiment were Dr. E. C. Anderson of the University of Chicago and Dr. A. V. Grosse, president of the Temple University Research Institute.

Their experimentation established that radiocarbon was generated some 30,000 feet up in the atmosphere and that a "steady state condition" has existed on earth for a long time. Dr. Libby estimated then that there must be present somewhere on earth a total of thirty-three metric tons of radio-carbon.

This amount, he explained, is several million times, the amount of the isotope likely to be manufactured by the Atomic Energy Commission "over a period of many years."

Originating in the air, the carbon atoms produce carbon dioxide, the essential food for plants, hence are radioactive, Dr. Libby reasoned. Since animals and humans eat plants, they also are radioactive as are all things not isolated from the atmospheric life cycle.

15

Dr. Libby defined life as plants' assimilation of carbon dioxide from the air and the returning of carbon dioxide to the air by the animals that have eaten the plants.

"One then expects," he explained, "that death terminates the flow of radio‑carbon atoms into the organism to replace those which are steadily disintegrating to form nitrogen."

Since this disintegration rate is immutable, he added, the elapsed time since death can be measured by determination of the radiocarbon "assay" and comparing it with the general worldwide assay which was obtained at the time of death. The word "assay" is equivalent to the measurement of radioactivity.

He made the point that materials must contain exactly the same carbon atoms they had at time of death, and that their cosmic ray intensity must have remained constant over tens of thousands of years.

Dr. Libby reported that Dr. James Arnold of the University of Chicago collaborated during the last two years to "prove" the method. Using the special Geiger counter, they measured samples of wood that had their origins in antiquity. They were thus able to check for accuracy because a tree's age can be determined by its number of rings.

Wood samples collected from all parts of the world included fragments from a floor of a room in a palace of the "Syro-Hittite" period in northwestern Syria, dated at 625 to 725 BC; a California redwood tree felled in 1874; and wood from "a mummiform coffin from Egypt dated in the Ptolemaic period, 332 to 30 BC. Two samples came from the tomb of Egyptian kings, whose coffins were dated 4,600 years ago.

Measurements were lumped and they agree with one another. In further investigations some eleven different major dating questions were selected and prominent scientists in each case were invited to collaborate. The venture was supported financially by the Viking Fund of New York.

Dr. Libby said that within the next year a "decisive" test of the method would be obtained through the other inquiries. Some twenty-five "unknown" samples have been measured with satisfactory results.

"We have reason to believe," Dr. Libby said, "that ages up to 15,000 to 20,000 years can be measured with some accuracy by the present method, and we hope to go to 25,000 years."

Dr. Helmut de Terra, who is associated with the Viking Fund, said at the symposium that he had supplied many samples to Dr. Libby that had been collected from the core of the Pyramid of the Sun at Teotihuacan in the Valley of Mexico, said to be one of the oldest pyramids in this hemisphere. The pyramid's age had never been known but estimates had ranged up to 15,000 years. The Geiger counter set the correct age as 2,951 years.

Wider information of scientific validity is looked for which may develop a historical record of the earth's physical formation, and in that sense make the prehistoric period historic.

—September 6, 1949

Jamestown Fort, "Birthplace" of America in 1607, Is Found

By JOHN NOBLE WILFORD

On May 13, 1607, the first successful English settlement in what is now the United States got off to an unsteady start at a swampy island in a broad tidal river. The 104 settlers of the Virginia Company named the place James Fort, later Jamestown, after the sovereign for whom more than half would die in a few wretched months.

This was the colony where Capt. John Smith made his name in history and Pocahontas, a young daughter of the powerful Indian chief Powhatan, became a legend befriending the beleaguered English and supposedly rescuing Captain Smith from death. Looking back from the 19th century, the orator Edward Everett spoke of the Jamestown fort in tidewater Virginia as the place "where the first germs of the mighty republic . . . were planted."

All traces of the original fort, however, had been lost for more than two centuries. The log fort burned a year after it was built. Buried remains of the wall, along with any weapons, ceramics and other artifacts, were thought to have been washed away by changing currents of the James River. Anything of the first structures at the first permanent English settlement in America seemed beyond recovery.

But archeologists announced yesterday that the lost had been found. In recent excavations, they uncovered stains of decayed wood where they said logs of the palisade wall had stood in the ground.

From this they established the three-sided fort's footprint, determining the outlines of a rounded bastion at one of the three corners and the angle at which the walls were joined. Archeologists said these observations conformed exactly with contemporary written descriptions of the original fort.

Archeologists also reported finding traces of two buildings within the fort, evidence of glass-making and copper-working industries there and thousands "of other artifacts, including swords, armor, a smoking pipe, jewelry, ceramics and coins.

A grave at the site held the well-preserved skeleton of a white man in his early 20s who probably died of a gunshot wound, a fate raising intriguing questions about murder or mutiny or other tempests.

18

In making the official announcement at Jamestown Island, Gov. George Allen of Virginia declared, "We have discovered America's birthplace—the original fort."

Dr. William Kelso, the director of archeology of the Association for the Preservation of Virginia Antiquities, which owns the site and supported the excavations, said in a telephone interview that the artifacts should afford historians a clearer view of "a very early Colonial society" in the time of Queen Elizabeth I and her successor, James I.

Among the most interesting finds, Dr. Kelso said, were dozens of sky-blue glass beads apparently manufactured by the colony for trade with the Indians. "John Smith got himself out of hot spots by trading these beads," he said.

The project, which is also supported by the National Geographic Society, is expected to excavate most of the site of the one-acre fort and surrounding town ruins by the 400th anniversary of Jamestown, in 2007.

"Any time we get early Colonial artifacts, it's important," said Dr. James Axtell, a historian at the College of William and Mary in Williamsburg, Va., an expert on early encounters between Europeans and American Indians.

"There's also the symbolic significance here. Jamestown represents the founding of Virginia and of English America, even though other places are better examples of successful colonization."

Indeed, if John Rolfe, who married Pocahontas, had not developed a hybrid tobacco in 1613 that won a ready market in London and the first economic boost for the colony, historians say, Jamestown would have become just another Colonial failure.

An earlier English settlement, in 1584 at Roanoke Island, in present-day North Carolina, was a disastrous failure. In 1562, French Huguenots had no more luck with their Charlesfort, the site of which on the South Carolina coast was discovered earlier this year.

In 1565, the Spanish founded St. Augustine in what is now Florida and which is the oldest permanent European settlement in North America outside Mexico.

Jamestown's early tribulations were many. As Samuel Eliot Morison wrote in *The Oxford History of the American People*, the colonists "made the usual mistake of first-comers in America by settling on a low, swampy island." They also put down among a mighty Indian confederacy, which made its first attack on the fort on May 26, 1607.

So because of malaria, starvation and death at the hands of Indians, only 38 of the original 104 settlers lived through the first year. High mortality and bad management led to rumblings of mutiny, which provoked harsh punishment. Supply ships from England were few and inadequate; one relief ship became wrecked on Bermuda, an event that inspired Shakespeare's *The Tempest.*

"It's incredible that the colony survived," said Dr. Axtell, the historian at William and Mary. "There was 80 percent mortality in the first 15 years. They had no way to make money in the first years. They were a poorly organized community and ill supplied. And their relations with the native people were probably the poorest of any place."

An account of the settlement written in 1610 described the original fort as triangular in shape with a 360-foot wall on the riverside and the other two sides 300 feet long. Two sides of the fort form a 46.5-degree angle, identical to the number given in the written account.

The palisades were made of vertical planks and posts of oak and walnut. From the new excavations, it now appears that only 20 percent of the original fort may have been lost to the river. When the fort burned in 1608, it was replaced with a five-sided fort enclosing about four acres.

The excavators were surprised by the number of military artifacts at the site, said Dr. Kelso of the Antiquities Association. In a trash pile buried near one building, they found a helmet and metal breast plate of the type Captain Smith is depicted wearing in a contemporary engraving.

Archeologists also found a smoking pipe, marked with the letter *S.* "I'm not saying it's his," Dr. Kelso said, referring to Smith. "But who knows?"

The most recent discovery was of a young man's skeleton in a grave inside the fort. Forensic experts from the National Museum of Natural History in Washington examined the skeleton on Wednesday.

From the outline of wood decay and presence of nails, the scientists said, the man had been buried in a wooden coffin. The shape of the nose, teeth and cheekbones indicated that he was a European. Archeologists suspected that he was buried in the first two or three years of the colony.

Dr. Douglas W. Owsley, a forensic anthropologist at the Smithsonian Institution, said the man had definitely met a violent end. Both the tibia and fibula of the right lower leg were fractured, with a metal ball resting at the fracture site. There was a hole in one of the shoulder blades, which could be another gunshot wound.

Since the Indians did not have guns at that time, was this evidence of a murder among disgruntled settlers, an accident or an execution? That will be a problem for historians, who must comb through colony documents searching for references to men dying of gunshot wounds.

Dr. Owsley said there was an account of a man who had been killed as the result of a mutiny but also an account of an Indian who had seized a gun from a settler and shot him.

Anthropologists are more interested in studying the skeleton and any others found at the fort for insights into the nutrition, diseases and other physical conditions of early 17th-century people living in the harsh Colonial circumstances.

The fort site has so far yielded nothing to embellish the familiar story of Captain Smith and Pocahontas. Many historians suspect that elements of the story border on myth.

In an interview, Dr. Axtell of William and Mary said historical documents had failed to support the notion that Captain Smith and Pocahontas had a love affair. But she did go out of her way, often at great risk, to bring food to the English and warn them of impending attacks.

"I read her as a kind of traitor to her people, particularly to her father, Powhatan," the historian said.

Moreover, according to revisionist history, when Captain Smith was captured by Indians, he was held hostage and exchanged for goods, not freed by Pocahontas.

"The story that she saved him was added in much later editions of Smith's writings, as an afterthought," Dr. Kelso said.

—September 13, 1996

Bones Under Parking Lot
Belonged to Richard III

By JOHN F. BURNS

Until it was discovered beneath a city parking lot last fall, the skeleton had lain unmarked, and unmourned, for more than 500 years. Friars fearful of the men who slew him in battle buried the man in haste, naked and anonymous, without a winding sheet, rings or personal adornments of any kind, in a space so cramped his cloven skull was jammed upright and askew against the head of his shallow grave.

On Monday, confirming what many historians and archaeologists had suspected, a team of experts at the University of Leicester concluded on the basis of DNA and other evidence that the skeletal remains were those of King Richard III, for centuries the most reviled of English monarchs. But the conclusion, said to have been reached "beyond any reasonable doubt," promised to achieve much more than an end to the oblivion that has been Richard's fate since his death on Aug. 22, 1485, at the Battle of Bosworth Field, 20 miles from this ancient city in the sheep country of England's East Midlands.

Among those who found his remains, there is a passionate belief that new attention drawn to Richard by the discovery will inspire a reappraisal that could rehabilitate the medieval king and show him to be a man with a strong sympathy for the rights of the common man, who was deeply wronged by his vengeful Tudor successors. Far from the villainous character memorialized in English histories, films and novels, far from Shakespeare's damning representation of him as the limping, withered, haunted murderer of his two princely nephews, Richard III can become the subject of a new age of scholarship and popular reappraisal, these enthusiasts believe.

"I think he wanted to be found, he was ready to be found, and we found him, and now we can begin to tell the true story of who he was," said Philippa Langley, a writer who has been a longtime and fervent member of the Richard III Society, an organization that has worked for decades to bring what it sees as justice to an unjustly vilified man. "Now," Ms. Langley added, "we can rebury him with honor, and we can rebury him as a king."

Other members of the team at the University of Leicester pointed to Ms. Langley as the inspiration behind the project, responsible for raising much of the estimated $250,000—with major contributions from unnamed Americans—it cost to carry out the exhumation and the research that led to confirmation that indeed Richard had been found.

Ms. Langley's account was that her research for a play about the king had led her to a hunch that Richard's body would be found beneath the parking lot, in a corner of the buried ruins of the Greyfriars Priory, where John Rouse, a medieval historian writing in Latin within a few years after Richard's death, had recorded him as having been buried. Other unverified accounts said the king's body had been thrown by a mob into the River Soar, a mile or more from the priory.

Richard Taylor, the University of Leicester official who served as a coordinator for the project, said the last piece of the scientific puzzle fell into place with DNA findings that became available on Sunday, five months after the skeletal remains were uncovered. At that point, he said, members of the team knew that they had achieved something historic.

"We knew then, beyond reasonable doubt, that this was Richard III," Mr. Taylor said. "We're certain now, as certain as you can be of anything in life."

The team's leading geneticist, Turi King, said at a news conference that DNA samples from two modern-day descendants of Richard III's family had provided a match with samples taken from the skeleton found in the priory ruins. Kevin Schurer, a historian and demographer, tracked down two living descendants of Anne of York, Richard III's sister, one of them a London-based, Canadian-born furniture maker, Michael Ibsen, 55, and the other a second cousin of Mr. Ibsen's who has requested anonymity.

Dr. King said tests conducted at three laboratories in England and France had found that the descendants' mitochondrial DNA, a genetic element inherited through the maternal line of descent, matched that extracted from the parking lot skeleton. She said all three samples belonged to a type of mitochondrial DNA that is carried by only 1 to 2 percent of the English population, a rare enough group to satisfy the project team, pending more work on the samples, that a match had been found.

When she studied the results for the first time, she said, she "went very quiet, then did a little dance around the laboratory."

Even before the DNA findings came in, team members said, evidence pointed conclusively at the remains as being those of the king. These included confirmation that the body was that of a slight, slender man in his late 20s or early 30s—Richard was 32 at his death—and an analysis of his bones that showed that his high-protein diet had been rich in meat and marine fish, characteristic of a privileged life in the 15th century.

Also strongly indicative, they said, was the radiocarbon dating of two rib bones that showed that they were those of somebody who died between 1455 and 1540. In addition, team members said, the remains showed an array of injuries consistent with historical accounts of the fatal blows Richard III suffered on the battlefield, and other blows he was likely to have sustained after death from vengeful soldiers of the army of Henry Tudor, the Bosworth victor, who succeeded Richard as King Henry VII.

The fatal wound, researchers said, was almost certainly a large skull fracture behind the left ear that was consistent with a crushing blow from a halberd, a medieval weapon with an axlike head on a long pole—the kind of blow that was described by some who witnessed Richard's death. The team also identified nine other wounds, including what appeared to be dagger blows to the cheek, jaw and lower back, possibly inflicted after death.

But perhaps the most conclusive evidence from the skeletal remains was the deep curvature of the upper spine that the research team said showed the remains to be those of a sufferer of a form of scoliosis, a disease that causes the hunchback appearance, with a raised right shoulder, that was represented in Shakespeare's play as Richard III's most pronounced and unappealing feature.

The sense of an important historical watershed was underscored when reporters were escorted to a viewing of the skeletal remains, laid out in a locked room in the university's library, lying on a black velvet cushion inside a glass case. Two members of the university's chaplaincy's staff, one of them in the black-and-red robes of a Roman Catholic priest, sat beside the remains as reporters filed silently by, cautioned by university staff to behave with the "dignity" owed to a king.

Members of the Richard III Society have said in the past that they believed he should be reburied, once found, alongside other British monarchs in Westminster Abbey in London, the traditional venue for most royal weddings and burials. But in Leicester, officials said that plans were in hand to bury the bones early next year in the city's Anglican cathedral, barely 200 yards from where the skeleton was found, with a visitors' center dedicated to Richard to be opened in the cathedral grounds at the same time.

—February 4, 2013

CHAPTER 2

Astronomy:
Of Time and the Stars

THE BIG BANG AND BEFORE

Star Birth Sudden, Lemaître Asserts

By WALDEMAR KAEMPFFERT

S trange things happened to the universe when the physicists of the British Association for the Advancement of Science gathered today to hear such celebrities as Abbé Georges Lemaître, Professor Willem de Sitter and Sir Arthur Eddington discuss the new cosmogony.

The universe was pushed in and pulled out, blown up like a bubble and reduced to a mathematical point.

Complicated equations chalked on blackboards were evidently beautiful harmonies which these sensitively attuned mathematicians heard as a modernistic music of the spheres.

The reason for this mathematical juggling is to be found in the strange impasse to which a development of the Einstein theory of relativity has brought science. This development would make the earth, sun and stars a thousand times older than the universe itself. To extricate themselves from this uncomfortable position, relativists at Leicester wrote another chapter in the new mathematical book of Genesis.

Abbé Lemaître is especially interested in the problem presented. He is the original inventor of the closed and expanding universe. He rose to explain his latest conceptions, designed to reconcile the ages of old stars and the infant universe of which they are a part.

Lemaître Offers a Solution

"This conflict can be settled," said the Abbé, "if the mechanism of an expanding universe is provided with some means of substituting for the slow evolution of the stars a more rapid evolution."

So he created an entirely new universe. We start with a cloud of matter. It expands as a whole, slows down to pass over what Abbé Lemaître calls the equilibrium radius, and then continues to expand forever at a faster and faster rate.

But while the universe as a whole is thus being blown up, parts of the cloud of matter that are denser than the rest condense or shrink. In this process of contraction "diffused matter suddenly agglutinates into stars."

In this new theory "stars and nebulae are born together in an astronomical instant; sudden evolution takes the place of the slow evolution of the stars."

Hence the Milky Way, in which part of the solar system swims, and hundreds of millions of clusters that lie far beyond, are under this theory all a result of the collapses of matter in a universe being blown up like a bubble.

Abbé Lemaître finds some confirmation of his new theory in the work of Dr. Edwin P. Hubble of the Mount Wilson Observatory in California, but wants more.

Eddington Brings Gasps

Sir Arthur Eddington, who likes his paradoxes, has less respect for the so-called hard facts. He rose to say: "I hope it will not shock the experimental physicists too much if I say that we do not accept their observations unless they are confirmed by theory."

The mathematicians gasped a little and Professor de Sitter protested mildly, but a moment later did a little startling on his own account by remarking that the "universe is only a hypothesis, like the atom." M. de Sitter refused to be disconcerted by the discrepancy between the infancy of the universe and the hoary age of the stars.

"Out in California," he illustrated, "are giant redwood trees 2,000 years old. California as a political region is only a few generations old. We see nothing strange in this because we know when California received its name and became an entity. We have to deduce what we can about California and of the universe. Hence these possible universes which we devise, hoping to find one that fits the observed facts."

Question of Beginning

M. de Sitter sees no difficulty in adopting a point of view in which the beginning of the universe is no more remarkable a phenomenon than the crossing of the meridian by the sun.

On this supposition, the stars can be of any age. If the galaxies are now rushing away from us and from one another, there must have been a time when they interpenetrated. This interpenetration, with frequent collisions, occurred ten thousand million years ago.

In all this there was an impressive philosophic humility. He went on:

"Nature works without mathematics. Mathematics is simply a device to find out, if possible, what nature is doing. If we flounder to darkness it is not nature's fault but mathematicians'. In the end we know little. We are convinced that our sense impressions originate in the outer world and that there is actually an outer world.

"In order to understand the outer world we must give it some mathematical form, which means that we must measure it in some way. It was in this way that mathematicians reached the conclusion that space is curved. So far as we can be certain of anything, we are certain that this is a four-dimensional universe.

Expansion Is Seen

"Moreover, we are forced to the conclusion that this universe is expanding, both because our equations tell us so and because the outer nebulae are rushing away at a rate of 12,000 miles a second.

"To conclude that the universe is expanding is the easiest way for the mathematician to account for what is actually seen.

"All these mathematicians refuse to be bound by what usually passes for common sense. That is, they allow for tricks the mind plays in making its inferences from what we see, hear and feel."

Dr. G. C. McVittie considered this phase of the question in one part of his discussion of the expanding universe. He said:

"If our universe is flying to pieces, why is it that the sun and planets seem to be exceedingly stable and permanent? The answer is that the observer, using the instruments and methods of today, necessarily concludes that the planetary systems are fixed and unchanging in size, while the system of nebulae expands.

Workings of the Mind

"The nature of our minds is such that we instinctively endow our immediate surroundings with an element of permanency, and relegate to distant nebulae evidence of instability of the universe in which we live."

Professor Miles Walker told the association that the expenditure of huge sums for public works as a remedy for unemployment was unsound.

"The main object," he said, "should be to create as much wealth as possible with the least expenditure of money, and instead of employing as many as possible on some job to employ as few as possible on some job to employ as few as possible. There is no limit to the number of men who can be employed because there is no limit to the work to be done."

Professor Walker's solution is to produce the things people need.

"Provide them with these and they can do without money and employment," he said. "Employment is always desirable because of the moral restraint it imposes."

He admits it may be difficult to engage millions in producing things they need.

Long View Is Urged

"We must take a rather longer view so as to comprise in terms of wealth things not so immediately valuable but which will yield value in years to come," he added.

Hence, he would turn the idle population loose on the construction of underground railways, bridges, extensions of electric power systems, electrification of railways and the substitution of the Diesel-electric for the steam locomotive.

In an address on the "Significance of Genetics in Evolution," Dr. C. C. Hurst showed that new species of plants and animals can now be produced experimentally. To be sure, it is not yet known why one species is different from another, but work done with X-rays proves that ultimately the biologist must deal with atoms and electrons. Cosmic rays, he said, might be a primary force in evolution.

"The time is at hand," Dr. Hurst predicted, "when natural selection will be largely displaced by human selection, and man himself will be able to control evolution and even determine his own destiny."

—September 13, 1933

Peering Back in Time, Astronomers Glimpse Galaxies Aborning

By JOHN NOBLE WILFORD

In the beginning, there was light, the bursting dawn of cosmic creation. As the fireball of the Big Bang expanded and cooled, though, the light faded away and darkness fell on the young universe, by then little more than a formless sea of dark matter flecked with traces of primordial hydrogen and helium and broken only by scattered but fateful ripples.

The universe had entered what astronomers characterize as its dark age. Over a period from 300,000 years to about half a billion years after the Big Bang the darkness persisted, denying to astronomers any direct evidence of this time of momentous transformation. Somehow this epoch gave birth to the architecture of the modern universe and conditions that would eventually lead to planets and life.

But what happened in the dark age and how the light came on again is an enduring and frustrating mystery. If there was no light or other detectable radiation from the dark age, what was there to see? Besides, no telescope could see back that far. Sir Martin Rees, a cosmologist at Cambridge University in England, calls this the "most important unsolved problem in astronomy."

In spite of the obstacles, by looking deeper and deeper into space and thus into time with new telescopes in Earth orbit and on the ground, astronomers are making progress in their search for the first light. They are at last catching sight of some of the earlier, if not necessarily the first, galaxies—each observation seems to put their formation back earlier in time. Astronomers are also gathering clues to what makes up most of the universe, the mysterious dark matter, and how it and other dark-age conditions gave birth to the first stars and the first galaxies. The new knowledge is enabling scientists to test their theories and conduct increasingly detailed computer simulations of cosmic history, including the dark age.

The dark age ended when light returned to the firmament with the emergence of the first stars and then vast communities of stars gravitationally bound in galaxies. "Suddenly, the universe lit up like a Christmas tree," said Dr. Abraham Loeb, a theoretical astrophysicist at the Harvard-Smithsonian Center for Astrophysics in Cambridge, Mass. "Somewhere in the dark age, the transition from the universe's initial simple state to its current complex state occurred. It would be philosophically exciting to glimpse that moment."

A few years ago, Dr. Loeb recalled, he was one of a handful of scientists studying the dark age and early star formation. But a workshop on the subject at the University of Maryland last week drew several dozen of the top observational and theoretical astronomers. "Now this has become a very hot topic," he said.

Two weeks ago, the National Aeronautics and Space Administration released pictures taken by the Hubble Space Telescope of what astronomers think are the most distant galaxies ever studied. These galaxies probably formed in the first billion years of cosmic history, or more than 12 billion years ago. Such estimates are rough because they depend on cosmological models and chronological yardsticks based on the universe's being about 13 billion years old, which is controversial and subject to revision.

"We have really made a step into new territory," said Dr. Rodger I. Thompson, a University of Arizona astronomer who directed the observations with the Earth-orbiting Hubble telescope's near-infrared camera and spectrometer.

Only an infrared camera can detect the faintest, most distant galaxies. Dust in deep space obscures most visible light, more so than radiation in the longer infrared wavelengths. And it is a phenomenon of the expanding universe that the light emitted by a galaxy gets stretched out to the redder and even near-infrared wavelengths. At distances of vanishing visibility, the infrared galactic light becomes the cosmic version of the Cheshire cat's grin.

The more distant galaxies have a higher expansion velocity and thus a higher red shift. Since the age of the universe is still uncertain, astronomers express cosmic distance, and thus age, in terms of an object's red shift.

Not so long ago, it was all astronomers could do to see objects at distances at a red shift of 1, encompassing vistas extending the recent half of the age of the universe. When they picked up some red shift-3 sightings, it was the equivalent of aviation's breaking the sound barrier. In recent months, astronomers have broken the red shift-5 barrier, when the universe was only one-tenth its present age.

"We are seeing things at much higher red shifts than we once thought possible," said Dr. John C. Mather, an astronomer at the Goddard Space Flight Center in Greenbelt, Md.

Focused on a tiny patch of sky, the Hubble's infrared camera made a 36-hour exposure and saw more than 100 extremely faint galaxies that were absent in visible-light pictures. At least some of these galaxies are thought to have the highest red shifts ever measured. Precise measurements may have to await the development of more powerful telescopes.

"This is an exciting step into the infrared view of the universe," said Dr. Alan M. Dressler, an astronomer at the Carnegie Observatories in Pasadena, Calif. "That's where we're going to be looking for galaxy formation."

The first big break in distant observations came in 1995, when Dr. Robert Williams, director of the Space Telescope Science Institute in Baltimore, directed astronomers to concentrate the Hubble on a narrow sector of the sky, a region virtually devoid of light as seen from the ground. The telescope's visible-light camera took a 10-day-long exposure, long enough to resolve objects at then-record distances.

The Hubble Deep Field, as the resulting multicolor picture is called, gave astronomers a remarkable view of more than 3,000 galaxies, some at distances beyond red shift 4. They were astonished to find such a dense population of galaxies so far away and so early in time. It meant that astronomers had a way to go to reach the time of galactic genesis, the frontier to the dark age.

At about the same time, a team of astronomers led by Dr. Charles C. Steidel of the California Institute of Technology in Pasadena developed a novel technique for detecting faraway galaxies using the world's most powerful telescopes at Keck Observatory in Hawaii. By now, the astronomers have confirmed the presence of 440 high-red shift galaxies existing just a few billion years after the Big Bang. No objects at such distances had ever been examined in such close detail.

Dr. Steidel's observations are made by the ultraviolet "dropout" technique. Ultraviolet light from a remote galaxy tends to be absorbed by hydrogen within the galaxy itself and between there and Earth. The more distant a galaxy, the more hydrogen its light encounters in the intervening space, and thus the more its ultraviolet emissions are dimmed. By seeking objects that show up in green and red filters but not in ultraviolet, the astronomers detected galaxies with red shifts from 2.2 to 4.

The Hubble Deep Field and Dr. Steidel's snapshots have given theorists much to think about. The early galaxies appear to be smaller, more concentrated and less regular than today's galaxies. And they are "as strongly clustered in the early universe as they are today," Dr. Steidel said.

Small, irregular galaxies, cosmologists say, fit their models in which structure in the universe evolved from the bottom up, and not from the top down. That is, galaxies presumably started as small collections of stars and these merged to form much larger objects, instead of stars starting out in extraordinarily vast congregations and then fragmenting into galaxies.

"The small, young objects we find are still in the process of formation," Dr. Loeb said, explaining why the findings seem to support a growing consensus in favor of the bottom-up interpretation.

The observations also suggest that cosmologists are on the right track with their ideas about the nature and amount of cosmic mass. Years of studying the rotations of galaxies near and far have convinced scientists that there is much more to the universe than what they can see. Galaxies would have disintegrated long ago if they were not held together by more gravity than ordinary matter can account for. Most of the universe's mass, maybe as much as 99 percent, must be in some mysterious "dark" form.

In *Before the Beginning*, published this year by Perseus Books, Dr. Rees of Cambridge writes, "Clearly we will never understand galaxies properly until we understand the dominant stuff whose gravity binds them together."

The favored hypothesis predicts the existence of cold dark matter, exotic and as yet unknown subatomic elementary particles that emerged from the Big Bang. They are so small and slow that they virtually never interact with each other or with ordinary matter except through gravity and thus produce no detectable radiations. Yet they could have supplied the decisive gravitational force to marshal ordinary matter into galaxies and clusters of galaxies.

It all began, according to theory, with tiny ripples in the cosmic microwave background radiation, the detectable afterglow of the Big Bang that is the source of much knowledge about the universe's first 300,000 years. These slight fluctuations in radiation, confirmed and plotted by COBE, the Cosmic Background Explorer spacecraft, in the early 1990s, are caused by irregularities in the density of matter in the Big Bang. Such clumps of dark matter would have attracted ordinary matter to create larger clumps and set in motion agglomerations leading to stars and galaxies.

Computer-generated simulations based on the COBE findings and many other physical data and assumptions have showed that a universe dominated by hot dark matter, an alternative idea invoking known fast-moving particles called "neutrinos," should have resulted in a top-down creation in which much larger conglomerations of matter would form first and then break up into galaxies. This scenario, however would result in a much longer dark age than now appears to be the case. Cold dark matter, on the other hand, should have formed galaxies on small scales and at a relatively early time. Which is just what the new observations are revealing.

In February, astronomers announced the first detection of a red shift-5 object, a young galaxy that existed when the universe was only 8 percent of its present age. They had crossed the frontier into the first billion years of the universe, with the boundary to the dark age still not in sight.

Working at Keck Observatory, a team led by Dr. Arjun Dey of Johns Hopkins University in Baltimore measured a red shift of 5.34 for the early galaxy, meaning that it was shining 820 million years after the Big Bang (also assuming an age of 13 billion years for the universe).

"Now that we know what to look for, I'm sure this record will be broken in a matter of months," Dr. Dey remarked at the time.

Sure enough, in May astronomers at the University of Hawaii and Cambridge University reported sighting a more distant galaxy with a 5.64 red shift, pushing back galactic discovery about 60 million years. One of the discoverers, Dr. Esther M. Hu, said astronomers were on the verge of seeing galaxies back to about 500 million years after the Big Bang.

Dr. Loeb noted two intriguing clues that light had returned to the universe at least that early, perhaps at red shifts beyond 10. Both came from observations of light from quasars, cores of distant galaxies and the brightest objects in the universe. Because the quasar light passes through clouds of hydrogen gas before it reaches Earth, it serves as a probe of the cosmic environment out to great distances.

The fact that the quasar light is not absorbed by the gas, Dr. Loeb said, is evidence that nearly all of the hydrogen atoms in deep cosmic space have been ionized, their electrons stripped away by ultraviolet radiation. Such radiation could have come only from a large population of stars at red shifts beyond that of the quasars. Other quasar studies yielded the second clue: some of the distant gas clouds contained not only hydrogen but heavy elements like carbon, which are produced only by the nuclear furnaces inside stars and scattered through space when stars explode at the end of their lives. So there had been stars living and dying long before these early galaxies.

"We are going to fainter and fainter objects and getting smarter and smarter each month," said Dr. Dressler of the Carnegie Observatories. "But we've pushed just about as far as we can expect from today's telescopes, even the Keck or the Hubble."

Astronomers are in the early stages of planning a more powerful replacement for the orbiting Hubble telescope. Called the Next Generation Space Telescope, it is expected to be launched a million miles into space in 2007 and be able to detect infrared emissions from objects 400 times fainter than those being studied with any of today's ground-based and orbiting infrared telescopes. Its mirror will be nearly three times larger than Hubble's.

With the new telescope thus able to gather so much more light, Dr. Mather of the Goddard Space Center said, astronomers in exposure times as short as two weeks should "see as far as you can imagine seeing." They should be able to make out galaxies with red shifts of 10 or more.

That might be just enough to observe the first stars and galaxies and understand more about how the early universe passed its long night of darkness organizing itself to bring form and light everywhere.

—October 20, 1998

What Happened Before the Big Bang?

By DENNIS OVERBYE

L ike baseball, in which three strikes make an out, three outs on a side make an inning, nine innings make a regular game, the universe makes its own time. There is no outside timekeeper. Space and time are part of the universe, not the other way around, thinkers since Augustine have said, and that is one of the central and haunting lessons of Einstein's general theory of relativity.

In explaining gravity as the "bending" of space-time geometry, Einstein's theory predicted the expansion of the universe, the primal fact of 20th-century astronomy. By imagining the expansion going backward, like a film in reverse, cosmologists have traced the history of the universe credibly back to a millionth of a second after the Big Bang that began it all.

But to ask what happened before the Big Bang is a little bit like asking who was on base before the first pitch was thrown out in a game, say between the Yankees and the Red Sox. There was no "then" then.

Still, this has not stopped some theorists with infinity in their eyes from trying to imagine how the universe made its "quantum leap from eternity into time," as the physicist Dr. Sidney Coleman of Harvard once put it.

Some physicists speculate that on the other side of the looking glass of Time Zero is another universe going backward in time. Others suggest that creation as we know it is punctuated by an eternal dance of clashing island universes.

In their so-called quantum cosmology, Dr. Stephen Hawking, the Cambridge University physicist and author, and his collaborators envision the universe as a kind of self-contained entity, a crystalline melt of all possibilities existing in "imaginary time."

All these will remain just fancy ideas until physicists have married Einstein's gravity to the paradoxical quantum laws that describe behavior of subatomic particles. Such a theory of quantum gravity, scientists agree, is needed to describe the universe when it was so small and dense that even space and time become fuzzy and discontinuous. "Our clocks and our rulers break," as Dr. Andrei Linde, a Stanford cosmologist, likes to put it.

At the moment there are two pretenders to the throne of that ultimate theory. One is string theory, the putative "theory of everything," which posits that the ultimate constituents of nature are tiny vibrating strings rather than points. String theorists have scored some striking successes in the study of black holes, in which matter has been compressed to catastrophic densities similar to the Big Bang, but they have made little progress with the Big Bang itself.

String's lesser-known rival, called "loop quantum gravity," is the result of applying quantum strictures directly to Einstein's equations. This theory makes no pretensions to explaining anything but gravity and space-time. But recently Dr. Martin Bojowald of the Max Planck Institute for Gravitational Physics in Golm, Germany, found that using the theory he could follow the evolution of the universe back past the alleged beginning point. Instead of having a "zero moment" of infinite density—a so-called singularity—the universe instead behaved as if it were contracting from an earlier phase, according to the theory, he said. As if the Big Bang were a big bounce.

In their dreams, theorists of both stripes hope that they will discover that they have been exploring the Janus faces of a single idea, yet unknown, but which might explain how time, space and everything else can be built out of nothing. A prescription for, as the physicist Dr. John Archibald Wheeler of Princeton puts it, "law without law."

Dr. Wheeler himself, the pre-eminent poet-adventurer in physics, has put forth his own proposal. According to quantum theory's famous uncertainty principle, the properties of a subatomic particle like its momentum or position remain in abeyance, in a sort of fog of possibility until something measures it or hits it.

Likewise he has wondered out loud if the universe bootstraps itself into being by the accumulation of billions upon billions of quantum interactions— the universe stepping on its own feet, microscopically, and bumbling itself awake. It's a notion he once called "genesis by observership," but now calls "it from bit" to emphasize a proposed connection between quantum mechanics and information theory.

One implication of quantum genesis, if it is correct, is that the notion of the creation of the universe as something far away and long ago must go. "The past is theory," Dr. Wheeler once wrote. "It has no existence except in the records of the present. We are all participators, at the microscopic level, in making that past as well as the present and the future."

If the creation of the universe happens outside time, then it must happen all the time. The Big Bang is here and now, the foundation of every moment.

And you are there.

—November 11, 2003

THE SOLAR SYSTEM

The Satellites of Mars

The official report of Rear Admiral John Rodgers, Superintendent of the Naval Observatory in this city, in regard to the recent discovery of the satellites of Mars, was received at the Navy Department this morning. He says, from observations made by Prof. Asaph Hall, of the Observatory, with the 26-inch refractor, the planet Mars has been found to have at least two satellites.

Prof. Hall finds, on examining his observing book, that the satellite which was first discovered, and which he supposed was seen for the first time on Aug. 16 at 11h. 42m., had been, in fact, observed on Aug. 11, at 14h. 40m.; but as he had no opportunity at that date to wait for the planet's motions, he failed to recognize the object as a satellite. It was, however, recognized and observed as such on Aug. 16, and has been observed, on the 17th, 18th, and 19th of August. This satellite has an apparent distance from the center of Mars of 82 seconds, and its time of revolution around the planet is 30 hours. Its magnitude Prof. Hall estimates as the 13th or 14th. The plane of its orbit has now a considerable inclining to the line of sight from the earth to Mars. At its elongations its angles of positions are 72 degrees and 252 degrees. The second satellite was discovered Aug. 17, at 16h. It appears to be quite as bright as the first one and at its elongations has nearly the same angles of position which correspond to the equator of Mars. Its apparent distance at the elongations and its periodic time are not yet known. The following are the preliminary elements of the outer satellite, as calculated by Prof. Simon Newcomb: Major axis of orbit, 82 seconds; angles of major axis, 70 degrees and 250 degrees; minor axis, 28 seconds; passage of satellite through western apsis, Aug. 19, 16h. 40m. The period of the inner satellites is so short, probably less than eight hours, that it cannot be fixed.

—August 22, 1877

The New Solar Theory

While the acknowledged masters of experimental physics in this City, such as Prof. Rood, of Columbia College, and Prof. Draper, of the University of the City of New-York, are cautious about expressing an opinion of Mr. J. Norman Lockyer's alleged discoveries respecting the constitution of the solar atmosphere, it is very evident, from the general drift of their conversation, that, while they concede the probability that the so-called elements are not simples, they are not prepared to concede that Mr. Lockyer furnished any sufficient demonstration. This hypothesis, which has taken a new significance in view of the doctrine that all forces are but modes of one primal force, has been conditionally accepted by physicists since the days of Sir Isaac Newton, and formed the basis of the speculations of medieval science. But as Prof. Draper very cautiously observed in the course of an interview last evening, it is one thing to accept this doctrine in a vague and general way as something that may possibly be proved some time, and quite another thing to bring sure optical demonstration that such is the case.

Prof. Draper was found yesterday in his laboratory on the third floor of his residence on Madison-avenue. He laughed heartily when his attention was called to the double-leaded telegram from London that appeared in a morning journal, and called attention to a passage in *Comptes Rendus* (journal of the Academy of Sciences, France) for 1878, page 674. The passage occupies the space of four lines, and runs as follows:

> En raisonnant d'après les analogies fournies par la manière d'agir des composés connus, j'ai mis en évidence que indépendamment du calcium, beaucoup de corps considérés commes éléments, sont aussi des corps composés.

Such, word for word, is the note in *Comptes Rendus*, to which the Secretary of the Academy briefly adds that Mr. Lockyer will furnish the facts in a few days, and this note, when translated into English, will be found to convey all the real information upon which the great hydrogen craze was based. While Prof. Draper would not wish to be understood as disparaging the work of a spectroscopist with whom he has had some controversy, he believes that Mr. Lockyer ought to publish his verifying facts immediately, if he has amassed sufficient data to furnish unquestionable proof of his hypothesis, and will wait patiently for the official transcript of his paper in the *Transactions of the Royal Society* before expressing a decided opinion. From such meagre memoranda of that paper as he has been able to obtain he is, however, constrained to maintain that such proofs have not yet been adduced as would justify the scientific world in accepting Mr. Lockyer's claims to the discovery that hydrogen is the Protean basis of many of the so-called elements; and he is particularly cautious about subscribing to the doctrine of dissociation in the photosphere and association of the 14 elements in the color-sphere of the sun as the true explanation of the different spectra presented by the two spheres, ingenious as that doctrine is. . . .

—January 14, 1879

Canals of Mars Called an
Optical Illusion

One of those expert accountants who would make a name for himself as an efficiency engineer has estimated that if it were possible to forward telegraphic messages from the earth to Mars, and if it had been possible to send them in years long past—before telegraphy had been invented—there would still be more than sixty years to come before the planet's sublimated population would receive news of the fighting in Belgium. The reference is not to the fighting which is going on now, but of that which ended at Waterloo almost exactly one hundred years ago.

And now, six years and more since the 1909 opposition of Mars, there has only recently been published a report by the Section for the Observation of Mars of the British Astronomical Association, under the direction of E. M. Antoniadi, announcing that the famous canals of Mars are an optical illusion. The entire number of canals on Mars listed in 1908 amounted to 585.

"The alleged existence of a geometrical network of canals on Mars has received a lasting and unanswerable confutation," is the sweeping conclusion at which the Director of the British Astronomical Association has arrived, after prolonged study of the results of his personal observations of the planet, using the great Meudon refractor, the most powerful telescope in the Old World. . . .

Mr. Antoniadi designed for his work of observation at Meudon, in France, a sliding shed which rolled about on wheels. The length (33 feet) of the rails enabled the telescope to remain at such a distance from the structure as to receive none of the waves of heat emitted by the latter. He found that a contrivance of that kind was superior to a dome or any other arrangement, for the isolated body of the reflector, exposed to the sky, cools down very rapidly, especially if a draught be produced in the tube by opening the door of the mirror for a quarter of an hour before starting work.

While observing Mr. Antoniadi discovered that it was disadvantageous to keep the eye which does no work open and troublesome to keep it closed. Hence the advantage of a pair of old spectacles, in the use of which the inactive eye remains open against a black glass while the other looks in front of the eye-piece through the empty elliptical frame. It was such a pair of old spectacles that Mr. Antoniadi wore while making those observations which, with such force of scientific argument, he now asserts prove the theory of canals on the God of War, the ruddy or red planet, to be a myth only.

In working with smaller instruments he himself had, like other observers, obtained frequent glimpses of narrow, straight lines, but in the Meudon instrument these lines were seen by Mr. Antoniadi, only when the definition was bad, and the image of the planet "flaring." With good seeing, a complex natural structure of the so-called "continental" regions of the planet was revealed, a variety of irregular bands and shadings replacing the sharp, narrow lines drawn by Giovanni Schiaparelli, the Italian astronomer.

The modern studies of Mars which have aroused so much interest, says Professor Simon Newcomb, LL.D., D.Sc., began with the work of Schiaparelli in 1877. Accepting the word "ocean" used by the older observers to designate the widely extended darker regions on the planet, and holding that they were really bodies of water, he found that they were connected by comparatively narrow streaks. In accordance with the adopted system of nomenclature, he termed these streaks "canale," a word of which the proper rendering into English would be channels. But the word was actually translated into both English and French as "canal," thus connoting artificiality in the supposed waterways, which were attributed to the inhabitants of the planet. The fact that they were many miles in breadth, and that it was absurd to call them canals, did not prevent this term from being so extensively used that it is now scarcely possible to do away with it. . . .

A second series of observations was made by Schiaparelli at the opposition (or when Mars was 180 degrees from the sun) of 1879, when the planet was further away, but was better situated as to altitude above the horizon. Schiaparelli now found a number of additional channels, which were much finer than those he had previously drawn. The great interest attaching to their seemingly artificial character gave an impetus to telescopic study of the planet which has continued to the present time. New canals were added, especially at the Lowell Observatory, until the entire number listed in 1908 amounted to more than 585. The general character of this complex system of lines is described by Lowell as a network covering the whole face of the planet, light and dark regions alike, and connecting at either end with the respective polar caps there. . . .

At their juncture are small dark pinheads of spots. The lines vary in size between themselves, but each maintains its own width throughout. But the more difficult of these objects are only seen occasionally and are variable in definiteness. One remarkable feature of these objects is their occasional "gemination," some of the canals appearing as if doubled. "This was first noted by Schiaparelli, and has been confirmed," asserts Professor Newcomb, "so far as observations can confirm it."

And now Mr. Antoniadi pronounces the geometrical lines, and also the doubling of the lines, mere optical illusions, and presents a number of his own drawings side by side with those made by Schiaparelli and Lowell in support of his contention. He notes that the markings which Schiaparelli only glimpsed with his modest 8½-inch refractor were held quite steadily in the 32¾-inch refractor at Meudon. . . .

"Now, if we consider," says Mr. Antoniadi,

"(a) That Mars was defined in 1909 with the Meudon 32¾-in., as if he were from four to sixteen times nearer the earth than in Schiaparelli's modest 8½-in., refractor;

"(b) That all the irregularities of the above markings were held quite steadily in the large telescope while the straight lines were only glimpsed at Milan;

"(c) That these irregular markings obey rigorously the laws of perspective, while this is too frequently not the case with the lines;

"(d) That a real narrow, dark, planetary line, like Cassini's division in front of Saturn, is naturally, owing to reduced diffraction, much broader and far more conspicuous in a large telescope than a small one;

"(e) That minute irregular detail, utterly beyond the reach of Schiaparelli's instruments, and confined by photography, was held steadily in the 32¾-in., when no trace of lines could be seen at all,

"We reach the conclusion, which leaves no room for doubt, that the natural appearances revealed by the great French telescope give a much more truthful representation of the details on the planet than the rude spider's webs of the Italian astronomer." . . .

—January 23, 1916

Viking Robot Sets Down Safely on Mars and Sends Back Pictures of Rocky Plain

By JOHN NOBLE WILFORD

An explorer from Earth, the robot craft *Viking 1*, made the first successful landing on Mars today and transmitted spectacular photographs of a rocky, wind-scoured desert plain, the site for the first direct search for life on another world.

The squat, three-legged *Viking* landing craft came to rest, upright and intact, on the Chryse Plain of Mars at 7:53 a.m. Eastern daylight time after a voyage of 11 months and nearly half a billion miles. The final and most suspenseful step, the craft's descent to the surface from its mother ship in Mars orbit, took 3 hours 13 minutes.

Then, Touchdown

Responding to automatic computer commands, the lander's rockets fired, its parachute unfurled, protective shielding broke away, more rockets were fired— and then, touchdown. It was 19 minutes, because of the great distance between Mars and Earth, now more than 212 million miles, before confirmation of the safe landing reached the control rooms here [in Pasadena, Calif.] at the Jet Propulsion Laboratory.

"Touchdown!" announced Richard Bender, one of the flight controllers. "We have touchdown. We have several indications of touchdown."

It was an emotional moment for the scientists and engineers of the $1 billion *Viking* project, many of whom had spent eight years preparing for this day.

Applause and Amazement

There was applause in the control room and throughout the laboratory. There were broad smiles and moist eyes. There were soft expressions of numbed amazement at what they had wrought.

With the *Viking* landing begins the first surface exploration of Mars (two Soviet landings failed to produce usable data). The planet has fascinated man for centuries and been the object of legend and endless scientific speculation.

In eight days, if all continues to go according to plan, a mechanical arm on the lander is to reach out and scoop up soil samples for chemical and biological analysis by onboard instruments. This will mark the beginning of the mission's search for signs of possible life on Mars.

Though Mars is no longer seriously thought of as an abode of "super civilizations," as it was as recently as the early 20th century, scientists generally believe that it, of all the planets in solar system, could be the most hospitable to some forms of life.

But in addition to its life-detection laboratory, the *Viking* lander has instruments to study Martian soil chemistry, weather and atmosphere, seismic activity and geology. As such, it is the most sophisticated vehicle ever dispatched to a neighboring world.

At a news conference after the landing and the transmission of two clear black-and-white photographs of the spacecraft and its surroundings, James S. Martin Jr., the project manager, said: "This has got to be the happiest time of my life. It's incredible to me that it all worked so perfectly."

Of the *Viking*'s first two pictures, Dr. Thomas A. Mutch of Brown University, the geologist in charge of the imaging team, spoke with unabashed excitement, "It just couldn't be better."

The photographs disclosed that the 1,300-pound lander had settled on a plain strewn with rocks. Some were dark, others light. Many of them had sharp edges as if they had undergone very little erosion, but others were more rounded. The geologists on the Viking team refrained from giving more detailed interpretations of what they were seeing.

Sand and Craters

In a 300-degree panorama of *Viking*'s strange new surroundings there could be seen what appeared to be sand dunes, a depression that might be an eroded crater, bright patches of rocks partly buried by drifting sand, some low ridges and, out on the horizon about two miles away, a ridge that scientists said could be the rim of a large-impact crater.

Dr. Carl Sagan of Cornell University, an astronomer on the imaging team, remarked, "Even in a place chosen for its blandness, for safety reasons, Mars is an extraordinarily interesting place."

To Alan B. Binder of Science Applications, Inc., another member of the team, the landing region seemed "very reminiscent of fairly heavily eroded lava fields one sees in Arizona and northern Mexico."

Other scientists tended to agree, after hurried examination of the pictures, that the primary material in the area was most likely volcanic in origin, but that it had been modified by meteoric impacts, the winds of the thin Martian atmosphere and possibly water erosion.

A Bright Sky

Project scientists were surprised by the brightness of the sky. It was late afternoon on a midsummer day at the landing site. (The Martian day is 24 hours 37 minutes long, and the Martian year lasts 25 months.) A shadow in one picture suggested that a cloud, formed by the small amounts of water vapor known to be in the Martian atmosphere, might have passed over the lander. In the distant sky there were several horizontal layers of wispy clouds.

The photographs were taken by one of the lander's two identical cameras, beginning within a minute after the touchdown. They were transmitted in the form of electronic data to the orbiting vehicle, more than 12,000 miles above the Martian surface, and then relayed to a tracking antenna on Earth.

The first picture began taking shape on television screens in the control center at 8:54 a.m. It took 22 minutes for the picture to be completed, line by vertical scan line. The photograph covered a 50-degree-wide area, as close as two feet from the spacecraft at the bottom and six feet away at the top of the screen.

Selecting Site

Between now and the time when the mechanical arm on the lander scoops up soil samples, Viking is to transmit several more pictures, including some in color and others in stereoscopic form. These will be used by scientists in selecting a promising place for the scoop to collect its samples.

The first photograph also showed that the lander's foot pad had made little or no penetration in the firm soil. Other data indicated that it had set down gently at a velocity of 5.5 miles an hour, as planned, and that all systems aboard appeared to be functioning normally.

The exact location of the landing site has yet to be determined. But A. Thomas Young, the mission director, said that since all steps in the descent sequence appear to have gone precisely according to plan, it is assumed that the vehicle put down close to the center of its target area, an ellipse 62 miles wide and

130 miles long. It is on the western slopes of Chryse Planitia, the "gold plain" of Mars, at 22.4 degrees north latitude and 47.5 degrees west longitude.

The *Viking 1* spacecraft—the combined orbiter and lander vehicles—had gone into orbit of the planet on June 19 and spent the last four weeks photographing the surface in search of a relatively flat and safe place to land. . . .

It took 36 minutes to complete the panoramic view, a nearly complete circumference of the landing area. Often the view reduced Dr. Mutch to such unscientific expressions as "just lovely, just lovely" and "this is just an incredible scene." At the sight of one field of boulders, he said with a grin, "As we sophisticated geologists say, those are dark rocks."

Pictures Are Analyzed

A more detailed analysis of the photographs is already under way by geologists, physicists, meteorologists, biologists and astronomers. The quality of the pictures is also being enhanced by computer processing of the returned electronic signals.

During the transmission, President Ford telephoned from the White House to congratulate the *Viking* team. He spoke to Dr. James C. Fletcher, the administrator of the National Aeronautics and Space Administration, and Mr. Martin, both of whom were in the control room.

Calling it "a job well done," Mr. Ford reminded them that their success came on the seventh anniversary of the Apollo 11 landing on the moon, the first time men set foot on another world. The president had declared July 20, 1976, Space Exploration Day.

As soon as the president hung up, Dr. Mutch continued his first-impression commentary on the incoming photographs. He noted that shadows cast by rocks on Mars were softer than those on the moon, which he attributed to the diffusion of sunlight by the thin Martian atmosphere; the moon is airless.

At one point, his excitement rising, Dr. Mutch remarked, "You just wish you could be standing there, walking across that terrain."

Several other scientists said that the first pictures were much sharper and clearer than they had expected. They had also feared that they would lose radio contact with the lander before the completion of the panorama.

But the orbiting vehicle remained in range of Earth just long enough to complete the transmission before passing behind Mars. . . .

At a news conference later, project scientists said that they had seen nothing in the photographs suggesting the presence of any forms of life, but they had not really expected any visible signs. Nor did they see any readily identifiable evidence of water, on or below the surface, having once eroded the area.

Orbital photographs of the regions nearby strongly suggested that floods of water once caused deep channels in the surface and left islands shaped like teardrops. The landing site was thought to be an area where sediment from those ancient floods could have been deposited. And instruments on the orbiting craft had detected significant amounts of water vapor in the atmosphere over the low-lying Chryse region.

The principal constituent of the Martian atmosphere, which is only 1 percent as dense as the Earth's, is carbon dioxide. Small amounts of argon, a rare inert gas, were detected in the atmosphere during the lander's descent. More substantial amounts of argon had been predicted by earlier Soviet findings and by *Viking* orbital observations.

The presence of sharp-edged rocks raised some scientific eyebrows, for it indicated that erosional forces (either wind or water) may not be very recent or strong. Dr. Sagan suggested three possible interpretations—either the rocks were recently fractured by some catastrophic event, or they were formed long ago but buried early and recently exhumed, or the erosional processes "are relatively feeble in this area."

Winds of more than 200 miles an hour have been observed on Mars, creating global dust storms. But with the rarefied atmosphere, a 200-mile wind on Mars would be comparable to a 20-mile wind on earth.

In many ways, Dr. Mutch reflected, the most interesting object in the photographs was the lander itself, bristling with booms and antennas, its three legs solidly planted on the plain.

"When you go to an unknown place you look around for a friend," he commented, "and the lander is a good friend, a very good friend."

Its descent to the surface began when it separated from the orbiting vehicle at 4:40 a.m. The "go" command for separation had been radioed to the spacecraft computer a couple of hours beforehand. The command was a single coded word KVUGNG.

From then on, the spacecraft was on its own, operated by commands that had been programmed in its computer over the weekend.

Pyrotechnic devices were fired to release the lander from the orbiter. Three sets of springs gently pushed it away at a rate of one foot a second. The firing of eight small rockets on the edge of the lander's aeroshell, its dome-shaped protective covering, oriented the vehicle and braked it so that Martian gravity could pull it downward out of orbit and toward the surface.

Facing Heat Shield

After a long coasting period and until just before reaching the upper fringes of the Martian atmosphere, the lander was reoriented so that its heat shield faced the angle of attack. The shield's corklike coating material was designed to protect against entry heat of up to 2,700 degrees Fahrenheit.

Atmospheric drag and lift further slowed down the craft's descent and kept it from plunging in too fast or too steeply. At an altitude of 19,400 feet, a 53-foot-wide parachute deployed, braking the descent to about 100 miles an hour. At 4,600 feet, the lander's three throttable rockets fired to ease the vehicle to a soft landing.

For the first 12 seconds it was on the surface the lander radioed engineering instructions—"how-am-I" data, as the flight controllers call it—to report on its condition. There were no anomalies, Mr. Young, the mission director, reported.

Soviet Failures

Three attempts by Soviet spacecraft to land on Mars had not been so fortunate. One crashed on landing, another ceased communication after only 20 seconds on the surface, and the third failed during the descent.

Viking 2, identical to *Viking 1* in every way, is scheduled to enter Mars' orbit on Aug. 7 and go for a second landing, possibly as early as Sept. 4. It may be aimed to land in the more rugged and scientifically intriguing northern latitudes below the Martian polar cap.

But for the moment all attention here at control center was on *Viking 1* and the first pictures from the surface of Mars.

Dr. Noel W. Hinners, NASA's associate administrator for space science, admitted: "I had tears in my eyes for the first time since—well, I guess, since I got married. It's fantastic."

—July 21, 1976

Pluto's Not a Planet?
Only in New York

By KENNETH CHANG

As she walked past a display of photos of planets at the Rose Center for Earth and Space, Pamela Curtice of Atlanta scrunched her brow, perplexed. There didn't seem to be enough planets.

She started counting on her fingers, trying to remember the mnemonic her son had learned in school years ago.

> My Very Educated Mother Just Served Us Nine Pizzas.
> Mercury, Venus, Earth, Mars, Jupiter, Saturn, Uranus, Neptune.

"I had to go through the whole thing to figure out which one was missing," she said.

Pluto.

Pluto was not there.

"Now I know my mother just served us nine," Mrs. Curtice said. "Nine nothings."

Quietly, and apparently uniquely among major scientific institutions, the American Museum of Natural History cast Pluto out of the pantheon of planets when it opened the Rose Center last February. Nowhere does the center describe Pluto as a planet, but nowhere do its exhibits declare "Pluto is not a planet," either.

"We're not that confrontational about it," said Dr. Neil deGrasse Tyson, director of the museum's Hayden Planetarium. "You actually have to pay attention to make note of this."

Still, the move is surprising, because the museum appears to have unilaterally demoted Pluto, reassigning it as one of more than 300 icy bodies orbiting beyond Neptune, in a region called the Kuiper Belt (pronounced KY-per).

"Pluto is noticeable by its difficulty to find," Dr. Richard P. Binzel, a professor of planetary science at the Massachusetts Institute of Technology, said of the Rose exhibits. "They went too far in demoting Pluto, way beyond what the mainstream astronomers think."

Dr. S. Alan Stern, director of Southwest Research Institute's space studies department in Boulder, Colo., also dislikes the change. "They are a minority viewpoint," he said. "It's absurd. The astronomical community has settled this issue. There is no issue."

The International Astronomical Union, the pre-eminent society of astronomers, still calls Pluto a planet, one of nine of the solar system. Even a proposal in 1999 to list Pluto as both a planet and a member of the Kuiper Belt drew fierce protest from people who felt that the additional "minor planet" designation would diminish Pluto's stature.

The proposal was abandoned, and the astronomical union, which is based in Paris, released a statement reaffirming that Pluto was and is a planet. "This process was explicitly designed to not change Pluto's status as a planet," the organization said.

But even some astronomers defending Pluto admit that were it discovered today, it might not be awarded planethood, because it is so small—only about 1,400 miles wide—and so different from the other planets.

While the international union's debate stirred considerable astronomical passion, the Rose Center's Plutoless planet display has not generated much controversy or consternation.

"I learned it one way for the first 50 years," said Mrs. Curtice's husband, William. "I'll learn it another way now, I guess."

Jane Levenson, an "explainer" at the center, says that perhaps one out of every 10 visitors asks her about the missing planet. She tells them about the debate over Pluto's status and says "a decision had to be made" as the museum was assembling the new exhibits.

"Children in particular ask," she said. "Children say, 'Did they forget about Pluto?' Some even say, 'Did you forget my friend Pluto?'"

Ilisse Familia, a sixth grader from the Good Shepherd School in Manhattan, was surprised when she heard the museum no longer counted Pluto among the planets. "No wonder I couldn't find Pluto," she said. "It's kind of weird."

As a planet, Pluto has always been an oddball. Its composition is like a comet's. Its elliptical orbit is tilted 17 degrees from the orbits of the other planets. Pluto was discovered on Feb. 18, 1930, by Clyde W. Tombaugh, and astronomers initially estimated it to be as large as Earth. They have since learned it is much smaller, smaller than Earth's Moon.

But Pluto continued to be called a planet, because there was nothing else to call it. Then, in 1992, astronomers found the first Kuiper Belt object. Now they have found hundreds of additional chunks of rock and ice beyond Neptune, including about 70 that share orbits similar to Pluto's, the so-called Plutinos.

"We're much more subtle, but not deviously subtle," Dr. Tyson, the planetarium director, said of the Hayden exhibits. "We decided to organize the information for the visitor in such a way that Pluto's classification would become self-evident."

The exhibits refer to the inner four "terrestrial planets"—Mercury, Venus, Earth, Mars—and the four gas giant planets—Jupiter, Saturn, Uranus and Neptune. Pluto, a small ball of rock and ice, does not fall into either group. "Pluto does not have a family except for the icy bodies in the outer solar system," Dr. Tyson said. "So we simply group it with the Kuiper Belt. In a sense, we're sidestepping the definitional problem altogether."

A display describing the solar system includes this carefully worded sentence: "Beyond the outer planets is the Kuiper Belt of comets, a disk of small, icy worlds including Pluto."

A diagram of the planets shows eight, not nine, rings around the Sun.

Other planetariums have not followed the Rose Center's lead. The entryway to the Adler Planetarium in Chicago includes bronze plaques of only eight planets, but that is because it opened just before Pluto was officially named. Inside, the exhibits include Pluto among the planets.

The Denver Museum of Nature and Science is building a $45 million, 30,000-square-foot space science center, scheduled to open in 2003. Those exhibits will also still count nine planets in the solar system. "We're sticking with Pluto," said Dr. Laura Danly, curator of space sciences at the Denver museum. "We like Pluto as a planet."

But, she also said, "I think there is no right or wrong on this issue. It's a moving target right now, no pun intended, what is and is not a planet."

Planet, in the original Greek word, meant "wanderer," referring to the dots of light that moved across the night sky. When the 16th-century astronomer Nicolaus Copernicus realized that the universe did not revolve around the Earth, Earth became another planet circling the Sun.

For Dr. Tyson, the redefining of Pluto has historical precedent. In 1801, astronomers combing the large gap between Mars and Jupiter discovered Ceres, and for a short while, Ceres was a planet. Then another large rock was found in the same region. And another. Soon it became apparent there was a ring of rocky bodies between Mars and Jupiter. Since astronomers did not want to call all of them planets, they renamed them asteroids.

Just as Ceres, which turned out to be about 580 miles wide, was reassigned from planet to asteroid, Pluto should join the Kuiper Belt objects, Dr. Tyson said. "It's entirely analogous to the asteroid belt," he said, "except there's a 60-year delay between the discovery of the first and second objects."

The new view of Pluto would recast it "from puniest planet to king of the Kuiper Belt," Dr. Tyson said. "And I think it's happier that way. I'm convinced our approach will prevail. It makes too much scientific sense and too much pedagogical sense."

—January 22, 2001

Curiosity Rover's Quest for Clues on Mars

By KENNETH CHANG

More than 3.5 billion years ago, a meteor slammed into Mars near its equator, carving a 96-mile depression now known as Gale Crater.

That was unremarkable. Back then, Mars, Earth and other bodies in the inner solar system were regularly pummeled by space rocks, leaving crater scars large and small.

What was remarkable was what happened after the impact.

Even though planetary scientists disagree on exactly what that was, they can clearly see the result: a mountain rising more than three miles from the floor of Gale.

More remarkable still, the mountain is layer upon layer of sedimentary rock.

The layered rock drew the attention of the scientists who chose Gale as the destination for NASA's *Curiosity* rover, a mobile laboratory the size of a Mini Cooper.

Now, more than two years after arriving on Mars, *Curiosity* is climbing the mountain.

In sedimentary rock, each layer encases the geological conditions of the time it formed, each a page from the book of Mars' history. As *Curiosity* traverses the layers, scientists working on the $2.5 billion mission hope to read the story of how young Mars, apparently once much warmer and wetter, turned dry and cold in what John P. Grotzinger, the project scientist, calls "the great desiccation event."

Dr. Grotzinger remembers the first time he heard about Gale. "I looked at it, and immediately I'm like, 'This is a fantastic site,'" he said. "What's that mountain in the middle?"

Officially, the name is Aeolis Mons, but mission scientists call it Mount Sharp in homage to Robert P. Sharp, a prominent geologist and Mars expert at the California Institute of Technology who died in 2004.

On Earth, mountains rise out of volcanic eruptions or are pushed upward by plate tectonics, the collision of pieces of the planet's crust.

Mars lacks plate tectonics, and volcanoes do not spew out of sedimentary rock. So how did this 18,000-foot mountain form?

In the late 1990s, NASA's *Mars Global Surveyor* spacecraft was sending back images of the Martian surface far sharper than those from earlier missions, like *Mariner* and *Viking*.

Kenneth S. Edgett and Michael C. Malin of Malin Space Science Systems, the San Diego company that built Global Surveyor's camera, saw fine layered deposits at many places on Mars, including Gale. In 2000, they offered the hypothesis that they were sedimentary, cemented into rock.

Indeed, Dr. Edgett said, it appeared that Gale Crater had been fully buried with sediment and that later winds excavated most of it, leaving the mountain in the middle.

Imagine carving out of an expanse as large as 1.5 Delawares—a mound as tall, from base to peak, as Mount McKinley in Alaska, the tallest mountain in North America at 20,237 feet.

Dr. Edgett asserts that that is plausible on Mars. He points to other Martian craters of similar size that remain partly buried. "There are places where this did happen, so it's not ridiculous to think this is what happened at Gale," he said.

Still, in 2007 Gale had been discarded from the list of potential landing sites for *Curiosity*, because observations from orbit did not show strong evidence for water-bearing minerals in the rocks. NASA's Mars mantra for the past two decades has been "Follow the water," because water is an essential ingredient for life.

Dr. Grotzinger asked Ralph E. Milliken, then a postdoc in his research group at Caltech, to take a closer look at Gale. With data from an instrument on NASA's Mars Reconnaissance Orbiter that can identify minerals in the rocks below, Dr. Milliken showed the presence of clays at the base of Mount Sharp as well as other minerals that most likely formed in the presence of water.

"The fact we have this mountain, and it's not all the same stuff—the mineralogy is changing from one layer to the next—that gives us the hope that maybe those minerals are recording the interaction of the water and the atmosphere and the rocks," said Dr. Milliken, now a geologist at Brown.

Were water conditions there becoming more acidic? Was there oxygen in the water? "That's something we can assess with the rover on the ground," Dr. Milliken said.

Since its landing on Mars in August 2012, *Curiosity* took a detour to explore a section named Yellowknife Bay and discovered geological signs that Gale was once habitable, perhaps a freshwater lake.

After that, the rover drove to Mount Sharp, with only brief stops for science. To date, the rover, operated by NASA's Jet Propulsion Laboratory in Pasadena, Calif., has driven more than six miles, taken more than 104,000 pictures and fired more than 188,000 shots from a laser instrument that vaporizes rock and dirt to identify what they are made of.

In September, *Curiosity* drilled its first hole in an outcrop of Mount Sharp and identified the iron mineral hematite in a rock. That was the first confirmation on the ground for a Gale mineral that had been first identified from orbit.

When *Curiosity* reaches rocks containing clays, which form in waters with a neutral pH, that will be the most promising place to look for organic molecules, the carbon compounds that could serve as the building blocks of life, particularly if the rover can maneuver into a spot shielded from radiation. (It does not have instruments that directly test for life, past or present.)

The orbiter also detected magnesium sulfate salts, which Dr. Milliken described as possibly similar to Epsom salts.

That layer appears to be roughly as old as sulfates that NASA's older *Opportunity* rover discovered on the other side of Mars. If Mount Sharp sulfates turn out to be the same, that could reflect global changes in the Martian climate. Or they could be different, suggesting broad regional variations in Martian conditions.

"We're finally beginning the scientific exploration of Mount Sharp," Dr. Milliken said. "That was the goal."

Along the way, *Curiosity* may also turn up clues to the origins of Mount Sharp. While Dr. Edgett thinks Gale Crater filled to the brim before winds excavated the mountain, others, like Edwin S. Kite, a postdoctoral researcher at Princeton who is moving to the University of Chicago as a professor, think the mountain formed as a mound, with winds blowing layers of sand together that then were cemented by transient water. "Can you build up a pile like that without necessarily filling up the whole bowl with water?" Dr. Kite said. "Perhaps just a little bit of snow melt as the pile grows up."

He said the layers of Mount Sharp dip outward at the edges, as in an accumulating mound; they are not flat, as would be expected if they were lake sediments subsequently eroded by wind.

Dr. Grotzinger thinks that both could have happened: that Gale Crater partly filled, then emptied to form the lower half of Mount Sharp, and a different process formed the upper portion. A sharp divide between the upper and lower parts of the mountain is suggestive.

On Monday, during a NASA telephone news conference, Dr. Grotzinger and other members of the science team described new data suggesting long-lived lakes in the crater. The deposits at Yellowknife Bay could have been part of an ancient lake filled by streams flowing from the crater rim. As *Curiosity* drove toward Mount Sharp, it appeared to be traveling down a stack of accumulated deltas— angled layers where river sediment emptied into a standing body of water—and yet it was heading uphill. That pattern could have occurred if the water level were rising over time, and Mount Sharp was not there yet.

That does not mean Gale was continually filled with water, but it suggests repeated wet episodes. "We don't imagine that this environment was a single lake that stood for millions of years," Dr. Grotzinger said, "but rather a system of alluvial fans, deltas and lakes and dry deserts that alternated probably for millions if not tens of millions of years as a connected system."

Ashwin Vasavada, the deputy project scientist, said that to explain the episodes of a lake-filled Gale crater, "the climate system must have been loaded with water."

But answers will remain elusive. "We're not going to solve this one with the rover," Dr. Edgett said. "We're not going to solve this one with our orbiter data. We're going to be scratching our heads a hundred years from now. Unless we could send some people there."

As successful as the NASA Mars rovers have been, their work is limited and slow. *Curiosity*'s top speed is not quite a tenth of a mile per hour. What might be obvious at a glance to a human geologist, who can quickly crack open a rock to peer at the minerals inside, could take days or weeks of examination by *Curiosity*.

"I'd like to think it would take only a few months," Dr. Edgett said of solving Mount Sharp's mysteries, "with a few people on the ground."

—December 8, 2014

THE UNIVERSE AND BEYOND

Universe Multiplied a Thousand Times by Harvard Astronomer's Calculations

Dr. Harlow Shapley, an astronomer who recently came to Harvard from the Mount Wilson Observatory in California, announces that he has made certain discoveries that reveal the universe to be a thousand times greater than scientists have conceived it.

By so doing he has relegated the earth to a place one thousand times less important than it has heretofore occupied, and instead of its being in the "centre of things," as has been understood heretofore, he now estimates it to be something like 360,000,000,000,000,000 miles from the centre of the universe.

By triangulation, taking the distance between sun and Earth as a base for measurements, vast distances in the past have been recorded until such lines have been extended hundreds of light-years, and even to the limit of measurement, the Pleiades.

These same Pleiades are scarcely in the front yard of Dr. Shapley's galaxy, which he has measured and found to be in length about 300,000 light-years from end to end. It is a super-Milky Way.

A light-year, the distance a beam of light will travel in one year, is 6,000,000,000,000 miles. It takes but eight minutes for light to come to the earth from the sun, 93,000,000 miles away.

The young astronomer has proved to his satisfaction by various calculations that the sun, the little speck of light around which a tiny shadow called the earth revolves, is 60,000 light-years from the centre of the universe.

"Personally I am glad to see man sink into such physical nothingness," says the scientist, "and it is wholesome for human beings to realize of what small importance they are in comparison with the universe."

At the Mount Wilson Observatory Dr. Shapley spent a great deal of time studying globular clusters of stars, and principally one called "Messier 13," which is 36,000 light-years from the earth. Small stars in this group are 100 times as large as our sun, while the larger and more brilliant he described as 1,000 times as large.

—May 31, 1921

What Man's New 200-Inch Eye Will See

By SIR JAMES H. JEANS

Night after night, for hundreds of thousands of years, primitive man watched the stars rising in the east, circling round the Pole and setting in the west, providing evidence, had he been able to read it aright, of the existence of a universe incomparably grander than the earth on which he dwelt. Yet until quite recently he had no means of studying this universe.

A man's eye admits only so much light as falls on a circle a fifth of an inch in diameter. He will be unable to see an object at all unless this amount of light produces a distinct impression on his brain. There are something like 30,000 million stars surrounding the sun, but our unaided eyes can see at most 6,000 of these: we are too blind to see more than one star in 5,000,000.

Galileo's Telescope

A deaf man can improve his hearing by collecting waves of sound in an ear-trumpet and projecting them into his eardrum. In the same way the seeing power of the human eye can be increased by collecting the rays of light which fall on a large area and bending them so that they all pass through the pupil of his eye onto his retina. This is the principle of the telescope. Roger Bacon propounded it in the thirteenth century, explaining how a lens of glass could be shaped "to make the stars appear as near as we please." Yet not until 1608 was the first working telescope constructed, probably by Lippershey, a spectacle-maker of Magdeburg.

The next year Galileo constructed the tiny instrument whose discoveries were to revolutionize human thought. This had an aperture of two and one-quarter inches, which means that it collected all the rays of light falling on a circle of two and one-quarter inches diameter and bent them so that they fell onto the retina of the observer. This increased the amount of light falling on the retina a hundredfold.

The results of thus increasing the capacity of the human eye were almost indescribable. Sixty-six years earlier Copernicus had shown how the complicated tracks of the planets in the sky could be explained by supposing that they, together with the earth—the majestic abode of man and the supposed center of the universe—circled around the sun like moths around a candle flame. The idea was not new: Aristarchus of Samos had advanced it five centuries before Christ, as had many others in the intervening 2,000 years. Copernicus had merely produced powerful new arguments in support of an old conception. But this conception was so devastating in its implications, and so difficult to reconcile with the superficial appearance of things, that it met with well-nigh universal condemnation.

Then Galileo, turning his telescope onto Jupiter, saw four small bodies circling around a big central body—a perfect model of the solar system as described by Copernicus. Turning it onto Venus, he observed "phases" like those of the moon, the appearance of the planet varying from a crescent to a full circle. This was of overwhelming significance, since the supposed absence of such phases had been repeatedly urged as proof that the earth and Venus could not revolve around the sun.

Even these discoveries did not satisfy the skeptics; too much, including human self-esteem, was at stake, and, many years later, the old Ptolemaic doctrine of sun and planets revolving about a fixed central earth was being taught both at Harvard and at Yale, concurrently with the new ideas. Nevertheless, of all the many factors which ultimately dethroned man from his self-arrogated position at the center of the universe, the most potent was probably the little two-and-one-quarter-inch telescope of Galileo.

The Work of Herschel

Every subsequent increase of telescopic power revealed new worlds to conquer and provided the means of conquest. A landmark in the advance was Sir William Herschel's construction, in 1789, of a telescope with an aperture of four feet. This gathered in 500 times as much light as Galileo's telescope and 50,000 times as much as the unaided eye. What Galileo's telescope had done for the solar system, this did for the huge family of stars—the "galactic" system, bounded by the Milky Way—to which our sun belongs. Man, the explorer of the universe, had left the solar system far behind and was now voyaging in depths of space so profound that light took thousands of years to travel from them to earth; he was viewing objects not as they then were, but as they had been before the birth of Christ, before the fall of Troy, or before civilization had come to mankind at all.

The Spiral Nebulae

And just as Galileo's telescope had established the Copernican structure of the solar system by discovering the similar systems of Jupiter and Saturn, so Herschel confirmed his own view of the structure of the galactic system by discovering other systems of similar shape—"island universes," he called them. In 1845 Lord Rosse's great telescope of six-foot aperture revealed the spiral structure of many of these systems; we now call them "spiral nebulae."

The culmination of power was reached in 1920 in the present 100-inch telescope at Mount Wilson. By now the retina of the human eye, which can only retain light for a fraction of a second, had been practically superseded by the photographic plate which records cumulatively all the light received for hours or even for successive nights. The 100-inch telescope not only gathers in 250,000 times as much light as the human eye, but projects it onto the far more sensitive photographic plate.

An example of its tremendous power is provided by Dr. Hubble's study, with its aid, of the faintest of spiral nebulae. He finds that these are faint merely because of their great distance; they are so remote that light, traveling 186,000 miles every second, takes 340,000,000 years to travel from them to us. We see them not as they are, but as they were inconceivable ages before man appeared on earth. A wireless signal will girdle the earth in a seventh of a second; or we could send it out into space and receive a reply from Mars in a few minutes if there were any one to dispatch it to us, but we could not conceivably get a reply from these nebulae in less than 280,000,000 years. By the time our remote descendants received it, mankind would be a thousand times as old as now, and we who had sent the message out would be regarded as the earliest forefathers of the race. Such are the depths of space which the 100-inch telescope brings within our purview. Even the nearest of these nebulae are nearly 1,000,000 years away, but their size is so gigantic that even at this distance they show a great wealth of detail. The 100-inch telescope resolves their outer regions into distinct spots of light which we recognize as stars because many of them exhibit the special characteristics of stars much nearer home.

And now comes the news that a new telescope of 200-inch aperture is to be erected near Pasadena on a height not far from Mount Wilson. In aperture this represents as great an advance over the 100-inch telescope of 1920 as the latter did over Herschel's telescope of 130 years earlier. Just as Herschel might have wondered what new discoveries a doubling of his aperture might produce, so we may wonder what will result from the doubling of the aperture which is actually about to take place. The new telescope will gather in four times as much light as the old, with the result that objects will be visible up to double the distance at which similar objects are now visible.

Outside the sphere which the old telescope has explored lies a sphere of double radius awaiting exploration by the new. What will it contain? And what shall we find in the sphere which has already been explored when it is searched afresh with a fourfold more powerful instrument?

The 100-inch telescope hardly shows any type of object that Herschel had not seen through his 4-foot telescope. The universe appears to be uniform, and the traveler in far-off regions does not encounter new species of objects, but merely further examples of the species which abound nearer home. For example, the 100-inch telescope shows about 2,000,000 nebulae, so that the new telescope ought to show about 16,000,000, but the new 14,000,000 will in all probability prove very similar to the old 2,000,000. Yet this mere multiplication of specimens is not altogether valueless. If for no other reason, the new museum of nebulae is likely to contain eight times as many freaks and oddities as the old, and a few of these are often of more service than innumerable examples of normal formations.

To mention one instance, the solar system is now thought to have come into being as the result of a very unusual occurrence. In their blind motion through space two stars, so it is supposed, chanced to pass so close to one another that gigantic tides were raised, tides so high that the crests of the tidal waves broke off like drops of spray and subsequently condensed into planets. We may never hope to see such an event in progress; it is far too rare and in any case the stars appear far too small for us to be able to study it in detail even if it occurred. But the giant nebulae ought at times to exhibit the same phenomenon on a far grander scale. Indeed, the 100-inch telescope discloses five or six objects which may be examples of this process. The new telescope ought to show about fifty such, and with fifty specimens we may begin a systematic study, and hope really to learn something. Thus, although it is beyond all bounds of probability that we shall ever witness the birth of planets similar to our earth, a study of nebulae in the new telescope may disclose the general mechanism of its occurrence.

Let us pass to the other end of the scale. The star Proxima Centauri, our nearest neighbor, so far as we at present know, in the whole sky, is a quite recent discovery. It emits only a ten-thousandth part of the light of the sun and is too faint to be seen at all in a small telescope. Still fainter stars exist, but the vast majority of these are too distant to be seen by any telescope on earth. Even the new 200-inch telescope will not show many, but it ought to show enough to make a systematic study possible. So far we have only been able to study the bright stars; a whole universe of faint stars awaits the new telescope.

Stars of Powdered Atoms

It may be asked why we should trouble about those dying cinders. One reason is that the substance of some, at least, of them is in a state which we cannot reproduce on earth. A mass of sugar can be packed into far less space by powdering it; after the process it contains just as many atoms as before, but they fit closer together. But we have hardly yet started the true powdering process. The atoms have so far remained intact, each consisting of a positive charge of electricity holding a lot of negative charges at arm's length. Powder the atoms themselves and there is hardly any limit to the closeness with which you can pack the constituents; you could pack the Statue of Liberty in a teacup.

Now some of the very faint stars consist of atoms in this completely powdered state; at the center of "Van Maanen's star" a mass equal to the Statue of Liberty in weight probably occupies rather less space than a teacup. A systematic study of matter in this state may be expected to throw light on the ultimate structure of the atom. The powdering process just described can only occur at a temperature of hundreds of millions of degrees; so that the physicist must necessarily ask help of the astronomer for its study.

And yet the main service of the new telescope will probably be neither in the remote depths of space which it alone can reach, nor yet with the dim objects close at hand which it alone can discern. If a prophecy is permissible, it is likely to be in the vast middle reaches.

What Is Behind It All

As the 100-inch telescope discovered no new types of astronomical object, the 200-inch is not very likely to do so. But the main problem of present-day astronomy is not so much to discover new species of objects as to understand existing species; interpretation has become more important than discovery. It is reasonably certain that the known types form an evolutionary sequence; stars and other objects differ primarily because they are at different stages of their long journey from birth to death. And one great problem before astronomy is to trace out this journey—to string the beads in their right order, so to speak.

Man is no longer content to stare through a telescope as at a rare show; the dumb attitude of astonished wonder has passed and he is beginning to ask insistently what it all means. What is the unifying concept behind it all and what, in particular, is the relation of our puny planet and our ephemeral existence to the terrifying grandeur of the universe outside? The mathematicians have given the best answers they can, but the telescopic power so far available has been inadequate either to confirm or to disprove their answers. Even a little increase of power would be of great value; that to be embodied in the new instrument will be valuable beyond words.

In its main lines the answer to the problem is already beginning to emerge; it is, in brief, that the universe as a whole is transforming its substance into radiation. Matter, which appears to be indestructible on earth, is not so in the stars; there it is "burning up" so completely that neither ashes nor smoke are left—only radiation. Where two tons of matter existed in the past, only one remains today; the other ton has been annihilated, or rather is transformed into radiation which is destined to wander round and round space to the end of time. Recent investigations indicate that direct physical evidence of this annihilation of matter is probably to be found in the cosmic radiation which Professor Millikan and his colleagues at Pasadena are now studying. But we have to explain why, as matter transforms itself into radiation, it should assume in turn all the various forms and shapes disclosed by our telescopes. To this end we must study the interaction of the process of annihilation with others such as changes in the rate of spin, changes in temperature and density, and so on. At every step direct observational confirmation will be needed, and this the new telescope ought to provide—at any rate within limits.

Light on Man's Destiny

Yet, looked at from this point of view, it is clear that the new telescope does not mark the end of an era; is it not rather a beginning? For the first 99.9 percent of his life on earth mankind was without telescopes, and paid the price by falling into the crude error of imagining himself the central fact, as well as the central point, of the universe. Only in the last 300 years of his long life of 300,000 years, has he begun to open his eyes to the wide world around him; here, he now realizes, is material which must throw light on his beginnings, meaning and destiny. He has learned to mistrust such explanations as originate in his own inner consciousness or are created by his own desires.

The first necessary step toward obtaining a true answer is to collect all the objective information which science can provide. And if mankind is not prepared to accept anything less than the best and surest of such information, then clearly the services of astronomy to man are only at their outset. We have not yet even probed to the furthest limits of space. There are reasons for thinking that such limits exist, so that an increase in telescopic power will always carry us into new regions of space. At present our astronomic explorers may be discovering America, but they have not yet circumnavigated the globe. And even when they are armed with instruments capable of sweeping the whole of space their task will only be beginning; the great problems of interpretation will remain. At present no limit can be set either to the magnitude of the task or to the services which astronomy can render to man, services of which Galileo's exploration of the solar system figures as the prototype.

—January 6, 1929

If Theory Is Right, Most of Universe Is Still "Missing"

By WILLIAM J. BROAD

A cosmic mystery of immense proportions, once seemingly on the verge of solution, has deepened and left astronomers and astrophysicists more baffled than ever. The crux of the riddle is that the vast majority of the mass of the universe seems to be missing. Or, more accurately, it is invisible to the most powerful telescopes on earth or in the heavens, which simply cannot detect all the mass that ought to exist in even nearby galaxies.

A major explanation for the invisible mass—that it is made up of unseen subatomic particles known as neutrinos—has now largely been discredited, according to a recent article in the British journal *Nature*.

The increasingly unsettling news has caused a crisis among cosmologists, according to Dr. Jeremiah P. Ostriker, an astrophysicist at Princeton University. "The discrepancy between what was expected and what has been observed has grown over the years, and we're straining harder and harder to fill the gap," he said.

According to astrophysicists, calculations show that the sum of all the known dust, planets, comets, asteroids, stars, pulsars and quasars now accounts for about 1 percent of the matter that theory says ought to make up the universe—that is, unless there is a flaw in current understanding of the laws of nature.

"It remains one of our great, great problems," said Dr. Frank Wilczek, an astrophysicist at the Institute for Theoretical Physics at the University of California at Santa Barbara.

The depth of the mystery is illustrated by the odd suspects now being put forward to account for the invisible mass: Giant slush balls, swarms of black holes, cosmic rocks and exotic new types of hypothetical subatomic particles with names like photinos and winos. "Why now further strain credulity?" asked John Maddox, *Nature*'s editor, in a recent editorial. "Would it not be simpler if cosmologists abandoned their belief that there is a missing mass?"

The answer given by Dr. Maddox and other scientists is that the evidence for some kind of "dark matter" has been growing steadily for nearly a half century. Indeed, the case is now so compelling that scientists around the globe are redoubling their efforts to track it down.

Effects of Mysterious Powers

The first hint of the riddle came in the 1930s when an astronomer noticed that some galaxies within a dense galactic cluster were moving much faster than expected, suggesting that some kind of invisible force was at work. Decades later astronomers came up with an even more startling discovery: The stars on the outer rims of galaxies were rotating around galactic centers at speeds much faster than expected.

Both observations suggested the presence of invisible matter. The reason, according to Dr. Wilczek, can be seen by visualizing the same kind of discovery in our own solar system. It has been observed for centuries that the outer planets move more slowly around the Sun than the inner ones. This is because the strength of the Sun's gravitational pull increases as planets get closer to it.

If outer planets like Pluto had been observed to be orbiting much faster than they actually do, astronomers would have deduced that there was a great mass at work in addition to the Sun, exerting a powerful gravitational pull from somewhere beyond the edge of the solar system.

"In the same way," Dr. Wilczek said, "there seems to be a massive halo of dark matter surrounding every galaxy. What has been found consistently is that the mass of the universe measured gravitationally is always bigger than what you observe."

How much bigger? Based on calculations from gravitational tugs, there seems to be about 10 times more matter than has been observed. "The problem," Dr. Ostriker noted, "is not missing matter but missing light. Something is out there. We just don't know what it is."

Astrophysicists say the problem is even worse than that. Even the unseen matter implied by the powerful gravitational tugs brings the total known mass of the universe to only about 10 percent of what is believed, on the basis of theory, to be out there. More than 90 percent is still missing.

The Ever-Expanding Universe

Belief in this ghostly matter is based in part on philosophy, astrophysicists say. If there is not more matter, the universe will go on expanding forever, pushed ever outward by the residual power of the Big Bang—that moment in the primordial past when the universe was born in a hot explosion of dazzling brilliance. If, on the other hand, there is a critical amount of additional matter out there somewhere, the universe's overall gravitational pull will cause it to eventually slam back together.

According to Dr. David Schramm, an astrophysicist at the University of Chicago, new theories of cosmology are giving this view a firm basis in fact. Further, he said, the added "missing mass" must be something other than everyday matter.

This odd twist has been deduced by measuring the spectral lines in starlight and calculating the total amount of elements like deuterium in the universe, he said. Only a certain amount of it could have been synthesized during the Big Bang, and it would occur in a certain ratio to the rest of the normal matter in the universe.

The rest of the invisible mass, he said, is probably not in the form of baryons—the generic name for the protons and neutrons that make up ordinary matter. It is something different, something that might be wholly outside the realm of human experience.

Number of Suspects Grows

As the mystery of the invisible mass has grown over the years, so have the number of candidates put forward to account for it.

According to Dr. Ostriker, the first ones were low-mass stars, brown dwarfs and intergalactic dust and gas. The problem is that astronomers over the past decade have developed sophisticated ways to detect such low-luminosity matter, and so far not enough has been found to make a dent in the overall problem.

Another candidate is swarms of black holes—stars so dense that even light cannot escape their powerful gravitational pull. Dr. Ostriker said the search for these black holes is intensifying. Each black hole would need to have a few million solar masses, and there would need to be millions of them scattered around the edges of galaxies.

"We're looking for gravitational lens and other effects that might be caused by them," he said. "It's speculative, but in this realm everything is speculative."

The Case Against Neutrinos

The main candidate since the mid-1970s has been neutrinos, a ghostly subatomic particle that does not react with normal matter. Astrophysicists say neutrinos are appealing candidates precisely because they are not baryons—that is, not normal matter. Another reason for their allure is that unlike black holes or slush balls, they are clearly one of the main forms of matter in the universe. A single human body at any one instant holds a few hundred million of them, according to Dr. Wilczek.

The problem with neutrinos is that they have no mass—at least, they did not until Russian researchers in the mid-1970s reported that they did. That discovery immediately made them a key candidate for the missing mass. Unfortunately, no other researchers around the globe have been able to detect mass in neutrinos.

Further, according to the Aug. 23 issue of *Nature*, a universe in which neutrinos had mass would look quite different from the one we see about us. "It's probably the end of neutrinos in the standard cosmology," said Dr. Simon D. M. White, an author of the article, in a telephone interview from his office at the Steward Observatory at the University of Arizona. "We tried to see if you could fix things up, but it was extremely difficult to find anything that worked."

For instance, Dr. White and Dr. Piet Hut, his coauthor, found on the basis of computer simulations that if neutrinos had mass, all the galaxies in the universe would still be in the process of forming, rather than being largely complete.

Odd Lineup of Suspects

Thus, to date, all the major suspects have either been discredited or are still under investigation. Nothing so far fits the description for the invisible matter.

The search, according to Dr. Schramm, is thus increasingly turning to different types of hypothetical nonbaryonic particles with names like gravitinos, photinos and axions.

Teasing these odd particles into the light of day might take the construction of the biggest and costliest project in the history of pure science, a multibillion-dollar atom smasher that American physicists want to build in the 1990s. It would stretch 60 to 120 miles in circumference.

So too, new suspects will be sought in the heavens as astronomers during the next decade put powerful telescopes into orbit that will, for the first time, be able to peer to the edge of the universe. Astronomers say strange new candidates, more bizarre than black holes, might be discovered.

A final way out of the conundrum, scientists say, would be some kind of breakthrough in their understanding of the forces of universe, for instance, a new theory of gravity that would explain why stars on the rims of galaxies move so quickly. But that solution, they say, is the most radical of all. They would much rather discover the stars or particles that have so far eluded their grasp.

—September 11, 1984

In Key Repair, Telescope Gets Better Eyesight

By JOHN NOBLE WILFORD

T he myopic Hubble Space Telescope, which has had an embarrassingly out-of-focus view of the heavens for three and a half years, got two new sets of expensive prescription eyeglasses today. The repair work continued a string of such successes by shuttle astronauts and raised hopes of impressive astronomical discoveries in years to come.

Astronomers considered today's accomplishment the most important of the mission.

On Monday night and early today, in the mission's third of five planned space walks, astronauts of the shuttle *Endeavour* removed an old camera and inserted an upgraded 620-pound, $100 million replacement model with special mirrors to compensate for the telescope's flawed vision. This camera is considered so essential because it is responsible for half of Hubble's observations.

And late tonight, in the fourth space walk, another team of astronauts installed the other piece of equipment designed to clarify Hubble's vision for its other scientific instruments: a faint-object camera and two spectrographs.

Initial engineering tests this morning indicated that the new camera had been correctly installed; it was signaling that the shutter, sensing electronics and other systems were performing normally. But it should take at least six weeks of fine-tuning after the mission before the camera is ready for the ultimate picture-taking tests to see whether Hubble has been made clear-sighted.

With the replacements made today and the new solar-power panels and gyroscopes installed on the first two space walks, elated officials of the National Aeronautics and Space Administration said the *Endeavour* astronauts had ful-filled the minimum criteria for mission success, as defined before the launching last Thursday.

The electricity-generating solar panels and the gyroscopes for pointing the telescope at celestial targets were essential for Hubble's long-term operation. And the new camera should enable astronomers to study some of the most distant cosmic objects and determine the rate at which the universe is expanding.

At a news conference after the space walk early today, Joe Rothenberg, the Hubble project director at the Goddard Space Flight Center in Greenbelt, Md., said: "We have met the mission success criteria. But I don't think anyone would be satisfied to leave Hubble the way it is now."

Dr. Edward J. Weiler, the program's chief scientist, was even more cautious about claiming success until the crucial post-flight testing could be completed. "Today we took a great leap," he said. "We really have to wait six to eight weeks before we know if we have a total success."

Commenting on the crew's accomplishments, Col. Richard O. Covey of the Air Force, the shuttle commander, radioed to Mission Control, "We've been incredibly lucky so far."

The plan is for a final space walk on Thursday to complete the assigned tasks, including replacing electronics controls for the solar panels, installing an ultraviolet-light-detector switch and improvising a protective cap over the broken shell of a magnetic sensing unit. Astronauts found the break when they were inserting a replacement for this particular device, which had experienced erratic behavior.

Then the $1.6 billion telescope is to be released from the shuttle's cargo bay, back on its own and presumably fit for at least three or four more years, until the next scheduled visit by shuttle astronauts. The Hubble, with its modular construction affording easy access to most vital components, is the first spacecraft designed to be maintained by regular shuttle servicing missions over a lifetime of at least 15 years. All of its scientific instruments, and many of its operational components, were designed to be replaced with improved models or units for making different kinds of observations.

Flawed From the Start

The Hubble telescope is 43 feet long and 14 feet wide, about the size of a bus. Soon after it was launched in April 1990, astronomers discovered that its 94½-inch primary mirror was improperly ground and polished, leaving it with a focusing defect. Light collected by the mirror, instead of being focused into a sharp point, is spread over a larger area in a fuzzy halo. The result was blurry pictures and an inability to observe especially distant objects.

Since an upgraded version of the wide-field planetary camera was already being developed, engineers at the Jet Propulsion Laboratory in Pasadena, Calif., devised a way to insert four small precisely ground mirrors, about the size of dimes, that should remove the blur by intercepting and focusing incoming light collected by the primary mirror.

In addition, as part of its original design, the new camera has much more sensitive electronic detectors and should be able to make observations in ultraviolet light, as well as visible light.

"Look at That Baby"

Two space-walking astronauts, Dr. Story Musgrave and Dr. Jeffrey A. Hoffman, had no trouble removing the existing camera and inserting the new one. Standing on a platform at the end of the shuttle's mechanical arm, Dr. Hoffman pulled the old instrument out of its module on the side of the telescope. He then moved over in the cargo bay to pull the replacement camera out of its storage locker.

"Ohhh, look at that baby! It's a beautiful, spanking new wifpic," Dr. Hoffman exclaimed, using the project's nickname for the wide-field planetary camera. "We'll see some nice pictures with that."

With Dr. Musgrave assisting, Dr. Hoffman gently slid the wedge-shaped camera along guide rails inside the compartment. It was a tight squeeze. He had to be careful not to jiggle a protruding mirror.

Larry Simmons, manager for the camera development at the Jet Propulsion Laboratory, said early tests verified normal performance by the camera's electronics.

Acutely aware of the primary mirror's unexpected flaw, Mr. Simmons said, engineers had each component of the new camera tested by at least two independent methods and then had the entire system checked and checked again. Unfocused light, like that produced by the flawed mirror, was beamed at the corrective mirrors, and they made the appropriate compensations, producing clear images.

$50 Million Fix

In tonight's repair job, Dr. Kathryn C. Thornton and Lieut. Col. Tom Akers installed a $50 million device called the Corrective Optics Space Telescope Axial Replacement, or Costar. It is a 670-pound component the size of a refrigerator, which was designed for this mission to slide into a compartment in the aft end of the telescope.

Costar replaced the Hubble's high-speed photometer, its least-used science instrument. It has a set of 10 corrective mirrors attached to mechanical arms, and these are to be extended to intercept and correct light going to the faint-object camera and two spectrographs that astronomers use to determine the composition, temperatures and other physical properties of stars and galaxies.

The one problem that Dr. Musgrave and Dr. Hoffman encountered on their space walk occurred when they inserted two new magnetic sensing units, part of the system for rotating the big telescope in pointing toward observation targets. The existing units showed signs of serious degradation. Since these black boxes were not meant to be replaced, the two new ones were placed piggyback over the old ones.

Dr. Hoffman noticed that the aluminum sides of one of the old units were loose. Engineers decided that on the final space walk the astronauts should try to use a pouch that had protected one of the replacement units during flight as sort of a shower cap for the defective box. They were concerned that insulation foam inside the magnetometer might contaminate other components, including the telescope itself.

—*December 8, 1993*

Astronomers Relish Chance to
See Rare Supernova Blow Itself to Bits

By MALCOLM W. BROWNE

Observatories around the world have turned their telescopes to the brightest stellar explosion to appear in the Northern Hemisphere in decades, a cosmic furnace of the kind that produces the raw materials of life itself.

The new supernova flared into view on March 28, when it was discovered by an amateur astronomer with a small telescope in Spain. Several days elapsed before his observation could be officially confirmed and for the news to circulate among professional astronomers at leading observatories.

But by yesterday astronomers in North America, Europe and Asia, using both ground-based telescopes and satellite observatories, were rushing to analyze the explosion and were hoping to reap a bonanza of information, said Dr. Jay M. Pasachoff of Williams College in Williamstown, Mass.

Supernova explosions fascinate scientists for many reasons. Supernovas are thought to be the source of the heavy elements found in the universe, including the carbon from which all living organisms are made. Supernovas also initiate huge shock waves that some theorists believe may play a role in initiating the formation of galaxies. Scientists hope eventually to detect gravity waves caused by supernovas as these massive stars collapse on themselves to form neutron stars, pulsars or even black holes.

Similar to Milky Way

The new supernova, named SN1993J, lies about 12 million light-years from the Earth and is in one of the spiral arms of a galaxy named M-81. The galaxy, which is similar in size and shape to the Milky Way, is visible through powerful binoculars between the constellation Ursa Major (also known as the Big Dipper) and the pole star Polaris. It is always high enough in the sky to be seen from the Northern Hemisphere.

Daniel Green of the Smithsonian Astrophysical Observatory in Cambridge, Mass., a clearing house that records astronomical discoveries, said SN1993J was spotted by Francisco Garcia Diez of Lugo, Spain, using a small telescope with a light-gathering mirror only 10 inches in diameter. Mr. Garcia, who belongs to a volunteer team combing the sky for supernovas, succeeded in finding the supernova where at least one major observatory with automatic supernova-hunting equipment failed.

Although the light from SN1993J is already fading, it is still visible to amateur telescopes, and in any case, it may brighten again in the next few weeks.

Massive stars explode frequently throughout the universe, but most of these blasts are so distant from the Earth, or so obscured by interstellar dust clouds, that they are difficult or impossible to detect, much less measure. SN1993J exploded relatively nearby, and the light and radiation it spewed into space 12 million years ago are intense enough even after their long voyage to analyze.

Dr. Alex Fillipenko of the University of California at Berkeley has coordinated observations of SN1993J since March 29. His team at Lick Observatory has confirmed the existence of the supernova with a three-meter-diameter telescope equipped with a spectrograph for analyzing its light.

Reports have also arrived from the German Rosat X-ray satellite and the Japanese Astro-D satellite that SN1993J is emitting X-rays, an important fact for astrophysicists. It appears that the new supernova is shrouded by a large dust cloud, which is absorbing intense radiation from the explosion and reemitting it in the form of X-rays.

"This is the brightest supernova explosion the Northern Hemisphere has seen since 1937, except for a somewhat brighter one in 1972 that was too low on the horizon for optimal observation," Dr. Fillipenko said.

By far the brightest supernova in four centuries appeared in February 1987, and even in its present faded form, that supernova, SN1987A, is an important object of scientific study. But SN1987A was visible only from the Southern Hemisphere. Moreover, every year sees new technological developments that mine more information from supernova blasts.

"There are many new instruments available for supernova study that simply didn't exist in 1980," said Dr. Stan E. Woosley, an astrophysicist at the University of California at Santa Cruz.

But at 12 million light-years, SN1993J is about 70 times more distant from the Earth than SN1987A, and the greater distance rules out some observations. In particular, scientists have neither detected, nor do they hope to detect, showers of neutrinoparticles coming from the new supernova. Theorists and astrophysicists were delighted when neutrino detectors in both Japan and the United States saw neutrino bursts at the moment SN1987A popped into view, thereby confirming a prediction arising from a leading theory about supernova physics. But the new supernova is too far away.

Two Types of Supernova

Astronomers say there are two main types of supernova. In the first type, a small, super-dense star called a "white dwarf" draws matter away from a nearby companion star, and when the white dwarf's mass grows to more than 1.4 times the mass of the Sun, a chain of fusion reactions leads to an explosion that blows the star apart.

The second type, to which the latest supernova is believed to belong, starts with a single star more than eight times the mass of the Sun. The star consumes its fuel in a series of fusion reactions and finally collapses under the force of its own gravity. An explosion then blows the star apart.

—April 7, 1993

The End of Everything

By DENNIS OVERBYE

In the decades that astronomers have debated the fate of the expanding universe—whether it will all end one day in a big crunch, or whether the galaxies will sail apart forever—aficionados of eternal expansion have always been braced by its seemingly endless possibilities for development and evolution. As the Yale cosmologist Dr. Beatrice Tinsley once wrote, "I think I am tied to the idea of expanding forever."

Life and intelligence could sustain themselves indefinitely in such a universe, even as the stars winked out and the galaxies were all swallowed by black holes, Dr. Freeman Dyson, a physicist at the Institute for Advanced Study, argued in a landmark paper in 1979. "If my view of the future is correct," he wrote, "it means that the world of physics and astronomy is also inexhaustible; no matter how far we go into the future, there will always be new things happening, new information coming in, new worlds to explore, a constantly expanding domain of life, consciousness, and memory."

Now, however, even Dr. Dyson admits that all bets are off. If recent astronomical observations are correct, the future of life and the universe will be far bleaker.

In the last four years astronomers have reported evidence that the expansion of the universe is not just continuing but is speeding up, under the influence of a mysterious "dark energy," an antigravity that seems to be embedded in space itself. If that is true and the universe goes on accelerating, astronomers say, rather than coasting gently into the night, distant galaxies will eventually be moving apart so quickly that they cannot communicate with one another. In effect, it would be like living in the middle of a black hole that kept getting emptier and colder.

In such a universe, some physicists say, the usual methods of formulating physics may not all apply. Instead of new worlds coming into view, old ones would constantly be disappearing over the horizon, lost from view forever.

Cosmological knowledge would be fragmented, with different observers doomed to seeing different pieces of the puzzle and no single observer able to know the fate of the whole universe or arrive at a theory of physics that was more than approximate.

"There would be a lot of things about the universe that we simply couldn't predict," said Dr. Thomas Banks, a physicist at the University of California at Santa Cruz.

And perhaps most important, starved finally of the energy even to complete a thought or a computation, the domain of life and intelligence would not expand, but constrict and eventually vanish like a dwindling echo into the silence of eternity. "I find the fate of a universe that is accelerating forever not very appealing," said Dr. Edward Witten, a theorist at the Institute for Advanced Study.

That is an understatement, in the view of Dr. Lawrence M. Krauss, an astrophysicist at Case Western Reserve University in Cleveland, who along with his colleague Dr. Glenn D. Starkman has recently tried to limn the possibilities of the far future. An accelerating universe "would be the worst possible universe, both for the quality and quantity of life," Dr. Krauss said, adding: "All our knowledge, civilization and culture are destined to be forgotten. There's no long-term future."

Einstein's Last Laugh

Until about four years ago, an overwhelming preponderance of astronomers subscribed to the view that the cosmic expansion was probably slowing down because of the collective gravity of the galaxies and everything else in the universe, the way a handful of stones tossed in the air gradually slow their ascent. The only question was whether the universe had enough gravitational oomph to stop expanding and bring itself back together in a "big crunch," or whether the galaxies would sail ever more slowly outward forever.

It was to measure that rate of slowing of this outward flight, and thus find the long-sought and elusive answer to the cosmic question, that two teams of astronomers started competing projects in the 1990s using distant exploding stars, supernovas, as cosmic beacons.

In 1998 the two teams announced that instead of the expected slowing, the galaxies actually seem to have speeded up over the last five or six billion years, as if some "dark energy" was pushing them outward.

"It's definitely the strangest experimental finding since I've been in physics," Dr. Witten said. "People find it difficult to accept. I've stopped expecting that the finding will be proved wrong, but it's an extremely uncomfortable result."

To astronomers this dark energy bears a haunting resemblance to an idea that Albert Einstein had back in 1917 and then abandoned, later calling it his biggest blunder. In that year he inserted a mathematical fudge factor that came to be known as the cosmological constant into his equations of general relativity in order to stabilize the universe against collapse; Einstein's constant acted as a kind of cosmic repulsion to balance the gravitational pull of the galaxies on one another.

Einstein gave up the cosmological constant after the American astronomer Edwin Hubble discovered that the universe was expanding and thus did not need stabilizing. But his fudge factor refused to die. It gained a new identity with the advent of quantum mechanics, the bizarre-sounding rules that govern the subatomic realm. According to those rules, empty space is not empty, but rather foaming with energy. Inserted into Einstein's equations, this energy would act like a cosmological constant, and try to blow the universe apart.

According to astronomers, the recently discovered dark energy now accounts for about two-thirds of the mass of the universe. But is this Einstein's old fudge factor, the cosmological constant, come home to roost—in which case the universe will accelerate eternally? Or is the presumed acceleration only temporary, driven by one of the many mysterious force fields, dubbed quintessence, allowed by various theories of high-energy physics?

Or is the acceleration even real?

"It's important to find out if the cosmological constant is really constant," said Dr. Witten.

Because the repulsive force resides in space itself, as the universe grows, the push from dark energy grows as well. "If dark energy is the cosmological constant then it is a property of the vacuum that will always be with us, getting more powerful as the universe gets bigger and the universe will expand forever," explained Dr. Adam Riess of the Space Telescope Science Institute in Baltimore. But if the dark energy is some form of quintessence, "then there may be more such fields which arise in the future, possibly of the opposite sign, and then all bets are off for the future of the universe."

Dr. Krauss said, "The good news is that we can't prove that this is the worst of all possible universes."

The Long Goodbye

It might seem strange or presumptuous for astronomers to try to describe events all the way to the end of time when physicists are still groping for a "theory of everything." But to Dr. Krauss, this is testimony to the power of ordinary physics.

"We can still put ultimate limits on things without even knowing the ultimate theory," he said. "We can put limits on things based on ordinary physics."

Dr. Dyson said his venture into eschatology was inspired partly by a 1977 paper on the future of an ever-expanding universe by Dr. J. N. Islam, now at the University of Chittagong in Bangladesh, in the *Quarterly Journal of the Royal Astronomical Society*. Dr. Dyson was also motivated, he wrote in his paper, to provide a counterpoint to a famously dour statement by Dr. Steven Weinberg, who wrote in his book *The First Three Minutes*, "The more the universe seems comprehensible, the more it also seems pointless."

Dr. Dyson wrote, "If Weinberg is speaking for the 20th century, I prefer the 18th."

If the present trend of acceleration continues this is the forecast:

In about two billion years Earth will become uninhabitable as a gradually warming Sun produces a runaway greenhouse effect. In five billion years the Sun will swell up and die, burning Earth to a crisp in the process. At about the same time the Milky Way will collide with its twin the Andromeda galaxy, now about two million light-years away and closing fast, spewing stars, gas and planets across intergalactic space.

Any civilization that managed to survive these events would face a future of increasing ignorance and darkness as the accelerating cosmic expansion rushes most of the universe away from us. "Our ability to know about the universe will decrease with time," said Dr. Krauss. "The longer you wait, the less you see, the opposite of what we always thought."

As he explains it, the disappearance of the universe is a gradual process. The faster a galaxy flies away from us, the dimmer and dimmer it will appear, as its light is "redshifted" to lower frequencies and energies, the way a police siren sounds lower when it is receding. When it reaches the speed of light, the galaxy will appear to "freeze," like a dancer caught in midair in a photograph, in accordance to Einstein's theory of relativity, and we will never see it get older, said Dr. Abraham Loeb, an astronomer at Harvard. Rather it will simply seem dimmer. The farther away an object is in the sky, he said, the younger it will appear as it fades out of sight. "There is a finite amount of information we can collect from the universe," Dr. Loeb said.

About 150 billion years from now almost all of the galaxies in the universe will be receding fast enough to be invisible from the Milky Way. The exceptions will be galaxies that are gravitationally bound to the cloud of galaxies, known as the Local

Group, to which the Milky Way belongs. Within this cloud, life would look much the same at first. There would be galaxies in the sky. "When you look at the night, the stars will still be there," said Dr. Krauss. "To the astronomer who wants to see beyond, the sky will be sadly empty. Lovers won't be disturbed—scientists will be."

But about 100 trillion years from now, when the interstellar gas and dust from which new stars condense is finally used up, new stars will cease to be born. From that time on, the sky will grow darker and darker. The galaxies themselves, astronomers say, will collapse in black holes within about 10^{30} years.

But even a black hole is not forever, as Dr. Stephen Hawking, the Cambridge University physicist and best-selling author, showed in path-breaking calculations back in 1973. Applying the principles of quantum mechanics to these dread-sounding objects, Dr. Hawking discovered that a black hole's surface, its so-called event horizon, would fluctuate and exude energy in the form of random bursts of particles and radiation, growing hotter and hotter until the black hole eventually exploded and vanished.

Black holes the mass of the sun would take 10^{64} years to explode. For black holes the mass of a galaxy, those fireworks would light up space-time 10^{98} years from now.

Against the Fall of Night

Will there be anything or anyone around to see these quantum fireworks?

Dr. Dyson argued in his 1979 paper that life and intelligence could survive the desert of darkness and cold in a universe that was expanding infinitely but ever more slowly by adopting ever slower and cooler forms of existence. Intelligence could reside, for example, in the pattern of electrically charged dust grains in an interstellar cloud, a situation described in the 1957 science fiction novel *The Black Cloud*, by the British astronomer Sir Fred Hoyle, who died in August.

As an organism like the black cloud cooled, he argued, it would think more slowly, but it would always metabolize energy even more slowly, so its appetite would always be less than its output. In fact, Dr. Dyson concluded, by making the amount of energy expended per thought smaller and smaller the cloud could have an infinite number of thoughts while consuming only a finite amount of energy.

But there was a hitch. Even just thinking requires energy and generates heat, which is why computers have fans. Dr. Dyson suggested that creatures would have to stop thinking and hibernate periodically to radiate away their heat.

In an accelerating universe, however, there is an additional source of heat that cannot be gotten rid of. The same calculations that predict black holes should explode also predict that in an accelerating universe space should be filled with so-called Hawking radiation. In effect, the horizon—the farthest distance we can see—looks mathematically like the surface of a black hole. The amount of this radiation is expected to be incredibly small—corresponding to a fraction of a billionth of a billionth of a billionth of a degree above absolute zero, but that is enough to doom sentient life.

"The Hawking radiation kills us because it gives a minimum temperature below which you cannot cool anything," said Dr. Krauss. Once an organism cools to that temperature, he explained, it would dissipate energy at some fixed rate. "Since there is a finite total energy, this means a finite lifetime."

Infinity on Trial

Although Dr. Dyson agrees with this gloomy view of life in an accelerating universe, he and Dr. Krauss and Dr. Starkman are still arguing about whether life is also doomed in a universe that is not accelerating, but just expanding and getting slower and colder.

Quantum theory, the Case Western authors point out, limits how finely the energy for new thoughts can be shaved. The theory decrees that energy is emitted and absorbed in tiny indivisible lumps called "quanta." Any computation must spend at least this much energy out of a limited supply. Each new thought is a step down an energy ladder with a finite number of steps. "So you can only have a finite number of thoughts," said Dr. Krauss.

"If you want to stare at your navel and not think any new thoughts, you won't dissipate energy," he explained. But that would be a boring way to spend eternity. If life is to involve more than the eternal reshuffling of the same data, he and Dr. Starkman say, it cannot be eternal.

Dr. Dyson, however, says this argument applies only to so-called digital life, in which there is a fixed number of quantum states. Creatures like the black cloud, which could grow along with the universe, he said, would have an increasing number of quantum states, and so there would always be more rungs of the ladder to step down. So the bottom need never be reached and life and thought could go on indefinitely.

But nobody knows whether such a life form can exist, said Dr. Krauss.

Compared with the sight of the World Trade Center towers collapsing or the plight of a sick child, this future extinction may seem a remote concern. Dr. Allan Sandage, an astronomer at Carnegie Observatories in Pasadena, Calif., who has spent his life investigating the expansion and fate of the universe, said: "Life on this earth is going to vanish in 4.5 billion years. I wouldn't get hung up on the fact that the lights are all going out in 30 billion years."

Dr. Dyson said he was still an optimist. It is too soon to start panicking, he counseled in an e-mail message. The observations could be wrong.

"At present all possibilities are open," he wrote. "The recent observations are important, not because they answer the big questions about the history of the universe, but because they give us new tools with which to explore the history."

Even in an accelerating universe, Dr. Dyson said, humans or their descendants might one day be able to rearrange the galaxies and save more of them from disappearing. Another glimmer of hope comes from the deadly and chilling Hawking radiation itself, said Dr. Raphael Bousso, from the Institute of Theoretical Physics at the University of California at Santa Barbara. Since that radiation is produced by unpredictable quantum fluctuations, he pointed out, if you wait long enough anything can appear in it, even a new universe. "Sooner or later one of those quantum fluctuations will look like a Big Bang," he said.

In that case there is the possibility of a future, if not for us, at least for something or somebody. In the fullness of time, after all, physics teaches that the improbable and even the seemingly impossible can become the inevitable. Nature is not done with us yet, nor, as Dr. Dyson indicates, are we necessarily done with nature.

We all die, and it is up to us to decide who and what to love, but, as Dr. Weinberg pointed out in a recent article in *The New York Review of Books*, there is a certain nobility in that prospect.

"Though aware that there is nothing in the universe that suggests any purpose for humanity," he wrote, "one way that we can find a purpose is to study the universe by the methods of science, without consoling ourselves with fairy tales about its future, or about our own."

—*January 1, 2002*

Hot on Trail of "Just Right" Far-Off Planet

By DENNIS OVERBYE

What does Goldilocks want?

At least four times in the last few years, astronomers have announced they have found planets orbiting other stars in the sweet spot known as the habitable zone—not too hot, not too cold—where water and thus perhaps life are possible. In short, a so-called Goldilocks planet fit to be inhabited by the biochemical likes of us.

None of these claims are without controversy, but astronomers who are making discoveries with NASA's *Kepler* spacecraft are meeting next week in California to review the first two years of their quest, which seems tantalizingly close to hitting pay dirt.

"Sooner or later, *Kepler* will find a lukewarm planet with a size making it probably Earthlike," said Geoffrey Marcy of the University of California, Berkeley, who spends his time tracking down candidates identified by *Kepler*. "We're no more than a year away" from such a discovery, he said.

Sara Seager, a planetary astronomer at the Massachusetts Institute of Technology, put it this way: "We are on the verge of being those people who will be remembered."

All this has brought to the fore a question long debated by geologists, chemists, paleontologists and cosmologists turned astrobiologists, namely: What does life really need to get going, flourish and evolve on some alien rock?

The answer depends of course on whom we expect to be living there. We might dream of green men with big eyes, ants with hive minds, or even cuddly octopuses as an antidote to cosmic loneliness, but what we are most likely to find, a growing number of scientists say, is alien pond slime.

Microbes can spring up anywhere that is wet and warm, astronomers say, although biologists are not so sanguine. But the emergence of large creatures, let alone intelligent ones, as evidenced by the history of the Earth, depends on a chain of events and accidents—from asteroid strikes to plate tectonics— that are unlikely to be repeated anytime soon. "If you reran Earth's history, how many times would you get animals?" asked Donald Brownlee, an astronomer at the University of Washington. He and a colleague, the paleontologist Peter Ward, made a case that we live on a lucky planet in their 1999 book, *Rare Earth.*

Single-cell life might be common, given the right simple conditions, explained Dr. Marcy in an e-mail. "But the steady, long-term evolution toward critters that play improv saxophone, write alliteration poems, and build heavy-lifting rocket boosters may depend on a prohibitive list of planetary prerequisites," he added.

Even warm and wet is a rare condition, however, occurring now on only one of the eight official planets in our solar system and three of the several dozen moons. Mars was once wet, but it is now a desert. And after billions of dollars spent exploring Mars and the remains of space probes littering the planet, we still do not know if a single microbe ever lived there.

But nobody really knows how rare or common are planets like Earth and its brand of life. . . . Some scientists deplore the emphasis on animals like us, saying it is hopelessly parochial and unimaginative—the scientific equivalent of the drunk searching for his car keys under a street light because that's where the light is.

"Animals are overgrown microbes," said Paul Falkowski, a biophysicist and biologist from Rutgers. "We are here to ferry microbes across the planet. Plants and animals are an afterthought of microbes." So, we should hardly be disappointed if we find our neighbors are microbes. After all, on Earth, microbes were the whole story for almost four billion years, paleontologists say, and now inhabit our intestines as well as every doorknob.

Dimitar Sasselov, an astronomer turned astrobiologist at the Harvard-Smithsonian Center for Astrophysics, said he was all for the existence of a microbial planet. "Don't assume microbes are simple," he said, noting that 99 percent of the genes in our bodies belong to microbes inhabiting us and without which we could not live.

Looking for Goldilocks

A blue-ribbon committee of chemists convened by the National Academy of Sciences concluded that there was only one ironclad requirement for life, besides energy: a place warm enough for chemical reactions to go on. So, determining how warm a planet's atmosphere keeps it—through assumptions, calculations or just plain guesses—has been crucial in reaching a verdict about its potential habitability.

This is how it has gone with the potential Goldilocks planets orbiting Gliese 581, a small cool red star about 20 light-years from here in the constellation Libra that has been at the center of exoplanet fantasies and speculation for the last few years. Depending on whom you talk to, it has five or six planets, three of which have at one time or another been claimed to be habitable.

The first in what would become a chain of potential Goldilocks planets, identified in 2007, was a presumably rocky ball about five times as massive as the Earth and orbiting only about seven million miles from Gliese 581, close enough within the small star's shrunken habitable zone to have a warm surface. "On the treasure map of the universe, one would be tempted to mark this planet with an X," Xavier Delfosse, one of the astronomers who discovered it, said at the time.

But before budding interstellar explorers could even begin conceiving of booking passage to Gliese 581c, as the planet is poetically called, other astronomers took a closer look and concluded that if the planet's geology and atmosphere resembled those of Earth, it would be a stifling greenhouse, no place to set solar sail for. Attention then shifted to a farther planet in the system, Gliese 581d, which had been dismissed as too cold. Could the same greenhouse effect that would torch the inner planet thaw the outer one and make it livable? The answer was yes, but only if it had "loads of carbon dioxide" and an atmosphere seven times thicker than Earth's, said Lisa Kaltenegger, a climate modeler at the Max Planck Institute for Astronomy in Heidelberg. Otherwise it would be freezing cold.

Meanwhile yet another planet was claimed for that system, smack between the other two, by a team led by R. Paul Butler of the Carnegie Institution and Steven S. Vogt of the University of California, Santa Cruz. "This is really the first Goldilocks planet," Dr. Butler said at a news conference last year organized by the National Science Foundation in Washington.

But the Geneva team that had discovered the earlier Gliese 581 planets could not find any evidence of the new planet's existence in their own data. For now, anyway, most astronomers have dismissed that planet. Pending the publication of new results by the Geneva team—one of the most prolific in the planet-hunting business—Dr. Butler said, "We are in a holding pattern."

In September, what some astronomers called the best and smallest Goldilocks candidate yet was announced by the Geneva team. About 3.6 times as massive as the Earth, it circles a faint orange star in Vela known as HD 85512 at a distance of some 24 million miles, about a quarter of the Earth's distance from the Sun. Dr. Kaltenegger and her colleagues calculated that this planet would be habitable if it had an Earth-type geology and at least 50 percent cloud cover. "So, so far we only have two great targets to search for atmospheric signatures of life," Dr. Kaltenegger wrote.

So goes the history of astrobiology, as well as its future.

The problem, as many astronomers point out, is getting any more information about these planets. "Astronomers are going to have to learn to live with ignorance," Dr. Seager said.

Some exoplanets, like the Gliese worlds, were discovered by the "wobble method"—looking at the motions they induce in their parent stars—which allows their masses and orbits to be measured. Other planets, like the ones identified by Kepler, are found by watching for the blinks when they pass in front of their stars; that also allows their sizes to be determined.

If, If, If

To date, none of the Goldilocks candidates have been observed to transit their stars, and thus none have been assigned both masses and sizes, which would allow astronomers to calculate their densities and compositions and find out if they are water worlds, rocks or gassy fluff balls.

Kepler fixes its gaze on a patch of stars in Cygnus that are hundreds if not thousands of light-years away—too far for any wobble detections that would assess the abundance of Earthlike planets in the galaxy or any other close scrutiny. We are liable to never know anymore about those planets than we know now, astronomers say. The brute reality, astronomers admit, is that even if there are thousands or millions of habitable planets in the galaxy, only a few hundred of them are within range of any telescope that will be built in the conceivable future.

Luckily there is some renewed hope for life on those nearby planets. David Charbonneau of the Harvard-Smithsonian Center for Astrophysics runs a project called MEarth that looks for planets around nearby stars. He pointed out that of the 300 stars within 25 light-years of here, 260 are red dwarfs like Gliese 581.

Until recently it was thought that habitable-zone planets around such stars would have to hug the star so closely that they would be tidally locked, like the Moon, keeping one face locked to the star and roasting, the other freezing.

But new studies have concluded that a proper atmosphere could spread the heat around.

Which is good. "These stars," Dr. Charbonneau said, "are our only hope for studying life in the universe in the coming decades."

In the original scheme of things, Kepler was to be succeeded by a space observatory called the Terrestrial Planet Finder, which would be big enough to find and study planets up to 100 light-years distant.

But plans for that telescope have collapsed, because of NASA's continuing fiscal woes and disagreements among astronomers, as well as the technological challenges involved.

Some astronomers hope that some of these functions can be performed by the James Webb Space Telescope—NASA's Hubble successor, overdue and over budget, now scheduled for a launch in 2018.

Equipped with a "starshade" that would blot out the glare of a planet's sun, the Webb could detect and study the pinpoint of light from an exoplanet itself.

But the starshade would be hostage to the same political and fiscal pressures that are threatening to decimate NASA's scientific programs. At best, scientists say, the search for life elsewhere has been postponed for decades.

"I'm beginning to despair that I will see it in my lifetime," said James Kasting of Pennsylvania State University.

Geology Is Destiny

Earth got lucky early. Fossil evidence suggests that microbial life was already inhabiting the Earth as early as 3.8 billion years ago—only 700 million years after the planet collapsed into existence, and a geological instant after the end of a rain of comets and asteroids that brought just the right amount of precious water in the form of ice from the outer solar system to what would otherwise be a dry planet, astronomers say. "The question of whether the Earth is unique because of its water abundance is perhaps the most interesting one in the arsenal of Rare Earth arguments," said Dr. Kasting, who explained that calculations showed that the planet could have easily had too much or too little water.

The planet has remained comfortable ever since thanks to a geological feedback process, by which weather, oceans and volcanoes act as a thermostat. Known as the carbonate silicate cycle, it regulates the amount of carbon dioxide in the atmosphere, where it acts like a greenhouse—trapping heat and keeping the planet temperate and mostly stable. Rain washes the gas out of the air and under the ocean; volcanoes disgorge it again from the underworld.

Without greenhouse gases and this cycle—which Dr. Brownlee called "this magic thing"—the Earth would have frozen into a snowball back in its early days when the Sun was only 70 percent as bright as it is now. Still, with all this magic, it took four billion years for animal life to appear on the Earth.

The seeds for animal life were sown sometime in the dim past when some bacterium learned to use sunlight to split water molecules and produce oxygen and sugar—photosynthesis, in short. The results began to kick in 2.4 billion years ago when the amount of oxygen in the atmosphere began to rise dramatically.

The Great Oxidation Event, as it is called in geology, "was clearly the biggest event in the history of the biosphere," said Dr. Ward from Washington. It culminated in what is known as the Cambrian explosion, about 550 million years ago, when multicellular creatures, that is to say, animals, appeared in sudden splendiferous profusion in the fossil record. We were off to the Darwinian races. Whatever happened to cause this flowering of species helped elevate Earth someplace special, say the Rare Earthers. Paleontologists argue about whether it could have been a spell of bad climate known as Snowball Earth, the breakup of a previous supercontinent, or something else.

In other words, alien planets that have been lucky enough to be habitable in the first place might have to be lucky again. "The big hurdle" for other planets, said Dr. Brownlee, is to have some event or series of events to trigger their own "Cambrian-like" explosions.

Eventually though, Earth's luck will run out. As the Sun ages it will get brighter, astronomers say, increasing the weathering and washing away of carbon dioxide. At the same time, as the interior of the Earth cools, volcanic activity will gradually subside, cutting off the replenishing of the greenhouse gas.

A billion years from now, Dr. Brownlee said, there will not be enough carbon dioxide left to support photosynthesis, that is to say, the oxygen we breathe.

And so much for us.

"Even Earth, wonderful and special as it is, will only have animal life for one billion years," Dr. Brownlee said.

—December 3, 2011

CHAPTER 3

Biology:
The Mechanisms of Life

Life Is Generated in Scientist's Tube

By WILLIAM L. LAURENCE

"Bottle babies," predicted by Aldous Huxley for the distant future in a "brave new world" where children will be born in test tubes, have been brought at least part way toward actuality by Dr. Gregory Pincus at the Harvard Biological Institute.

A report of his work, made here [in Washington] today at the annual meeting of the Federation of American Societies for Experimental Biology, told of several spectacular developments on the frontiers of the science of life.

Dr. Pincus took the female and the male elements of rabbits, and fertilized them outside the body in a test tube. This in itself was formerly impossible to do, but Dr. Pincus went an important step further.

Allowing the artificially fertilized ovum to reach a stage of early development in the test tube, Dr. Pincus removed it from the bottle and transplanted it into a living female rabbit.

The transplanted embryo developed and matured in the host-mother just as in nature. In due course of time normal rabbits were born, the world's first "semi-ectogenetic rabbits."

Extension of the Theory

As rabbits and men belong to the mammalian group, the work is viewed as pointing toward the possibility of human children being brought into the world by a "host-mother" not related by blood to the child.

It is reasoned that eventually women capable of having children whose health does not permit them to do so may "hire" other women to bear their children for them, children actually their own flesh and blood.

To one who desires to speculate at this point the Harvard experiment offers another possibility. Theoretically, at least, it may become possible for a woman so inclined, particularly in a country influenced by eugenic considerations, to bring into the world twelve children a year by "hiring" twelve "host-mothers" to bear their test-tube-conceived children for them.

Advocates of "race betterment" might urge such procedures for men and women of special aptitudes, physical, mental or spiritual.

"Eliminating" the Male

But the Harvard biologist reported on another advance in the experiments, "eliminating" the male in reproduction. He put a rabbit ovum in a test tube and fertilized it with a strong salt solution.

Finding that even this salt solution was not necessary, Dr. Pincus accomplished the same results by exposing the ovum for a few minutes to a high temperature of 45 degrees centigrade, or 113 degrees Fahrenheit.

The ovum, fertilized by a strong salt solution or exposure to high temperature, was then transplanted into a living host-mother rabbit.

The rabbit-embryo developed to the early embryonic stage as in the normal way. At the end of a week the experiment was stopped to allow the scientists to check on the results.

Further experiments will be carried on in this field with the object of bringing into the world the first mammalian creature having as its "father" a salt solution or a high temperature.

"Fatherless" Offspring Females

Dr. Pincus pointed out in an interview, however, that all rabbits produced from an ovum fertilized artificially by salts or heat would be all female because the process would lack the sex-determining "Y-chromosome" which is supplied by the male.

Even the early stages of activation of the rabbit ovum, Dr. Pincus reported, could now be brought about experimentally without the male, either by the administration of hormones from the anterior pituitary gland, or by electrical stimulation of the cervical sympathetic nerves.

Activation by electrical stimulus was achieved at the Harvard Laboratories by Dr. H. B. Friedgood and Professor Walter B. Cannon of the Harvard Medical School.

Functioning of Brain Areas

New light was thrown on the functions of different areas of the brain by experiments reported by Dr. C. R. Jacobsen, Dr. F. V. Taylor and Dr. G. M. Haslerud of Yale University's primate biology laboratory, before the central nervous system section of the American Physiological Society.

They removed completely the frontal and motor areas of the cerebral cortex of monkeys.

When the motor area was removed from an adult monkey the result was almost complete and permanent paralysis. The animal was unable to make any further voluntary movements. The same operation in infant monkeys, however, produced only a temporary paralysis. The animals soon were able to run, climb and manipulate objects.

Extirpation of the frontal association areas from adult animals caused total impairment of "recall" memory, although recognition memory was little disturbed. The results were essentially the same in the infants.

"It seems probable that this difference in recovery after motor and frontal area lesions arises from partial destruction of a dynamic system in the former instance, and complete removal in the latter," the report stated.

"Reversing Locomotion"

An experiment which produced "reversed locomotion" in salamanders was reported to the Physiological Society by Dr. Paul Weiss of the University of Chicago.

On the theory that a nerve becomes specified by its muscle and responds selectively to specific impulses from the brain intended for that muscle, he transposed right and left front legs of the salamanders so that the right leg was connected with the brain by the nerve fibers which had served the left leg, and vice versa.

Stating that "the movements exhibited by the transplanted limb in the various phases of locomotion should be expected to be exact mirror images of the ones that the normal limb would perform at the same instant," he said:

"This was actually found to be the case with eleven salamanders in which the developed forelimbs had been interchanged. These limbs moved always in the reverse direction.

"In the progressing animal they strode backward instead of forward and, although this seriously interfered with the locomotion of the body, the central pattern has never been changed and adjusted to the new conditions."

New Heart-Reviving Method

A new surgical technique to revive hearts which have nearly stopped beating because of coronary occlusion, or blocking of the coronary arteries, was described to a section of the Physiological Society by Dr. C. J. Wiggers of Western Reserve University.

He reported that the method had been successful in reviving dogs from five to seven minutes after their hearts had entered the stage known as fibrillation, after which heart action stops. He said that it should be of value when applied to human hearts which start fibrillating in the course of surgical operations.

Two or three years ago it was found that fibrillation could be stopped and strong action revived by the brief passage through the heart of an alternating current applied through padded electrodes. This was only true, however, provided the coronary occlusion was removed and fibrillation had not lasted more than two or three minutes.

Dr. Wiggers found, however, that by massaging the heart immediately before applying the electric shock—that is by compressing the ventricles with the hand about forty times a minute and each time forcing blood into the aorta—the expectancy period for revival could be tripled or quadrupled.

Dr. R. H. Cheney of Long Island University, reporting on an experiment with a group of college girls, said that caffeine increased both speed and accuracy, but taken in the form of coffee was only about half as effective as when minute amounts of the pure alkaloid were administered.

Extraction of a hitherto unknown chemical substance from the bacillus of leprosy was reported to the American Society of Biological Chemists by Dr. R. J. Anderson, Dr. J. A. Crowder, M. S. Newman and Dr. F. M. Stodola of Yale University.

The material, which has been named leprosin, is a snow-white amorphous powder. It can be resolved into a new acid of high molecular weight which has been named leprosinic acid.

A machine for milking rats and guinea pigs was described by Warren H. Cox Jr. and Arthur J. Mueller of Evansville, Ind. Comparison of rat's milk with cow's milk shows the former to be richer in fat and protein, while containing half as much sugar.

—March 27, 1936

Clue to Chemistry of Heredity Found

A scientific partnership between an American and a British biochemist at the Cavendish Laboratory in Cambridge has led to the unraveling of the structural pattern of a substance as important to biologists as uranium is to nuclear physicists. The substance is nucleic acid, the vital constituent of cells, the carrier of inherited characters and the fluid that links organic life with inorganic matter.

The form of nucleic acid under investigation is called DNA (desoxyribonucleic acid) and has been known since 1869.

But what nobody understood before the Cavendish Laboratory men considered the problem was how the molecules were grooved into each other like the strands of a wire hawser so they were able to pull inherited characters over from one generation to another.

Further Tests Slated

The two biochemists, James Dewey Watson, a former graduate student of the University of Chicago, and his British partner, Frances H. C. Crick, believe that in DNA they have at last found the clue to the chemistry of heredity. If further X-ray tests prove what has largely been demonstrated on paper, Drs. Watson and Crick will have made biochemical history.

Dr. Watson has now returned to the United States, where he intends to join Dr. Linus Pauling, of California, who has done most of the pioneer work on the problem.

[In Pasadena, Calif., Dr. Pauling said that the new Crick-Watson solution appeared to be somewhat better than the proposal for the structure of the nucleic acids worked out by Dr. Pauling and associates at the California Institute of Technology. The California solution was published in the February 1953 issue of the *Proceedings of the National Academy of Sciences*.]

Dr. Crick may leave Britain, too, when he has done some more work on the problem. Right now, he said, it "simply smells right" and confirms research in many institutions, particularly the Rockefeller Foundation in the United States and at King's College in London.

The acid DNA, Dr. Crick explained, is a "high polymer"—that is, its chemical components can be disentangled and rearranged in different ways.

DNA is the essential constituent of the miscroscopic life-threads called chromosomes that carry the genes of heredity like beads on a string.

In all life cells, including those of man, DNA is the substance that transmits inherited characters such as eye color, nose shape and certain types of blood and diseases. The transmission occurs at the vital moment of mitosis or cell division when a tangle of DNA containing chromosomes becomes thicker and the cell separates into two daughter cells.

Forming of Molecular Chain

Although DNA has never been synthesized, Drs. Watson and Crick knew it was composed of horizontal hook-ups of bases (sugars and phosphates) piled one above the other in chain-like formations. The problem was to find out how these giant molecules could be fitted together so they could duplicate themselves exactly.

By a method of scientific doodling with hand-drawn models of the molecules, Drs. Watson and Crick worked out which molecules could be joined together with regard to the fact that some molecules were more rigid than others and had critical angles of attachment. Some months ago they decided that the only possible interrelation of the molecules was in the form of two chains arranged in a double helix—like a spiral staircase, with the upper chain resembling the staircase handrail and the lower resembling the outside edge of the stairs.

New evidence for double DNA chains in helical form now has been obtained from the King's College Biophysics Department in London, where a group of workers extracted crystalline DNA from the thymus gland of a calf and bombarded it with X-rays.

The resulting X-ray diffraction photographs showed a whirlpool of light and shade that could be analyzed as the components of a double helix.

Dr. Crick emphasized that years of work still must be applied to the helical carriers of life's characteristics. But a working model to aid in the genetical studies of the future now has been laid out in blueprint form by Drs. Watson and Crick— or so most biochemists here believe.

Looks Good, Pauling Says

Reached by telephone in Pasadena, Dr. Pauling said last night that the Crick-Watson proposal for the structure of the nucleic acids "looks very good."

Dr. Pauling has just returned from London, where he talked with Dr. Crick and with Dr. Watson, who was formerly a student at California Institute of Technology.

Dr. Pauling said that he did not believe the problem of understanding "molecular genetics" had been finally solved, and that the shape of the molecules was a complicated matter. Both the California and the Crick-Watson explanations of the structure of the substances that control heredity are highly speculative, he remarked.

—June 13, 1953

50 Years Later, Rosalind Franklin's X-Ray Fuels Debate

By DENISE GRADY

Fifty years ago, a casual gesture at a laboratory in London became a defining moment in the history of science. James D. Watson was visiting King's College late one afternoon near the end of January 1953 when a researcher named Maurice Wilkins showed him an X-ray photograph of a molecule of DNA.

Describing the encounter years later in *The Double Helix*, Dr. Watson wrote, "The instant I saw the picture my mouth fell open and my pulse began to race."

The image was one of many by various researchers that hinted at a helix, but its singular clarity helped lead Dr. Watson and his colleague Francis Crick to the structure of DNA.

The scientist who took the picture was Dr. Rosalind Franklin, and though they cited other work she had done, Dr. Watson and Dr. Crick did not acknowledge the photograph itself, or additional work by her they had used, in their paper.

Some historians say that is of little importance because the two would have deduced the structure even without the image. Besides, they say, Dr. Franklin herself did not seem to recognize the picture's importance; she had put it aside.

But for others, over the years, Dr. Franklin has come to symbolize the plight of women in science, as men close ranks against them.

Dr. Franklin's X-ray image, labeled "Photograph 51," showed a distinctive X-shaped pattern. On the train back to Cambridge that night, Dr. Watson sketched on his newspaper the details he remembered from the picture, clues to the angles and spacing within the helix. By the time he got home, he had decided that its most likely structure was a double helix, and that he and Francis Crick should build a model to see if the pieces would fit.

The pressure was on: in America, Dr. Linus Pauling was at work, too, and he could not be far from cracking it.

A month later, the Watson-Crick team won the race: the pieces fit, they soon published a paper, and they went on to win the Nobel Prize, along with Dr. Wilkins.

But a question remains: had they used Dr. Franklin's data without her permission or knowledge, and without giving her adequate credit?

It was Dr. Franklin, not her colleague Dr. Wilkins, who created Photograph 51, an image of what was known as the B form of DNA. Dr. Wilkins, who did not get along with Dr. Franklin, showed Dr. Watson the picture without telling her. Dr. Watson and Dr. Crick also used a report Dr. Franklin had written, passed to them by Max Perutz, a colleague and member of a research oversight committee.

"Rosy, of course, did not directly give us her data," Dr. Watson wrote. "For that matter, no one at King's realized they were in our hands."

But Dr. Crick does not share this perception. Dr. Franklin "must have known we knew most of it," he said in an interview.

For one thing, Dr. Franklin had made her data public at a seminar both men had been invited to attend. Unfortunately, Dr. Watson misinterpreted her presentation and Dr. Crick was not there.

As a result, the photograph Dr. Wilkins showed Dr. Watson was "the thing that triggered this off," Dr. Crick said in an interview. But he added: "We could have got that information from earlier work. It would have made a big difference if I'd gone to that seminar."

"We never had a discussion afterward," he said of Dr. Franklin. "She never raised the issue."

Dr. Watson and Dr. Crick did acknowledge other work by Dr. Wilkins and Dr. Franklin in their paper, but left it to Dr. Wilkins to decide whether he and Dr. Franklin should share authorship, an offer he declined. Dr. Franklin and Dr. Wilkins themselves published much of their data on DNA, including the famous photograph, in the same issue of *Nature* as the Watson and Crick paper.

As for the 1962 Nobel Prize, Dr. Franklin could not have been included; she died in 1958 of ovarian cancer and the Nobel is never awarded posthumously. Whether she would have been included is not clear; the prize is not split more than three ways.

In any case, neither Dr. Watson nor Dr. Crick mentioned Dr. Franklin during his Nobel speech, though Dr. Wilkins did.

Dr. Watson's *Double Helix* has, if anything, contributed to the view that Dr. Franklin was unfairly deprived of credit she deserved. The book is peppered with snide comments about her looks and describes her as so snappish and fierce that Dr. Watson and Dr. Wilkins, who towered over her, supposedly feared she would hit them.

"Clearly Rosy had to go or be put in her place," Dr. Watson wrote in Chapter 2 and added, later, "The thought could not be avoided that the best home for a feminist was in another person's lab."

In a biography published in October, *Rosalind Franklin: The Dark Lady of DNA*, Brenda Maddox theorizes that Dr. Watson portrayed Dr. Franklin as hostile and unreasonable to justify using her data without telling her. Ms. Maddox writes that Dr. Watson added a "pious epilogue" praising Dr. Franklin only after numerous colleagues who read a draft expressed outrage at his depiction of her.

According to Ms. Maddox's biography, Dr. Watson and Dr. Crick eventually became friendly with Dr. Franklin. After her death, both men praised her generously in public forums.

"At no time did Rosalind, as far as is known, express any resentment at our having solved the structure," Dr. Crick said.

In a recent interview in which he discussed Dr. Franklin, Dr. Watson said: "She never felt she was robbed. People say, 'Well, she never knew we saw the B photo.' That was a question between Maurice and her."

—February 25, 2003

Scientist Reports First Cloning Ever of Adult Mammal

By GINA KOLATA

In a feat that may be the one bit of genetic engineering that has been anticipated and dreaded more than any other, researchers in Britain are reporting that they have cloned an adult mammal for the first time.

The group, led by Dr. Ian Wilmut, a 52-year-old embryologist at the Roslin Institute in Edinburgh, created a lamb using DNA from an adult sheep. The achievement shocked leading researchers who had said it could not be done. The researchers had assumed that the DNA of adult cells would not act like the DNA formed when a sperm's genes first mingle with those of an egg.

In theory, researchers said, such techniques could be used to take a cell from an adult human and use the DNA to create a genetically identical human— a time-delayed twin. That prospect raises the thorniest of ethical and philosophical questions.

Dr. Wilmut's experiment was simple, in retrospect. He took a mammary cell from an adult sheep and prepared its DNA so it would be accepted by an egg from another sheep. He then removed the egg's own DNA, replacing it with the DNA from the adult sheep by fusing the egg with the adult cell. The fused cells, carrying the adult DNA, began to grow and divide, just like a perfectly normal fertilized egg, to form an embryo.

Dr. Wilmut implanted the embryo into another ewe; in July, the ewe gave birth to a lamb, named Dolly. Though Dolly seems perfectly normal, DNA tests show that she is the clone of the adult ewe that supplied her DNA.

"What this will mostly be used for is to produce more health care products," Dr. Wilmut told the Press Association of Britain early today, the Reuters news agency reported.

"It will enable us to study genetic diseases for which there is presently no cure and track down the mechanisms that are involved. The next step is to use the cells in culture in the lab and target genetic changes into that culture."

Simple though it may be, the experiment, to be reported this coming Thursday in the British journal *Nature*, has startled biologists and ethicists. Dr. Wilmut said in a telephone interview last week that he planned to breed Dolly next fall to determine whether she was fertile. Dr. Wilmut said he was interested in the technique primarily as a tool in animal husbandry, but other scientists said it had opened doors to the unsettling prospect that humans could be cloned as well.

Dr. Lee Silver, a biology professor at Princeton University, said last week that the announcement had come just in time for him to revise his forthcoming book so the first chapter will no longer state that such cloning is impossible.

"It's unbelievable," Dr. Silver said. "It basically means that there are no limits. It means all of science fiction is true. They said it could never be done and now here it is, done before the year 2000."

Dr. Neal First, a professor of reproductive biology and animal biotechnology at the University of Wisconsin, who has been trying to clone cattle, said the ability to clone dairy cattle could have a bigger impact on the industry than the introduction of artificial insemination in the 1950s, a procedure that revolutionized dairy farming. Cloning could be used to make multiple copies of animals that are especially good at producing meat or milk or wool.

Although researchers have created genetically identical animals by dividing embryos very early in their development, Dr. Silver said, no one had cloned an animal from an adult until now. Earlier experiments, with frogs, have become a stock story in high school biology, but the experiments never produced cloned adult frogs. The frogs developed only to the tadpole stage before dying.

It was even worse with mammals. Researchers could swap DNA from one fertilized egg to another, but they could go no further. "They couldn't even put nuclei from late-stage mouse embryos into early mouse embryos," Dr. Silver said. The embryos failed to develop and died.

As a result, the researchers concluded that as cells developed, the proteins coating the DNA somehow masked all the important genes for embryo development. A skin cell may have all the genetic information that was present in the fertilized egg that produced the organism, for example, but almost all that information is pasted over. Now all the skin cell can do is be a skin cell.

Researchers could not even hope to strip off the proteins from an adult cell's DNA and replace them with proteins from an embryo's DNA. The DNA would shatter if anyone tried to strip it bare, Dr. Silver said.

Last year, Dr. Wilmut showed that he could clone DNA from sheep embryo cells, but even that was not taken as proof that the animal itself could be cloned. It could just be that the embryo cells had DNA that was unusually conducive to cloning, many thought.

Dr. Wilmut, however, hit on a clever strategy. He did not bother with the proteins that coat DNA, and instead focused on getting the DNA from an adult cell into a stage in its normal cycle of replication where it could take up residence in an egg.

DNA in growing cells goes through what is known as the cell cycle: it prepares itself to divide, then replicates itself and splits in two as the cell itself divides. The problem with earlier cloning attempts, Dr. Wilmut said, was that the DNA from the donor had been out of synchrony with that of the recipient cell. The solution, he discovered, was, in effect, to put the DNA from the adult cell to sleep, making it quiescent by depriving the adult cell of nutrients. When he then fused it with an egg cell from another sheep—after removing the egg cell's DNA—the donor DNA took over as though it belonged there.

Dr. Wilmut said in the telephone interview last week that the method could work for any animal and that he hoped to use it next to clone cattle. He said that he could use many types of cells from adults for cloning but that the easiest to use would be so-called stem cells, which give rise to a variety of other cells and are present throughout the body.

In his sheep experiment, he used mammary cells because a company that sponsored his work, PPL Therapeutics, is developing sheep that can be used to produce proteins that can be used as drugs in their milk, so it had sheep mammary cells readily available.

For Dr. Wilmut, the main interest of the experiment is to advance animal research. PPL, for example, wants to clone animals that can produce pharmacologically useful proteins, like the clotting factor needed by hemophiliacs. Scientists would grow cells in the laboratory, insert the genes for production of the desired protein, select those cells that most actively churned out the protein and use those cells to make cloned females. The cloned animals would produce immense amounts of the proteins in their milk, making the animals into living drug factories.

But that is only the beginning, Dr. Wilmut said. Researchers could use the same method to make animals with human diseases, like cystic fibrosis, and then test therapies on the cloned animals. Or they could use cloning to alter the proteins on the surfaces of pig organs, like the liver or heart, making the organs more like human organs. Then they could transplant those organs into humans.

Dr. First said the "exciting and astounding" cloning result could shake the dairy industry. It could allow the cloning of cows that are superproducers of milk, making 30,000 or even 40,000 pounds of milk a year. The average cow makes about 13,000 pounds of milk a year, he said.

"I think that if—and it's a very big if—cloning were highly efficient," Dr. First said last week, "then it could be a more significant revolution to the livestock industry than even artificial insemination."

Although Dr. Wilmut said he saw no intrinsic biological reason humans, too, could not be cloned, he dismissed the idea as being ethically unacceptable. Moreover, he said, it is illegal in Britain to clone people. "I would find it offensive" to clone a human being, Dr. Wilmut said, adding that he fervently hoped that no one would try it.

But others said that it was hard to imagine enforcing a ban on cloning people when cloning got more efficient. "I could see it going on surreptitiously," said Lori Andrews, a professor at Chicago-Kent College of Law who specializes in reproductive issues. For example, Professor Andrews said last week, in the early days of in vitro fertilization, Australia banned that practice. "So scientists moved to Singapore" and offered the procedure, she said. "I can imagine new crimes," she added.

People might be cloned without their knowledge or consent. After all, all that would be needed would be some cells. If there is a market for a sperm bank selling semen from Nobel laureates, how much better would it be to bear a child that would actually be a clone of a great thinker or, perhaps, a great beauty or great athlete?

"The genie is out of the bottle," said Dr. Ronald Munson, a medical ethicist at the University of Missouri in St. Louis. "This technology is not, in principle, policeable."

Dr. Munson called the possibilities incredible. For example, could researchers devise ways to add just the DNA of an adult cell, without fusing two living cells? If so, might it be possible to clone the dead?

"I had an idea for a story once," Dr. Munson said, in which a scientist obtains a spot of blood from the cross on which Jesus was crucified. He then uses it to clone a man who is Jesus Christ—or perhaps cannot be.

On a more practical note, Dr. Munson mused over the strange twist that science has taken.

"There's something ironic" about study, he said. "Here we have this incredible technical accomplishment, and what motivated it? The desire for more sheep milk of a certain type." It is, he said, "the theater of the absurd acted out by scientists."

In his interview with the Press Association, Britain's domestic news agency, Dr. Wilmut added early today: "We are aware that there is potential for misuse, and we have provided information to ethicists and the Human Embryology Authority. We believe that it is important that society decides how we want to use this technology and makes sure it prohibits what it wants to prohibit. It would be desperately sad if people started using this sort of technology with people."

—*February 23, 1997*

Pas de Deux of Sexuality Is Written in the Genes

By NICHOLAS WADE

When it comes to the matter of desire, evolution leaves little to chance. Human sexual behavior is not a free-form performance, biologists are finding, but is guided at every turn by genetic programs.

Desire between the sexes is not a matter of choice. Straight men, it seems, have neural circuits that prompt them to seek out women; gay men have those prompting them to seek other men. Women's brains may be organized to select men who seem likely to provide for them and their children. The deal is sealed with other neural programs that induce a burst of romantic love, followed by long-term attachment.

So much fuss, so intricate a dance, all to achieve success on the simple scale that is all evolution cares about, that of raising the greatest number of children to adulthood. Desire may seem the core of human sexual behavior, but it is just the central act in a long drama whose script is written quite substantially in the genes.

In the womb, the body of a developing fetus is female by default and becomes male if the male-determining gene known as SRY is present. This dominant gene, the Y chromosome's proudest and almost only possession, sidetracks the reproductive tissue from its ovarian fate and switches it into becoming testes. Hormones from the testes, chiefly testosterone, mold the body into male form.

In puberty, the reproductive systems are primed for action by the brain. Amazing electrical machine that it may be, the brain can also behave like a humble gland. In the hypothalamus, at the central base of the brain, lie a cluster of about 2,000 neurons that ignite puberty when they start to secrete pulses of gonadotropin-releasing hormone, which sets off a cascade of other hormones.

The trigger that stirs these neurons is still unknown, but probably the brain monitors internal signals as to whether the body is ready to reproduce and external cues as to whether circumstances are propitious for yielding to desire.

Several advances in the last decade have underlined the bizarre fact that the brain is a full-fledged sexual organ, in that the two sexes have profoundly different versions of it. This is the handiwork of testosterone, which masculinizes the brain as thoroughly as it does the rest of the body.

It is a misconception that the differences between men's and women's brains are small or erratic or found only in a few extreme cases, Dr. Larry Cahill of the University of California, Irvine, wrote last year in *Nature Reviews Neuroscience*. Widespread regions of the cortex, the brain's outer layer that performs much of its higher-level processing, are thicker in women. The hippocampus, where initial memories are formed, occupies a larger fraction of the female brain.

Techniques for imaging the brain have begun to show that men and women use their brains in different ways even when doing the same thing. In the case of the amygdala, a pair of organs that helps prioritize memories according to their emotional strength, women use the left amygdala for this purpose but men tend to use the right.

It is no surprise that the male and female versions of the human brain operate in distinct patterns, despite the heavy influence of culture. The male brain is sexually oriented toward women as an object of desire. The most direct evidence comes from a handful of cases, some of them circumcision accidents, in which boy babies have lost their penises and been reared as female. Despite every social inducement to the opposite, they grow up desiring women as partners, not men.

"If you can't make a male attracted to other males by cutting off his penis, how strong could any psychosocial effect be?" said J. Michael Bailey, an expert on sexual orientation at Northwestern University.

Presumably the masculinization of the brain shapes some neural circuit that makes women desirable. If so, this circuitry is wired differently in gay men. In experiments in which subjects are shown photographs of desirable men or women, straight men are aroused by women, gay men by men.

Such experiments do not show the same clear divide with women. Whether women describe themselves as straight or lesbian, "Their sexual arousal seems to be relatively indiscriminate—they get aroused by both male and female images," Dr. Bailey said. "I'm not even sure females have a sexual orientation. But they have sexual preferences. Women are very picky, and most choose to have sex with men."

Dr. Bailey believes that the systems for sexual orientation and arousal make men go out and find people to have sex with, whereas women are more focused on accepting or rejecting those who seek sex with them.

Similar differences between the sexes are seen by Marc Breedlove, a neuroscientist at Michigan State University. "Most males are quite stubborn in their ideas about which sex they want to pursue, while women seem more flexible," he said.

Sexual orientation, at least for men, seems to be settled before birth. "I think most of the scientists working on these questions are convinced that the antecedents of sexual orientation in males are happening early in life, probably before birth," Dr. Breedlove said, "whereas for females, some are probably born to become gay, but clearly some get there quite late in life."

Sexual behavior includes a lot more than sex. Helen Fisher, an anthropologist at Rutgers University, argues that three primary brain systems have evolved to direct reproductive behavior. One is the sex drive that motivates people to seek partners. A second is a program for romantic attraction that makes people fixate on specific partners. Third is a mechanism for long-term attachment that induces people to stay together long enough to complete their parental duties.

Romantic love, which in its intense early stage "can last 12–18 months," is a universal human phenomenon, Dr. Fisher wrote last year in *The Proceedings of the Royal Society*, and is likely to be a built-in feature of the brain. Brain imaging studies show that a particular area of the brain, one associated with the reward system, is activated when subjects contemplate a photo of their lover.

The best evidence for a long-term attachment process in mammals comes from studies of voles, a small mouselike rodent. A hormone called vasopressin, which is active in the brain, leads some voles to stay pair-bonded for life. People possess the same hormone, suggesting a similar mechanism could be at work in humans, though this has yet to be proved.

Researchers have devoted considerable effort to understanding homosexuality in men and women, both for its intrinsic interest and for the light it could shed on the more usual channels of desire. Studies of twins show that homosexuality, especially among men, is quite heritable, meaning there is a genetic component to it. But since gay men have about one-fifth as many children as straight men, any gene favoring homosexuality should quickly disappear from the population.

Such genes could be retained if gay men were unusually effective protectors of their nephews and nieces, helping genes just like theirs get into future generations. But gay men make no better uncles than straight men, according to a study by Dr. Bailey. So that leaves the possibility that being gay is a byproduct of a gene that persists because it enhances fertility in other family members. Some studies have found that gay men have more relatives than straight men, particularly on their mother's side.

But Dr. Bailey believes the effect, if real, would be more clear-cut. "Male homosexuality is evolutionarily maladaptive," he said, noting that the phrase means only that genes favoring homosexuality cannot be favored by evolution if fewer such genes reach the next generation.

A somewhat more straightforward clue to the origin of homosexuality is the fraternal birth order effect. Two Canadian researchers, Ray Blanchard and Anthony F. Bogaert, have shown that having older brothers substantially increases the chances that a man will be gay. Older sisters don't count, nor does it matter whether the brothers are in the house when the boy is reared.

The finding suggests that male homosexuality in these cases is caused by some event in the womb, such as "a maternal immune response to succeeding male pregnancies," Dr. Bogaert wrote last year in the *Proceedings of the National Academy of Sciences*. Antimale antibodies could perhaps interfere with the usual masculinization of the brain that occurs before birth, though no such antibodies have yet been detected.

The fraternal birth order effect is quite substantial. Some 15 percent of gay men can attribute their homosexuality to it, based on the assumption that 1 percent to 4 percent of men are gay, and each additional older brother increases the odds of same-sex attraction by 33 percent.

The effect supports the idea that the levels of circulating testosterone before birth are critical in determining sexual orientation. But testosterone in the fetus cannot be measured, and as adults, gay and straight men have the same levels of the hormone, giving no clue to prenatal exposure. So the hypothesis, though plausible, has not been proved.

A significant recent advance in understanding the basis of sexuality and desire has been the discovery that genes may have a direct effect on the sexual differentiation of the brain. Researchers had long assumed that steroid hormones like testosterone and estrogen did all the heavy lifting of shaping the male and female brains. But Arthur Arnold of the University of California, Los Angeles, has found that male and female neurons behave somewhat differently when kept in laboratory glassware. And last year Eric Vilain, also of UCLA, made the surprising finding that the SRY gene is active in certain cells of the brain, at least in mice. Its brain role is quite different from its testosterone-related activities, and women's neurons presumably perform that role by other means.

It so happens that an unusually large number of brain-related genes are situated on the X chromosome. The sudden emergence of the X and Y chromosomes in brain function has caught the attention of evolutionary biologists. Since men have only one X chromosome, natural selection can speedily promote any advantageous mutation that arises in one of the X's genes. So if those picky women should be looking for smartness in prospective male partners, that might explain why so many brain-related genes ended up on the X.

"It's popular among male academics to say that females preferred smarter guys," Dr. Arnold said. "Such genes will be quickly selected in males because new beneficial mutations will be quickly apparent."

Several profound consequences follow from the fact that men have only one copy of the many X-related brain genes and women two. One is that many neurological diseases are more common in men because women are unlikely to suffer mutations in both copies of a gene.

Another is that men, as a group, "will have more variable brain phenotypes," Dr. Arnold writes, because women's second copy of every gene dampens the effects of mutations that arise in the other.

Greater male variance means that although average IQ is identical in men and women, there are fewer average men and more at both extremes. Women's care in selecting mates, combined with the fast selection made possible by men's lack of backup copies of X-related genes, may have driven the divergence between male and female brains. The same factors could explain, some researchers believe, why the human brain has tripled in volume over just the last 2.5 million years.

Who can doubt it? It is indeed desire that makes the world go round.

—April 10, 2007

My Genome, Myself:
Seeking Clues in DNA

By AMY HARMON

The exploration of the human genome has long been relegated to elite scientists in research laboratories. But that is about to change. An infant industry is capitalizing on the plunging cost of genetic testing technology to offer any individual unprecedented—and unmediated—entrée to their own DNA.

For as little as $1,000 and a saliva sample, customers will be able to learn what is known so far about how the billions of bits in their biological code shape who they are. Three companies have already announced plans to market such services, one yesterday.

Offered the chance to be among the early testers, I agreed, but not without reservations. What if I learned I was likely to die young? Or that I might have passed on a rogue gene to my daughter? And more pragmatically, what if an insurance company or an employer used such information against me in the future?

But three weeks later, I was already somewhat addicted to the daily communion with my genes. (Recurring note to self: was this addiction genetic?)

For example, my hands hurt the other day. So naturally, I checked my DNA.

Was this the first sign that I had inherited the arthritis that gnarled my paternal grandmother's hard-working fingers? Logging onto my account at 23andMe, the start-up company that is now my genetic custodian, I typed my search into the "Genome Explorer" and hit return. I was, in essence, Googling my own DNA.

I had spent hours every day doing just that as new studies linking bits of DNA to diseases and aspects of appearance, temperament and behavior came out on an almost daily basis. At times, surfing my genome induced the same shock of recognition that comes when accidentally catching a glimpse of oneself in the mirror.

I had refused to drink milk growing up. Now, it turns out my DNA is devoid of the mutation that eases the digestion of milk after infancy, which became common in Europeans after the domestication of cows.

But it could also make me question my presumptions about myself. Apparently I lack the predisposition for good verbal memory, although I had always prided myself on my ability to recall quotations. Should I be recording more of my interviews? No, I decided; I remember what people say. DNA is not definitive.

I don't like brussels sprouts. Who knew it was genetic? But I have the snippet of DNA that gives me the ability to taste a compound that makes many vegetables taste bitter. I differ from people who are blind to bitter taste—who actually like brussels sprouts—by a single spelling change in our four-letter genetic alphabet: somewhere on human chromosome 7, I have a G where they have a C.

It is one of roughly 10 million tiny differences, known as single nucleotide polymorphisms, or SNPs (pronounced "snips") scattered across the 23 pairs of human chromosomes from which 23andMe takes its name. The company generated a list of my "genotypes"—ACs, CCs, CTs and so forth, based on which versions of every SNP I have on my collection of chromosome pairs.

For instance, I tragically lack the predisposition to eat fatty foods and not gain weight. But people who, like me, are GG at the SNP known to geneticists as rs3751812 are 6.3 pounds lighter, on average, than the AAs. Thanks, rs3751812!

And if an early finding is to be believed, my GG at rs6602024 means that I am an additional 10 pounds lighter than those whose genetic Boggle served up a different spelling. Good news, except that now I have only my slothful ways to blame for my inability to fit into my old jeans.

And although there is great controversy about the role that genes play in shaping intelligence, it was hard to resist looking up the SNPs that have been linked—however tenuously—to I.Q. Three went in my favor, three against. But I found hope in a study that appeared last week describing a SNP strongly linked with an increase in the I.Q. of breast-fed babies.

Babies with the CC or CG form of the SNP apparently benefit from a fatty acid found only in breast milk, while those with the GG form do not. My CC genotype meant that I had been eligible for the 6-point I.Q. boost when my mother breast-fed me. And because, by the laws of genetics, my daughter had to have inherited one of my Cs, she, too, would see the benefit of my having nursed her. Now where did I put those preschool applications?

I was not always so comfortable in my own genome. Before I spit into the vial, I called several major insurance companies to see if I was hurting my chances of getting coverage. They said no, but that is now, when almost no one has such information about their genetic make-up. In five years, if companies like 23andMe are at all successful, many more people presumably would. And isn't an individual's relative risk of disease precisely what insurance companies want to know?

Last month, alone in a room at 23andMe's headquarters in Mountain View, Calif., with my password for the first time, I wavered (genetic?) and walked down the hall to get lunch.

Once I looked at my results, I could never turn back. I had prepared for the worst of what I could learn this day. But what if something even worse came along tomorrow?

Some health care providers argue that the public is unprepared for such information and that it is irresponsible to provide it without an expert to help put it in context. And at times, as I worked up the courage to check on my risks of breast cancer and Alzheimer's, I could see their point.

One of the companies that plans to market personal DNA information, Navigenics, intends to provide a phone consultation with a genetic counselor along with the results. Its service would cost $2,500 and would initially provide data on 20 health conditions.

DeCODE Genetics and 23andMe will offer referrals. Although what they can tell you is limited right now, all three companies are hoping that people will be drawn by the prospect of instant updates on what is expected to be a flood of new findings.

I knew I would never be able to pass up the chance to fill in more pieces of my genetic puzzle.

But I had decided not to submit my daughter's DNA for testing—at least not yet—because I didn't want to regard anything about her as predestined. If she wants to play the piano, who cares if she lacks perfect pitch? If she wants to run the 100-meter dash, who cares if she lacks the sprinting gene? And did I really want to know—did she really want to know someday—what genes she got from which parent and which grandparent?

I, however, am not age three. Whatever was lurking in my genes had been there my entire life. Not looking would be like rejecting some fundamental part of myself.

Compelled to know (genetic?), I breezed through the warning screens on the site. There would be no definitive information, I read, and new discoveries might reverse whatever I was told. Even if I learned that my risk for developing a disease was high, there might well be nothing to do about it, and, besides, I should not regard this as a medical diagnosis. "If, after considering these points, you still wish to view your results," the screen read, "click here."

I clicked.

Like other testers of 23andMe's service, my first impulse was to look up the bits of genetic code associated with the diseases that scare me the most.

But in the bar charts that showed good genes in green and bad ones in red, I found a perverse sense of accomplishment. My risk of breast cancer was no higher than average, as was my chance of developing Alzheimer's. I was 23 percent less likely to get Type 2 diabetes than most people. And my chance of being paralyzed by multiple sclerosis, almost nil. I was three times more likely than the average person to get Crohn's disease, but my odds were still less than one in a hundred.

I was in remarkably good genetic health, and I hadn't even been to the gym in months!

Still, just studying my DNA had made me more acutely aware of the basic health risks we all face. I renounced my midafternoon M&M's.

And then I opened my "Gene Journal" for heart disease to find that I was 23 percent more likely than average to have a heart attack. "Healthy lifestyle choices play a major role in preventing the blockages that lead to heart attacks," it informed me.

Thanks, Gene Journal. Yet somehow even this banal advice resonated when the warning came from my own DNA.

Back in New York, I headed to the gym despite a looming story deadline and my daughter's still-unfinished preschool applications. At least I had more time. I had discovered a SNP that likely increased my life span.

But in what I have come to accept as the genomic law of averages, I soon found that I might well be sight impaired during those extra years. According to the five SNPs for macular degeneration I fed into the "Genome Explorer," I was nearly 100 times more likely to develop the disease than someone with the most favorable A-C-G-T combination.

And unlike the standard eat-right-and-exercise advice for heart health, there was not much I could do about it. Still, I found the knowledge of my potential future strangely comforting, even when it was not one I would wish for. At least my prospects for nimble fingers in old age were looking brighter. I didn't have the bad form of that arthritis SNP.

Maybe I was just typing too much.

—November 17, 2007

His Fertility Advance Draws Ire

By SABRINA TAVERNISE

To most people, the word "mitochondria" is only dimly familiar, the answer to a test question in some bygone high school biology class. But to Shoukhrat Mitalipov, the mysterious power producers inside every human cell are a lifelong obsession.

"My colleagues, they say I'm a 'mitochondriac,' that I only see this one thing," he said recently in his modest, clutter-free office at Oregon Health and Science University. He smiled. "Maybe they are right."

With a name that most Americans can't pronounce (it is Shoe-KHRAHT Mee-tuhl-EE-pov) and an accent that sounds like the villain's in a James Bond film, Dr. Mitalipov, 52, has shaken the field of genetics by perfecting a version of the world's tiniest surgery: removing the nucleus from a human egg and placing it into another. In doing so, this Soviet-born scientist has drawn the ire of bioethicists and the scrutiny of federal regulators.

The procedure is intended to help women conceive children without passing on genetic defects in their cellular mitochondria. Such mutations are rare, but they can cause severe problems, including neurological damage, heart failure and blindness. About one in 4,000 babies in the United States is born with an inherited mitochondrial disease; there is no treatment, and few live into adulthood.

Mitochondria have their own sets of genes, inherited solely from mothers, and women who carry mitochondrial mutations are understandably eager to not pass them to their children. Dr. Mitalipov's procedure would allow these women to bear children by placing the nucleus from the mother's egg into a donor egg whose nucleus has been removed. The defective mitochondria, which float outside the nucleus in the egg's cytoplasm, are left behind.

"It was a major breakthrough," said Douglas C. Wallace, a professor of pathology and laboratory medicine at the University of Pennsylvania. "He's an exceptionally talented person."

But the resulting baby would carry genetic material from three parents—the mother, the host egg's donor and the father—an outcome that ethicists have deplored.

That specter drew critics from all over the country to a hotel in suburban Maryland late last month, where Dr. Mitalipov tried to persuade a panel of experts convened by the Food and Drug Administration that the procedure, which he has pioneered in monkeys, was ready to test in people.

Some told the officials that the technique could introduce new genetic mutations into the human gene pool. Others warned that it could be used later for something ethically murkier—perhaps, said Marcy Darnovsky, executive director of the Center for Genetics and Society, "to engineer children with specific character traits."

Back in his office, Dr. Mitalipov waved off those warnings. Mitochondrial DNA comprises just 37 genes, which direct the production of enzymes and molecules that the cell needs for energy, he noted. They have nothing to do with traits like eye and hair color, which are encoded in the nucleus.

"There are always people trying to stir things up," said Dr. Mitalipov, an American citizen who grew up in what is now Kazakhstan. "Many of them made their careers by criticizing me."

The United States is not the only country weighing mitochondrial replacement. In Britain, the government has issued draft regulations that would govern clinical trials in people. If accepted into law by Parliament, such trials, which are now banned, would be allowed to go forward, although regulators would have to license any clinical application.

Dr. Mitalipov's fixation on mitochondria began in graduate school in Russia in the 1990s. After graduating from an agricultural institute—and a brief, unhappy stint as a manager on a collective farm—he began work on his doctoral thesis at the Research Center of Medical Genetics, a prestigious state-funded institution in Moscow. He focused on embryonic stem cells, which can be grown in the laboratory and turned into any type of cell in the body.

He noticed a strange thing. When stem cells were extracted from a mouse embryo and put in a petri dish, they stopped aging but remained healthy and growing, as if frozen in time. Somewhere in the cell, it seemed, was a clock that determined its life span.

The search for the clock took him to Utah State University for postdoctoral research in the mid-1990s. He developed an interest in cloning, a process in which the cellular clock is not only stopped but reset. Why, he wondered, do cloned animals have normal life spans?

The answer to the riddle of cellular aging was not to be found in the cell's nucleus, Dr. Mitalipov concluded, but in the surrounding cytoplasm. In the mitochondria.

"Everything was falling into place in my head," he said.

As researchers began to suspect defective mitochondria as a cause in more diseases, Dr. Mitalipov wondered whether replacing them might be possible.

Scientists already had experimented with combining genetic material from three people to make a baby. About 15 years ago, researchers in New Jersey injected a bit of cell fluid from donor eggs into the eggs of women who were having fertility problems. Those experiments, which came shortly after the cloning of Dolly the sheep, set off such an uproar that the F.D.A. eventually told researchers that they could not perform them without special permission.

Dr. Mitalipov persevered. At Oregon Health and Science University's National Primate Research Center, one of eight in the country, he spent years perfecting a way to create monkey eggs with donated mitochondria. He persuaded software developers to adapt a program that would allow real-time viewing of the necessary microsurgery. A special microscope was developed so that human hands, too blunt an instrument on their own, could conduct the operation with joysticks that look like upside-down flashlights.

"He's just a really practical guy," said Daniel M. Dorsa, senior vice president for research at the university. "He just nose-to-the-grindstone plowed through and figured out what it took."

Success came in 2008 in a darkened, hot laboratory room. On April 24, 2009, twin male rhesus monkeys, Mito and Tracker, were born with replaced mitochondria. Later, with some adjustments Dr. Mitalipov replicated the procedure in human eggs. Because of federal rules against genetic manipulation, the eggs were not allowed to mature.

His research has brought persistent criticism. "If these procedures are carried out, it crosses a very bright line," said Ms. Darnovsky of the genetics center.

She said that the current goal, mitochondrial replacement, may be narrow, but that Dr. Mitalipov's genetic techniques could lead to broader applications and eventually to a situation in which scientists or governments "compete to enhance future generations," such as producing soldiers who never need sleep.

Sheldon Krimsky, a bioethicist who attended the F.D.A. meeting on behalf of the Council for Responsible Genetics, argues that mitochondrial replacement is simply unnecessary. There are other options for women with mitochondrial defects to have healthy children, such as getting an egg from a donor, or having prenatal genetic diagnosis to find eggs with fewer mutations, he said.

"There's that genetic chauvinism that says unless my DNA is in the child, it will not be truly my child," he said.

Would-be parents, on the other hand, have been following Dr. Mitalipov's work with the intensity of the hungry waiting for food. When he came back from the meeting in Maryland, his inbox contained an avalanche of emails from women with mitochondrial mutations and other fertility problems.

Dr. Dorsa said the university still has not decided whether to formally ask the F.D.A. for permission to move forward with clinical trials.

Dr. Mitalipov, for his part, is determined.

"We are ready now to move on to the next stage," he said. "Not in 10 years, but in the next few years."

—March 18, 2014

CHAPTER 4

Earth Science: Plates, Poles and Oceans

THE FORCES THAT SHAPE THE PLANET

Theory of Continental Drift Is Tested in Icy Greenland

By KURT WEGENER

The expedition set on foot by the German Science Aid Association, organized and led by my brother Alfred, and brought to its conclusion by me after his death, was a spatially and mentally far-ramifying enterprise. First of all, it meant the first winter sojourn in this, one of the greatest of the earth's ice-wastes. Peary, Nansen and Quervain carried out their famous trips across Greenland in the summer, and of the winter climate there only vague ideas were current. During the summer the Greenland ice field is commonly under brilliant sunshine, and cloudiness is rare; solar radiation is so intense that at times the members of the expedition could work only by the light of the midnight sun, as otherwise they would have been severely snow-burned.

Our meteorological observations were carried on at the Eastern Station, on Scoresby Sound, at the expedition's headquarters on the west coast, and at the Central Station, on the inland ice, where Drs. Georgi, Sorge and Loewe passed the winter wholly cut off from the world. In addition, we had midway between the Western and the Central Station, at a distance of 125 miles from the coast, a weather kiosk with automatically registering instruments whose clocks—with a run of three months—were kept going by passing sled parties.

Year-Round Observations

With pride we reckon among our most important achievements that our expedition could take weather observations throughout the year and across the full width of Greenland—the first time this ever was done. Kites and balloons were sent up hundreds of times, pilot balloons rising as high as 10½ miles.

This work at times encountered great difficulties. Thus, in summer, Dr. Georgi had to place the bottle containing the gas for the balloons down an ice well several yards deep to obtain a tolerably even temperature. In winter, blizzards frequently made the daily reading of the instruments an extremely dangerous job, both Dr. Sorge and Dr. Georgi several times escaping death only by a close shave.

Important Results Obtained

The meteorological results, even as surveyable at present, are extremely important. Weather conditions over Greenland are shown to be far more complicated than had been assumed. The "Greenland High" (area of high barometric pressure), to which climatologists ascribe so much influence on the weather over the Atlantic and Central Europe, is structurally very different from what it had been supposed. Instead of extending only to a height of about 1¾ miles, it reaches up to 8¾ miles and even higher, thus up to the border of the stratosphere—a discovery of immense importance. Further, there does not prevail that constant clear weather which had been experienced over the inland ice in summer. Both Central and West Stations recorded strong temperature fluctuations and much fog and clouds.

In order to make possible comparison of the climatological data obtained at Central Station and West Station—whose winter house was at 3,116 feet altitude—with the weather factors at sea level, a supplementary meteorological station was set up on Umanak Island, on the west coast. We were thus able to obtain a complete meteorological profile for Greenland; and when all these data are worked up and interpreted we shall be in position to say what influence Greenland exerts on our weather and to formulate a judgment as to whether a transatlantic air route via Greenland is practicable.

Composition of Ice

Among the most important winter labors was the construction of shafts in the ice, serving for the investigation of temperatures and the composition of the inland ice. With the most primitive appliances—ice axes and a pail—Dr. Sorge during the polar night dug out a well 52.48 feet deep, while the West Station sank its shaft to a depth of 65.6 feet. To complete the latter, 3,500 pails of ice had to be hoisted to the surface. One man at the bottom would hack away the ice and fill it into the pail, another one would hoist this, while a third cleared away the contents. All that in darkest polar night, with primitive illumination, and the thermometer 54 degrees Fahrenheit below the freezing point. In the ice shafts thermometers were set up at every metre (3.28 feet) of depth and read at regular intervals.

At the Central Station Dr. Sorge ascertained that the upper layers of the ice are composed of annual "rings" 1.64 feet thick. The Greenland ice-cap would grow each year by that amount in height, were it not that the névé, turned at greater depth to ice, flows away to the sea through the innumerable glaciers.

During the reconnaissance of 1929 my brother Alfred had measuring rods sunk into the ice in order to measure melting and fresh accumulation of snow. Such snow gauges were also put up by the main expedition, 20 kilometers (12.42 miles) apart, all the way out to the Central Station and readings were taken by every sled party traveling that route. We found that at sea level melting amounts to 8.2 feet a year; at 3,116 feet above sea level (the altitude of the West Station's winter house) to only 3.28 feet; at 4,920 feet melting and annual accumulation balance each other, and at a height of 6,560 feet a year's fresh snow exceeds the year's melting by 3.28 feet.

Our expedition also carried out the first trigonometric altitude measurements in Greenland. The trigonometric method, employed everywhere in topographic surveys, gives the most reliable results, but prior to our expedition altitudes in Greenland had been determined only by means of the barometer. Since, however, the conditions of air pressure over Greenland had been insufficiently explored, barometric height measurements were defective. Now that we have trigonometrically surveyed thousands of square miles, as far as the Central Station, we have also at last gained knowledge of the real air pressure conditions over those icy wastes.

The sensational features of our expedition were measuring the thickness of the ice and gravity. This part of the expeditionary work my brother Alfred had specially at heart. Greenland, it is well known, is one of the largest islands and at the same time one of the oldest continental fragments. It is moving with a speed of slightly over 118 feet a year away from Europe toward the West. That sounds almost incredible, for only a few decades ago the general scientific assumption was that the crust of the earth was rigid and permanently immobile. According to Airy's "Sial" theory, however, the crust is no longer regarded as wholly rigid, but is presumed to have under the topmost hard rind a yielding layer on which the former floats, as it were.

In his celebrated theory of continental structure, Alfred Wegener put forward the view that originally the continents of the earth had formed one continuous land mass, were later sundered, and since then have been in continuing migration; and the Greenland expedition—that was my brother's deepest ambition—was to produce the final proof for his great theory.

An Ice-Filled Bowl

To the eye the interior of Greenland presents itself as a high plateau. The littoral mountains rise up to 6,520 feet. In the interior the ice-covered surface rises even to 9,850 feet. According to surface observation Greenland had been regarded as an ice-covered rock waste. My brother Alfred, however, held a view appearing fantastic at first: he was convinced that the rock bottom of Greenland sinks down deeply in the interior—that in structure Greenland was really like an ice-filled cup or bowl. Even before the expedition started he declared with much certainty that the depth of the ice in middle Greenland must lie between 8,200 and 9,850 feet; and these masses of ice, he held, had crushed in the land in the interior.

Here was the point where the thickness of the Greenland ice became important for geological history. For if Greenland turned out actually to have the configuration surmised by Alfred Wegener, and if it was further proved through gravity measurements—which also were among the expedition's tasks—that the continental fragment of Greenland is sinking or rising in relation to its environment, then new, and perhaps decisive, evidence would be produced for the theory of continental displacement.

The ice-thickness measurements taken by the expedition have confirmed my brother's inspired predictions in an astounding manner. Under the winter house at the West Station, the ice measured 459.2 feet in depth, thirty-eight and one-half miles inland already 2,900 to 4,920 feet, and at seventy-four and one-half miles from the coast the ice was 5,904 feet deep. In the same distance bedrock was found to drop from 2,952 to 984 feet. At the Central Station much greater thickness of ice was ascertained, and the last blast, with the last remaining ammunition, yielded the fabulous depth of 8,856 feet. My brother did not live to know of this splendid confirmation of his hypotheses.

Methods of Measurements

The ice measurements were made by a quite new method developed specially for the expedition by Dr. H. Mothes in the geophysical institute of Goettingen University, based on a sort of echo-sounding. Through a blast of dynamite there is produced an artificial "icequake" which, like any real earthquake, transmits its waves both along the surface and downward into the depths. These waves are recorded by a seismograph placed at some distance from the blasting point. The waves traveling along the surface naturally reach the instrument sooner; those sent downward keep traveling until they strike bedrock, by which they are reflected upward, and thus ultimately also reach the seismograph. From the difference in the times of arrival at the seismograph of the two kinds of waves, the distance of bedrock from the point of explosion, and thus also the thickness of the ice, can be computed.

In the field of anthropology and zoology, too, our expedition brought home rich booty. Dr. Peters, who worked at the East Station, Scoresby Sound, made a collection numbering thousands of specimens, and promising, even on a provisional stock-taking, important results.

Scoresby Sound has long passed for an animal paradise. Great herds of musk-ox, multitudes of the snow-hare and the polar fox are found on the steppes and heaths of the littoral.

Countless finds of skeletal remains and implements were dug up and brought home. When Scoresby Sound was discovered, there were no Eskimos there nor any traces of them. Only a few years ago the Danish Government settled West Greenland Eskimos in the Sound region and today they number about 710. It is therefore to be assumed that the Eskimos whose remains Dr. Peters recovered had never come in contact with whites.

A rich haul was made near the famous hot spring at Cape Tobin, which has a temperature of 143.6 degrees Fahrenheit. Among much else, Dr. Peters discovered there a small community of flies and spiders that must be so well adapted to their hot environment as to enable one to assume they are relics from the last warm interglacial period.

—January 17, 1932

U.S. Satellites Find Radiation Barrier

By JOHN W. FINNEY

United States satellites have detected a mysterious band of extremely intense radiation some 600 miles in space.

The radiation, 1,000 times more powerful than had been expected by scientists, raises a new obstacle to manned space flight. Scientists must now start redesigning future space ships to shield human passengers against the radiation.

The radiation is so intense that a space traveler would use up his weekly tolerance dose of radiation in one and a half hours. Scientists believe, however, that the radiation can be reduced to tolerable levels by enclosing the space travelers in a thin protective shield of lead.

The belt of intense radiation, which may stretch for several thousand miles into space, was discovered by the two Army *Explorer* satellites launched earlier this year. Both satellites were equipped with Geiger counters to measure the radiation, particularly from cosmic rays, in space.

The scientific findings obtained from the two satellites were described in detail for the first time today before scientists of the National Academy of Sciences and the American Physical Society. Later, the results were discussed at a news conference by scientists participating in the International Geophysical Year satellite program.

As a more encouraging sidelight, the satellites demonstrated that it was possible to develop space vehicles whose temperatures could be kept within tolerable limits for humans and confirmed that cosmic dust—or micrometeorites—was not so dense as to pose an undue hazard. The satellites also discovered that space a few hundred miles up was several times more dense than had been supposed.

By coincidence, the findings were disclosed as the first detailed information became available on the experiments performed by the Soviet Union's two satellites.

The findings of the Soviet satellites seemed to correspond closely to the scientific conclusions drawn from the United States satellite and rocket program, such as the higher-than-expected temperatures and density in space. Significantly new in the Soviet report was the effect of prolonged weightlessness on animals—an experiment not yet performed by the United States.

The second Soviet satellite, which was equipped to perform cosmic ray experiments, also apparently encountered the strange layer of radiation. The "mystifying" increase reported in the Soviet reports, however, was not so great as that found by the American satellites, nor did the Soviet scientists seem to have so complete a concept of the nature and probable source of the layer of radiation.

"Crash Program" Begun

The United States satellites encountered the intense belt of radiation as they neared the peak altitudes of their orbits. Aside from its implications for space travel, the discovery may shed new scientific light on how the earth's atmosphere is heated up and how aurora borealis (northern lights) is caused.

Satellite scientists were so startled and impressed by the findings that they are now drafting a "crash program" to launch a satellite within the next eight months that would be designed to probe the nature and intensity of the radiation.

The discovery of the high-intensity radiation was described by Dr. James A. Van Allen, head of the Department of Physics at the State University of Iowa, who developed the cosmic ray experiments for the two *Explorer* satellites.

At altitudes of 600 miles and higher, Dr. Van Allen reported, the satellites encountered unexpected radiation "1,000 times as intense as could be attributed to cosmic rays." The radiation became so intense, he said, that at times it "jammed" the Geiger counters so they did not put out any measurements.

The source of the intense radiation layer, he said, is believed to be ionized gas, probably hydrogen, shot out from the sun. In the electrified gas are fairly high-energy electrons, which produce X-rays as they bombard the satellite shell.

Dr. Van Allen offered this still preliminary theory about the radiation layer:

A reservoir of ionized gas, or plasma, apparently stretches from about 600 miles to perhaps 8,000 miles in space. The earth's magnetic field acts as an "umbrella" to hold the layer at least 600 miles away.

Gas "Drizzles" Away

The reservoir of ionized gas gradually "drizzles" away into the earth's atmosphere, particularly around the North and South Poles, as the electrons lose energy. Occasionally, the reservoir is replenished by new outbursts of ionized gas from the sun.

The electrons in the plasma rain on the satellite "like rain on a tin roof" and create X-rays, which go bouncing around inside the satellite. The X-rays are formed in the same manner as in an X-ray tube and apparently have about the same energy as medical X-rays.

Discussing the space-age implications of the satellite findings, Dr. Van Allen said that it was "likely that cosmic rays themselves are not a serious biological hazard, but this new radiation is something to worry about."

In a space ship, he said, the level of radiation could be reduced to about 10 percent by a shield of lead one millimeter thick. This would represent about 100 pounds of lead shielding for each space passenger.

Dr. Edward Manring, of the Air Force Cambridge Research Center, reported that a micrometeorite experiment performed by two satellites had shown that cosmic dust presented a "small hazard" for future space travel, although over an extended period the dust could erode some of the satellite surfaces. The density of cosmic dust discovered by the satellite was in "fair agreement" with previous estimates, he said.

Dr. G. F. Schilling and Dr. Theodore E. Sterne of the Smithsonian Astrophysical Observatory reported that visual and radio observations of the satellites' orbits had led to the conclusion that the atmosphere at a height of 220 miles had density of about two ounces a cubic mile—or fourteen times the density previously predicted. The means the atmosphere at 200 miles is 100,000,000,000 times lighter than at sea level.

—*May 2, 1958*

Mountain Birth Linked to Oceans

By WALTER SULLIVAN

Recent discoveries on the ocean floors are leading to a comprehensive theory on how the far-flung mountain ranges of the world were formed.

The theory explains why the ranges tend to fringe continents. It accounts for the timetable of their formation, from the crumpling of the Appalachians hundreds of millions of years ago to the present emerging mountains.

It offers an explanation for the fact that the world's highest mountains, the Himalayas, are built of ocean floor material.

The theory, known as "global plate tectonics," is part of a revolution in the earth sciences that many specialists in the field consider as important as the Darwinian revolution of a century ago.

While geologists for centuries have studied the folding and upthrusting of mountains from the Alps to the Cascades, there has never been a generally accepted explanation of what thrust them up or why it occurred when and where, in each case, it did so. This theory meets such a need.

It is a consequence of the discovery that the earth's crust is formed of great plates, both continental and oceanic, that move independently, colliding or over-riding one another.

Science of Tectonics

Tectonics is the science dealing with processes that have formed the mountains and other surface features of the earth.

Earlier this week a series of maps was made public showing how movements of the earth's crustal plates tore apart an ancient supercontinent some 200 million years ago and transported the fragments to the present locations of the continents.

According to the new theory, encounters between such drifting plates account for all of the world's major mountain systems.

While the theory has evolved in stages with many contributors, it has now been set forth in detail by two geologists.

They are Dr. John M. Bird, head of the geology department at the State University of New York in Albany, and Dr. John F. Dewey of Cambridge University in England.

Their presentations appear in recent issues of the *Journal of Geophysical Research* and the *Geological Society of America Bulletin*.

Elements of the theory are still in dispute. Not all specialists in the field agree with their hypothesis that the contraction of an ancient ocean, the so-called Appalachian Atlantic, led to the formation of the Appalachian Mountains.

Questions About Concepts

Dr. Bird, in turn, does not accept the widely held view that the oceanic plates are being driven apart by material welling up into the mid-ocean ridges.

He suspects, rather, that the plates are being carried on the shoulder of a layer of flowing rock, 100 miles thick, that is part of a deep churning activity within the earth.

The two geologists have sought, through detailed analysis of such encounters, to explain the complex rock structures of various mountain systems. Even in as relatively simple an encounter as that between South America and the Pacific Ocean floor the effects are manifold.

In such a process the ocean plate is warped downward where it meets the offshore edge of the continent. The ocean plate descends, forcing its way under the continental rock in a slope known as the Benioff Zone. This is a region of frequent earthquakes.

Where the plate first bends down there is a trench in the sea bottom and earthquakes near there are of shallow origin. However, the Benioff Zone, as it extends inland, slopes down several hundred miles, as shown by the pinpointing of its earthquakes.

At such depths pressures and temperatures become so great that some components of the oceanic plate melt and force their way to the surface in volcanic eruptions. The penetration of oceanic material also humps the rim of the continent.

Vast Oceanic Plates

There is upwelling of lava into mid-ocean ridges, Dr. Bird says, but it does not provide enough energy to push the plates.

Where an oceanic plate (the entire eastern Atlantic constitutes a single plate) has become thick, cool and heavy near its continental margin, in his view, its own weight may cause it to sink, pulling the rest of the plate slowly behind it.

An analogy, Dr. Bird said in a recent telephone interview, would be a towel that floats on water until one edge becomes soaked and sinks, pulling the rest behind it.

Dr. Bird and Dr. Dewey list four basic types of mountain building processes: a collision between continents, such as the one that formed the Himalayas when India drifted into Asia; a plate of ocean floor pushing under a continent, as along the West Coast of South America; ocean floor undercutting another plate of ocean floor, forming island arcs like those of Tonga and Marianas in the Western Pacific; and an island arc impinging on a continent, as may have occurred in New Guinea.

Effect of Volcanic Activity

It is this humping and volcanic activity that, it is now believed, has built the world's second highest mountain system. Similar volcanic activity is also evident in western Mexico and the Alaska Peninsula.

According to Dr. Dewey and Dr. Bird, it also occurred along the eastern shores of North America as the ancient Appalachian Atlantic shrank.

Off Newfoundland, where the oceanic plate penetrated under an extension of the continent (the continental shelf), volcanic activity helped build an island arc that, as the ocean was squeezed out of existence, was thrown against the continent.

At this stage, some 350 million years ago, the Appalachians may have been the loftiest mountains of the world having, like the Himalayas, been formed by continental collision. Then, at least 150 million years later, the Americas again broke away from Europe and Africa as hot lava welled up into the rupture zone.

Such a process of oceanic birth is now thought to be taking place in the Red Sea.

The process that gave birth to the modern Atlantic rumpled the landscapes of Africa and North America, forming highly symmetrical folds such as those of the Blue Ridge and Pocono Mountains.

When an island arc such as Japan forms off a continental coast enclosing an inland sea, oceanic sediments accumulate in the enclosed basin.

If the island arc is then driven against the continent, those sediments are crumpled. Or, in a collision of two continents, sediments on the intervening continental shelves are folded into mountain ranges.

Formation of Himalaya

The sediments that were heaped to form the Himalayas accumulated on an ocean floor during a 500-million-year period, according to specialists in that region.

There are indications, from submarine earthquakes, that the moving oceanic plate that pushed India against Asia, having been blocked by the impact, is beginning to burrow downward south of India.

Another front where great forces are contending lies between Africa and Eurasia. It appears that Africa has long been pushing north in a plate motion that has formed the Alps.

The most active zone is in the eastern Mediterranean, where, according to Dr. Dewey and Dr. Bird, "a collision of North Africa with Greece and Turkey seems inevitable."

The time is not imminent, the plate movements being in inches per year. However, the compression can already be blamed for the destructive earthquakes of Turkey and the Balkans. Similarly, the burrowing of the Pacific floor under South America has caused the quakes that in recent years have taken thousands of lives in Chile and Peru.

—September 12, 1970

The Earth Aroused:
Quakes, Volcanoes and Other Disasters

Record Made in Washington

The San Francisco earthquake registered itself most accurately on the seismograph at the Weather Bureau in this city. For the first time since the discovery of the instrument by which earthquake waves could be measured the experts of the bureau were able to see the vibrations as they recorded themselves.

It was not until a telephone message was sent to the bureau that attention was directed to the seismograph, when it was at once noted that the indicator was moving violently from one side to the other of the sheet of paper on which the record is marked. Prof. Charles F. Marvin, who has charge of the instrument, and his assistants, at once concentrated their work for the day on the phenomena of the earthquake.

The seismograph consists of a cylinder on which a sheet of paper six inches by three feet, coated with parafine [paraffin] and soot, is placed and over which a needle or stylus hangs. This stylus marks most delicately as the cylinder moves, the line resulting from the difference between the earth waves and a mass of matter that as far as possible by all means known to science is immovable.

There are a dozen similar instruments in use throughout the world and the scientists of the various countries provide for such observations, and there is an international bureau at Strasburg, Germany, where all records are compared and the results reported back to the various corresponding bureaus in various parts of the world.

There is no seismograph on the Pacific Coast similar to the one here, but there is one of a different type at the Lick Observatory and another maintained by the Canadian Government at Vancouver. The coast survey has one at Baldwin,

Kan., and there is one of the old type at Johns Hopkins University and one of the modern pattern at Albany recently installed.

The first disturbance on the seismograph here was noted at 8:19:20 o'clock. It appeared in minute wavering marks on the sooty paper. These kept on almost imperceptibly, increasing in size for several minutes, until at 8:25 they began to break into well-defined waves over a quarter of an inch long. From second to second they lengthened, and within an inch space as the cylinder moved on from the needle they gradually grew to be an inch and a half long. At 8:32 the waves had increased to be three inches long, and the needle passed off of one side of the sheet of paper.

For three minutes it remained off the sheet, and at 8:35 again came up on the bevel of the brass cylinder and began once more to mark the record of seismic disturbance. What movements the indicator made while off the paper can only be imagined. The waves may have been much longer during that period than at the time when the record was completed. The greatest length of the vibration or wave was four inches.

Under the old methods of fixing this basis of measurement a pendulum was used. This pendulum was sometimes made several hundred feet long. By means of levers this has been done away with, and a most acute and accurate means of measurement has been devised. The stylus makes a fine straight line around the cylinder when there is no disturbance.

The wave register continued for three hours, although the length of the waves had become again quite minute.

The Weather Bureau did not receive as usual the 5 o'clock observation which is taken every morning. This observation it is believed must have been taken before the earthquake occurred, but no dispatch could be sent reporting it. About noon a dispatch was received from Prof. McAdie, the observer in charge of the San Francisco Weather Station, saying he and his three assistants were safe and well. The station was in the Mills Building, which was in the heart of the destroyed district, and went down with other buildings.

The scientists rely on the staff of the Lick Observatory to make the observations of the earthquake phenomena that will give its history.

Prof. Willis Moore of the Weather Bureau said today that the eruption of Vesuvius was recorded during its entire period on the magnetic instruments of the research station at Mount Weather, a few miles from this city in Virginia. He stated, however, that this was not an earthquake record, but rather that

Vesuvius acted like a powerful wireless transmitter and Mount Weather as the receiver, Vesuvius with its eruptions there disturbing the electrical potentials which disturbances sent out electromagnetic waves that encircled the earth. The records at Mount Weather were clear and distinct.

—*April 19, 1906*

St. Helens: Scientists Report the Cause Is Finally Clear

By WALTER SULLIVAN

As the anniversary of the greatest explosion ever recorded in North America—that of Mount St. Helens—approaches, scientists have found a variety of clues that, they believe, pin down for the first time the rapid-fire sequence of events culminating in that deadly blast of flame-hot gas and ash.

The clues involve precisely timed seismic records, including one cut short last May 18 when the blast destroyed an observing site six miles northeast of the mountain.

Photographs taken from various angles and at different times during the eruption, some of them by visitors who had illegally penetrated the danger area (and survived) have been assembled. Using such pictures an avalanche specialist has mathematically run avalanches "backwards" up the mountain to determine their time of origin.

From these clues scientists have deduced that the triggering event on May 18 was a fairly severe earthquake (magnitude 5) that dislodged a gigantic bulge, formed in recent weeks on the mountain's north side. In the largest landslides ever recorded by American geologists, this bulge slipped down the mountain like a vast sliding door.

With this "door" open, rock that had become supercharged with highly compressed gas exploded, ejecting to the north a tremendous horizontal blast. The reconstruction has removed uncertainty over whether the earthquake caused the eruption or vice versa.

The scope and violence of the blast are reflected in its casualty list: more than 100 square miles of forest buried or blown down, 61 people killed or missing, and an estimated 7,000 elk, 12,000 deer, 300 bears and 5,000 coyotes.

Intensive observations are being made in search of symptoms that might warn of a repetition. Predictions of lesser eruptions are routinely being made with increasing success. In recent days the mountain has continued to shoot gas and steam into the air. It seems likely that activity will continue for a score of years, but probably without another lethal explosion.

It has been discovered that across the Pacific in Kamchatka, St. Helens has a twin, Mount Bezymianny, whose eruption in 1955 and subsequent development were strikingly similar to the recent history of St. Helens.

According to Soviet accounts one side blew out in a devastating explosion, followed by mudslides that swept down valleys radiating from the mountain. It is suspected that, as with St. Helens, the explosion was triggered by a great landslide, although it occurred in early morning with no witnesses.

After the blast a dome developed within the crater in successive stages, as has occurred here since last May. The dome within Bezymianny is now considerably larger than that presently inside St. Helens. Destructive eruptions of Bezymianny have occurred intermittently since 1955, the most recent a few years ago.

Dr. Stephen D. Malone, seismologist at the University of Washington, believes this means that St. Helens may continue to erupt for another 20 years or more. Dr. Malone and Dr. Barry Voight, an avalanche specialist at the University of California at Santa Barbara, have pieced together the timetable of events last May.

All efforts to find some sort of premonitory hint of the May 18 explosion have failed, possibly because of the special manner in which it was initiated by an earthquake. A smaller explosion had been expected because of recurrent earthquakes in previous weeks, but no indication of when it might take place had been identified. From evidence of past eruptions it had been assumed the blast would be upward and hence relatively harmless.

The quake that immediately preceded the eruption was no different from the many that had occurred in recent weeks, though unusually intense. Its only novel feature, according to Dr. Malone, was the eight-minute series of tremors that followed the quake, caused by subsequent landslides and avalanches.

Measuring Tremors and Swellings

According to Dr. Donald W. Peterson, scientist-in-charge for studies of the mountain by the United States Geological Survey, seismic tremors have proved the most useful indicators of impending eruptions. However, precise ranging on sites near the dome inside the crater is also making it possible to monitor the dome's growth before eruptions.

The chief site from which such measurements are made is Harry's Ridge, named for Harry Truman, keeper of the lodge that stood on the shore of Spirit Lake, then a scenic gem at the foot of the mountain. The blast filled the lake with debris and buried Mr. Truman and his lodge.

The lake is again filled with water but its surface is higher than the original lake. It is carpeted with logs, stripped of bark and branches, washed down from the surrounding slopes.

Whenever weather permits, Dr. Donald A. Swanson of the Geological Survey is landed on Harry's Ridge to take measurements from there to nine sites in and near the crater. Pulses of laser light are aimed at reflectors on each of the volcano sites. The travel time until their return is recorded electronically, indicating the distance to within a fraction of an inch.

On a recent day he made measurements while Dr. Peterson recorded the readings, including air temperature and pressure required for corrections. Because hydrofluoric acid in fumes within the crater erodes the reflectors, they cannot be left there. Each time measurements are made a helicopter lands someone who holds a reflector over a benchmark on the crater floor.

Into the Caldron

On this occasion there was so much steam swirling within the crater that Dr. Swanson asked the helicopter pilot to help him aim his laser by hovering over the man. Suddenly from within the caldron of steam a flickering pinpoint of red light became visible to those on Harry's Ridge and measurements could be made.

As the dome swells it pushes the crater floor outward, as the measurements show. During five days spanning an eruption on Dec. 25, the north side of the crater moved north 47 inches.

At three places on the volcano, devices have been installed to measure changes in tilt of its slopes, but such recordings in the most active area, on or near the dome, have not been practical because the tiltmeter, costing about $2,000, would soon be overrun by creeping lava.

According to Dr. Daniel Dzurisin, who has been studying ground deformation for the Geological Survey, expendable tiltmeters, costing about $100, have been developed by the California Institute of Technology. It is planned to place two of them closer to the scene of action and radio their readings back to the project headquarters in Vancouver.

Dangers of High Viscosity

Special caution is needed, for as the May 18 blast showed, during its active periods Mount St. Helens, like other volcanoes ringing the Pacific, can be explosive, particularly when its lava is rich in silica. The lava is then so stiff it clogs the volcano vents and allows enormous gas pressure to develop inside. Repeated analyses of the extruded lava at Mount St. Helens have provided no more than hints that its silica content is diminishing, making further explosions unlikely.

It is slow-flowing lava that accounts for the steep-sloped profiles of such volcanoes. They arise where slabs of seafloor are being overridden by an advancing continental plate.

One warning of an explosion might be a tapering off of gas emissions when the volcano's "safety valves" have become clogged. The problem, as explained by Dr. Thomas Casadevall, who has been conducting gas measurements, is that such a tapering off could have the opposite meaning. It could occur because the molten rock, or magma, that has risen from within the earth has discharged most of its gas. The emissions have tapered off steadily since last July.

The measurements are made on clear days by flying through the volcanic plume—admittedly a method subject to large errors—and by ground-based spectroscopic scanning, using a device designed to record pollutants from tall smokestacks.

No Place Was Safe

On the day before the explosion David Johnston of the Geological Survey made such measurements from the Timberline Parking Lot at the foot of the mountain. Emissions of sulfur dioxide had dropped from an estimated 40 tons a day to only 15 tons. The next day he observed from a "safer" distance—a lofty spot near Harry's Ridge. He was presumably swept away by the blast and his body has never been found.

Two probes on the crater floor record escaping hydrogen which, scientists believe, is released by the molten rock as it solidifies. They also suspect that a burst of hydrogen would occur if the carapace of solid rock capping the molten material were ruptured as prelude to an eruption.

Before the explosion a number of stations were established around the volcano for magnetic, seismic, tilt and gravity measurements. The latter, it was assumed, would indicate elevation changes or large-scale movements of subterranean material. Only small changes were recorded and the station 12 miles west of the mountain, used as a comparison site, was buried by the eruption.

Efforts to observe St. Helens in sufficient detail to diagnose its day-to-day developments have been severely handicapped by weather and by its explosive nature, requiring special caution. By contrast, Kilauea, the active Hawaiian volcano, is of the nonexplosive, midocean type in a fair-weather environment. Researchers can drive to it at any time of day or night to emplace or read instruments densely scattered around it.

Except for a few stations transmitting data by radio, the St. Helens measurements can only be made in daylight when the weather is clear. As a consequence there have been gaps of many days over the past year. Nevertheless no American volcano of the explosive type has ever been so intensively studied.

—April 28, 1981

Nothing's Easy for New Orleans Flood Control

By JON NORDHEIMER

Caught between the Mississippi and the long shoreline of Lake Pontchartrain, this low-lying city has long depended on levees and luck.

Now engineers say those are not enough to protect New Orleans, much of it below sea level, from a devastating flood that could threaten it if a storm surge from a powerful hurricane out of the Gulf of Mexico propelled a wall of water into the lake and the city.

That event could place vast sections under 20 feet or more of water, engineers and scientists say, with worst-case computer predictions showing death tolls in the tens of thousands with many more people trapped by high water that has no natural drainage outlets.

"There's no way to minimize the amount of devastation that could take place under such circumstances," warned Walter S. Maestri, director of emergency management of Jefferson Parish, a suburban region with 455,000 residents on the city's western and southern sides.

Perhaps the surest protection is building up the coastal marshes that lie between New Orleans and the sea and that have been eroding at high rates. But restoration will require time, a huge effort and prohibitive sums of money, perhaps $14 billion, according to a study by a panel from federal and state agencies, universities and business.

Engineers are considering other ways to protect the heart of the city and provide an island of refuge in the French Quarter and government centers. Though such approaches are less expensive, they come with their own problems. One plan involves walling off an area to keep out water. But where would the wall be built and who would benefit from it?

Many residents give little thought to such matters, counting on the knowledge that New Orleans has escaped hurricane disaster in the past.

The most nervous people are those paid to worry about such things, like Dr. Joseph N. Suhayda, director of the Louisiana Water Resources Research Institute at Louisiana State University. Like other coastal researchers, he has been using the latest geological and meteorological data to refine computer models of how different storms would damage the city.

On a bright spring day with fair skies and no trace of the sultry air that will dominate the weather in the months ahead, Dr. Suhayda and a few colleagues drove city streets 1,000 yards from levees that hold back Lake Pontchartrain.

At New York Avenue, near the lakefront campus of the University of New Orleans, the car stopped, and the engineer walked over and unfolded a wood measuring stick to its 25-foot length. He planted one end on the pavement and raised it until it was vertical. The other end poked into the sky well above a corner light pole, but it was still well beneath the level of a concrete wall that rose on top of a grassy slope 100 feet away.

"Behind that," Dr. Suhayda said, indicating the wall, "is a canal that runs into Lake Pontchartrain. Its surface is roughly about the same as the lake's surface."

In a hypothetical situation projected by his computers, Dr. Suhayda continued, a slow-moving Category 4 hurricane, with winds up to 155 miles an hour, or a Category 5 hurricane with even stronger winds could leave water 30 feet deep on this neighborhood street, which is more than five feet below sea level. Though Category 5 hurricanes are very rare, Camille in 1969 devastated Pass Christian, Miss., just 50 miles east of New Orleans, and killed scores of residents with winds that exceeded 200 miles an hour and a 35-foot storm surge.

In most areas vulnerable to hurricanes, the water would drain away quickly. That is not the case here.

So city planners and engineers continue to work on ways to improve an evacuation plan for the 1.3 million residents in the metropolitan region and to soften a storm's blow. Most long-term projects intended to blunt a hurricane involve slowing the loss of marshlands. One method calls for additional control gates to let the Mississippi pour sediment-rich water into surrounding lands, a process that would eventually raise or at least maintain their elevation.

Other proposals are to rebuild eroded offshore barrier islands, erect a wall of levees across much of the lower delta, plug the dredged containership channel that gives the Gulf of Mexico waters easy access to the vulnerable eastern shore of Lake Pontchartrain and help defend that shore with higher flood gates.

Perhaps the most unconventional is the "community haven" concept advanced by Dr. Suhayda and others in the belief that radical remedies may be necessary to soften a knockout punch by nature. More theory than an organized campaign, it envisions a two-story-high wall with flood gates at crucial intersections to seal off the southern part of the city from a bend in the river at the French Quarter to another one eight miles west.

If the rest of the city flooded, Dr. Suhayda said, the "island" between the wall and the river's levees could become a refuge for thousands of residents fleeing their homes, as well as preserving the cultural and government center.

But obtaining the money on the scale needed is far tougher than devising plans, especially if some skeptics dismiss the worst-case predictions as scare tactics to help finance university research or for further environmental intrusions on the coast.

Researchers, though, say they are not making up the city's potential peril, which arises from geology and history. As tight as a pimento in an olive, most of New Orleans is stuffed between the Mississippi and the lake, and it is settling as fast as the rest of the delta or faster, said Dr. Roy Dokka, a geologist at Louisiana State.

Although much agonizing has gone into problems of the river, and its metaphoric temperaments have become part of songs and folklore, it is the 300-square-mile lake that troubles him, Dr. Dokka said.

As New Orleans grew as a seaport, petrochemical hub, tourist destination and cultural phenomenon, neighboring marshes were drained and the levee system expanded to keep the water out of new suburbs and industrial parks, hastening the drying that led to sinking, and making a bad situation even worse.

A computerized cross-section of the city's topography recently created by L.S.U. scientists and engineers shows a shallow bowl-like profile. On the southern edge along the river, the rim rises about five feet above sea level for most of the French Quarter, a half-mile-wide sloping plain created by the river's natural earthen banks. A flood wall built by the Army Corps of Engineers to hold back a cresting river—which on normal days moves more than 300,000 cubic feet of water a second past the city at an average depth of 90 feet—raised the levee to a uniform height of 25 feet above sea level, or 10 feet above the average annual high water surface level of the river, when water can rush by at the rate of one million cubic feet of water, or more, a second.

Most of the city north of the French Quarter was reclaimed from a boggy morass. The lowest sections—residential areas and shops that sit on drained marshland at 5 to 10 feet below sea level—form a wide band near the lake's southern shore.

New Orleans International Airport to the west and industrial complexes and residential areas to the east are at sea level or below it.

Spider webs of city canals and wide ditches that measure 185 miles in length feed 22 pumping stations that lift water to a height where it can flow into the lake, over levees built more than a half-century ago that stand 15 to 17 feet above sea level.

Water cascading over the levee wall or flooding from swollen marshes at the lake's eastern and western ends is just one part of the nightmare, the experts say. Draining the city after the storm moves away may take weeks, they point out. The city would be trapped inside the levees, steeped in a worsening "witches' brew" of pollutants like sewage, landfill waste, chemicals and the bodies of drowned humans and animals.

Bourbon Street could remain under 10 feet of water, with water swirling above two-story houses in neighborhoods closer to the lake.

Dr. Ivor van Heerden, deputy director of the L.S.U. Hurricane Center, said that in a worst-case situation with incomplete evacuation "we could have up to 45,000 killed and 400,000 trapped on roofs, with 700,000 evacuees who would now be homeless."

Dr. van Heerden said it would take at least nine weeks to pump the city dry.

Pumping stations, which sometimes fail to keep up with heavy seasonal downpours in a city that receives nearly five feet of rain a year, would be inoperative under those conditions for days if not weeks, Dr. van Heerden said. The chemical stew that would be pumped into the lake and surrounding marshes would be an environmental disaster by itself, he added. He said that the cost in human misery might be incalculable but that the bills for insured damages, public works repairs and replacements and the economic effects might total $50 billion.

Despite the specter of such losses, it has been difficult to find enough money to build up the protective marshlands, said Jack C. Caldwell, head of the state's Natural Resources Department. Washington has been disinclined to earmark billions to protect the marshland and has resisted appeals from Baton Rouge to share revenues from offshore oil production with the state for that purpose, Mr. Caldwell said.

In the meantime, residents take their chances every hurricane season, which starts on June 1. According to his computer models, Dr. Suhayda said, the odds that the city will be hit by a cataclysmic storm in any given year are less than 1 in 100.

The American Red Cross is taking the threat seriously. It has declared it no longer will provide hurricane shelters in the New Orleans area, saying that placing staff there in a killer storm will represent too much risk for its employees, volunteers and the general public.

—April 30, 2002

Mercilessly Unpredictable, Quakes Defy Seismologists

By SANDRA BLAKESLEE

In one sense, the catastrophic earthquake off the coast of Sumatra on Sunday morning was no surprise. After decades of dissecting seismic fault zones, including the long seam between tectonic plates in the bed of the Indian Ocean, scientists have gotten good at describing where large quakes are likely to occur.

But the things people really want to know—exactly when and exactly where the next "big one" will strike—remain entirely unpredictable. Despite their best theories, despite more than $1 billion worth of spending on instruments on faults in California and Japan, scientists still cannot give timely warning to people in harm's way.

The best that seismologists can do today is predict probabilities of earthquakes' recurring in regions where they have happened before. The predictions are typically cast in terms of 10 to 30 years.

Such warnings are useful for deciding where to build dams, power stations, roads and pipelines in earthquake regions, said Dr. Lynn Sykes, a geophysicist at the Lamont-Doherty Earth Observatory in Palisades, N.Y. But they do not help ordinary people living in those areas.

The science of earthquake prediction has a long, sorry history, including good ideas that failed, disappointing experiments and an ever-expanding list of goofy claims.

In the 1970s, hopes ran high that a so-called seismic gap hypothesis would yield accurate predictions. Long fault zones were divided into smaller segments that were thought to rupture at somewhat regular intervals. By digging trenches along those segments and finding dates for past events, seismologists could infer the timing of future quakes.

But segments of the San Andreas Fault in California are not so well behaved, said Dr. Kerry Sieh, a seismologist at the California Institute of Technology who pioneered the technique. Some sections erupt every 50 to 350 years.

"We don't know why," Dr. Sieh said. "It may be because so many other faults interact. Or is it something fundamental about individual faults?"

Realizing he would never solve the prediction puzzle in California, Dr. Sieh turned his attention a decade ago to faults off the coast of Sumatra.

His focus is on a patch of ocean floor a few hundred miles south of Sunday's earthquake.

Both areas are part of a vast subduction zone, where a huge block of the earth's crust slides under the Indonesian archipelago.

By dating mortality patterns in coral reefs affected by fault motions and tsunamis, Dr. Sieh determined that large earthquakes occur regionally, in pairs, every 230 years or so. In 1797 there was a magnitude 8.2 quake; in 1833 a magnitude 8.7 or bigger quake occurred. There was a cluster in the 1500s and one in the 1300s.

"We are coming up on the beginning of the next cycle," Dr. Sieh said. "In the last four years, a number of smaller quakes have been flirting with these larger locked patches."

For example, in June 2000, a quake of magnitude 7.0 was felt in Singapore. A 7.4 temblor in 2002, under Simeulue Island, off the northwestern coast of Sumatra, may have been a foreshock for Sunday's quake.

A second very large earthquake in the same offshore area would not come as a surprise, Dr. Sieh said. But it could be decades away.

On land, quake predictions have never proved reliable. On Feb. 4, 1975, the Chinese government said it evacuated the town of Haiching based on earthquake precursors—changes in land elevation, groundwater level, swarms of small quakes and jittery horses, dogs and chickens. A quake of magnitude 7.3 struck two days later, and tens of thousands of lives were saved.

But later investigations suggest that the claim was based on ideology, not science. The quake, it turned out, had not been predicted so clearly. Tens of thousands of people were killed or injured.

A year later, a 7.6 earthquake struck Tangshan, China, without warning. An estimated 250,000 people died.

In January 1994, a "blind thrust," or previously unknown, fault ruptured beneath Los Angeles in a 6.7 quake that killed 61 people and did billions of dollars' worth of property damage. Exactly a year later, a similar quake, this one 6.9, killed more than 5,000 people in Kobe, Japan.

An experiment along a section of the San Andreas Fault near Parkfield in central California predicted that a moderate quake would occur no later than 1993. It came a decade late, rupturing in the "wrong" direction, on Sept. 28, 2004.

A scientist at the University of California, Los Angeles, Dr. Vladimir I. Keilis-Borok, predicted a 6.4 quake in Southern California between Jan. 5 and Sept. 5 of this year. It did not happen.

Such failure makes ample room for psychics and other soothsayers to explain quakes that do happen. Many signs—odd animal behavior, strange light in the sky, tidal forces—tend to be invoked after the fact.

Many seismologists now say earthquakes can never be predicted because the earth's crust is profoundly heterogeneous. Any small quake has some probability of cascading into a large event. Whether a small one grows into a large one or dissipates depends on myriad fine details of physical conditions in fractured earth.

Dr. John Rundle, a seismologist at the University of California, Davis, says the best hope for near-term prediction lies in a statistical method that identifies hot spots. Regions of concern are laid out in grids six miles on a side. Instruments pick up areas with frequent small quakes, around magnitude 3. If such activity increases, a quake is more likely within five years.

This method predicted the location of 12 out of 14 moderate California quakes over the last five years, Dr. Rundle said. Still, he added, "It only tells you where but not exactly when."

Once the ground starts shaking, it is possible to send signals to critical facilities like pipelines and power plants to shut down operations. Such systems are in place in Japan, including sirens to warn people in coastal areas prone to tsunamis.

The nations devastated by tsunamis on Sunday have no warning system, and experts said it would be some time before one could be put in place.

Such systems require integrated earthquake and tide and wave gauges and computer models that can quickly project where tsunamis may travel, said Dr. George D. Curtis, a tsunami expert affiliated with the University of Hawaii at Hilo.

Even more challenging, he said, is the need to create an efficient method for alerting hundreds of coastal communities, and educational efforts to be sure residents heed the alarm.

That is easier to do in advanced wealthy countries like the United States and Japan than in developing countries, where the vulnerability is enormous and resources limited.

"Even if you have the technical system in place," Dr. Curtis said, "you have to have people prepared to react immediately."

—December 28, 2004

A Bridge Built to Sway
When the Earth Shakes

By HENRY FOUNTAIN

Venture deep inside the new skyway of the San Francisco–Oakland Bay Bridge, and it becomes clear that the bridge's engineers have planned for the long term.

At intervals inside the elevated roadway's box girders—which have the closed-in feel of a submarine, if a submarine were made of concrete—are anchor blocks, called deadmen, cast into the structure. They are meant to be used decades from now, perhaps in the next century, when in their old age the concrete girders will start to sag. By running cables from deadman to deadman and tightening them, workers will be able to restore the girders to their original alignment.

The deadmen are one sign that the new eastern span of the Bay Bridge, which includes the skyway and a unique suspension bridge, is meant to last at least 150 years after its expected opening in 2013. (The existing eastern bridge, which is still in use, will then be torn down.)

But to make it to the 22nd century, the new span may at some point have to survive a major earthquake, like the one that destroyed much of San Francisco in 1906 or the one that partly severed the Bay Bridge in 1989. With two faults nearby that are capable of producing such large quakes, survival is no simple matter.

Say what you want about the project—and as the construction timeline has lengthened past a decade and costs have soared over $6 billion, plenty has been said—keeping the bridge intact in an earthquake has always been the engineers' chief goal.

And to meet that goal, they are going with the flow: designing flexible structures in which any potential damage would be limited to specific elements.

"We wanted to make this bridge flexible so that when the earthquake comes in, the flexibility of the system is such that it basically rides the earthquake," said its lead designer, Marwan Nader, a vice president at the engineering firm T. Y. Lin International.

That contrasts with another potential approach: making the bridge structures large enough, and rigid enough, to resist movement. "Massive and stiff structures would look absolutely ugly and be very, very expensive," said Frieder Seible, dean of the Jacobs School of Engineering at the University of California, San Diego, who tested many elements of the bridge design.

That design includes a 525-foot-tall suspension bridge tower made up of four steel shafts that should sway in a major earthquake, up to about five feet at the top. But the brunt of the force would be absorbed by connecting plates between the shafts, called shear links.

The bridge's concrete piers are designed to sway as well, limiting damage to areas with extra steel reinforcing. And at joints along the entire span there are 60-foot sliding steel tubes, called hinge pipe beams, with sacrificial sections of weaker steel that should help spare the rest of the structure as it moves in a quake.

"At the seismic displacement that we anticipate, there will be damage," Mr. Nader said. "But the damage is repairable and the bridge can be serviceable with no problems."

Emergency vehicles and personnel, at the least, should be able to use the bridge within hours of a major earthquake, after crews inspect the structure and make temporary fixes, like placing steel plates over certain joints. Given that the Bay Area's two major airports would be expected to be out of service after such a disaster, this bridge and the Benicia-Martinez Bridge, another seismically secure span about 20 miles to the northeast, would be "lifeline" structures to bring assistance to the stricken region from an Air Force base inland, said Bart Ney, a spokesman for the California Department of Transportation.

It was an earthquake that made this replacement span, which runs for 2.2 miles between Oakland and Yerba Buena Island in the middle of San Francisco Bay, necessary. The Loma Prieta quake of 1989, the first to occur along the San Andreas fault zone since the 1906 disaster, caused part of the existing steel-truss span to collapse, killing a motorist. The bridge was closed for a month. That quake, with a magnitude of 6.9, caused strong shaking that lasted about 15 seconds and movement far greater than the 1930s-vintage bridge had been designed to handle.

"When the bridge was subjected to those earthquake motions in 1989, it literally was stretched and, basically, one of the spans fell off," Mr. Nader said. Most experts agree that a stronger quake, most likely along the San Andreas or the Hayward fault, in the East Bay, could cause a total collapse of the old span.

There is a strong likelihood of a large earthquake in the Bay Area—about a 2-in-3 chance of magnitude 6.7 or larger before 2036, according to the United States Geological Survey and other institutions. But Mr. Nader and his colleagues were not so much concerned with magnitude measured at the epicenter as they were with ground motions at the bridge site. They planned for the largest motions expected to occur within 1,500 years.

After the 1989 quake, engineers determined that the bridge's western span—a double suspension bridge between San Francisco and Yerba Buena Island—could be made seismically safe, with some modifications. But the eastern truss bridge and causeway would eventually have to come down. (The Golden Gate Bridge was undamaged in the quake, but has been retrofitted to prepare it for a larger one.)

Among the eastern span's problems were that the foundations of the piers were sitting not on rock, but in mud that could shake like jelly in a quake, magnifying the motion.

The planning for the bridge's replacement was delayed by squabbles over the path the new bridge would take across the bay and the "look" of the span, with East Bay residents especially vocal about their desires.

"Folks would say that they feel that all the glamorous signature spans tend to be on the San Francisco side of the bay and that the East Bay gets the simple and utilitarian type of bridge," said Mr. Nader, who earned his doctorate at the University of California, Berkeley, and experienced the 1989 quake firsthand. "So they wanted a signature span."

They got one. Unlike more conventional suspension bridges, in which parallel cables are slung over towers and anchored at both ends in rock or concrete, the 2,047-foot suspension bridge has only a single tower and a single cable that is anchored to the road deck itself, looping from the eastern end to the western end and back again. (With a conventional design it would have been extremely difficult to create an anchorage on the eastern end, in the middle of the bay.)

The new bridge is the longest self-anchored suspension bridge in the world, and it is asymmetrical, with one side of the span longer than the other. (Mr. Nader says it looks like half a conventional suspension bridge.) The choice of such a design raised the cost of the project significantly. In a conventional suspension bridge, the road deck is added last, hung from suspender cables attached to the main cables. In a self-anchored design, the deck has to be built first.

"You have a kind of chicken-and-egg situation," Mr. Nader said. "You need the deck to carry the compression so that the cable anchors into it, but the deck can't carry itself until the cable is there to carry it. So you have to build a temporary system."

That system, called falsework, is basically a bridge to hold up the road deck until the cable is in place—an operation that began in late December and was expected to take up to six months. The falsework needs to be seismically secure as well, adding to the cost.

In all the discussions over a signature span, Mr. Nader said, there was deep interest in having only a single tower. But that created design problems. "In a single tower, there is a lack of redundancy," he said. "Just like a pole. If you have a pole and the pole starts shaking, all the damage will occur at the bottom."

The solution, fleshed out in conversations with the bridge's architects, was to split the tower into four shafts and tie them together with the shear links, which Mr. Nader had become familiar with during his Berkeley years through a professor who had tested them in certain kinds of building frames.

The links are of a special grade of steel that deforms more easily than other grades, and they are placed at specific points along the length of the tower, which affects how the shafts will move in a quake. "Based on where you place the shear links, you can tune the dynamic response of your tower," said Dr. Seible, of the University of California, San Diego.

Under normal conditions, the shear links help to stiffen the four shafts against wind and other loads. "But when you come to larger earthquake loads, these links start yielding," Mr. Nader said. "It's taking the energy that's being pumped into the tower."

Mr. Nader said he already knew which shear links would be most damaged in a major earthquake—those that are about two-thirds of the way up the tower. But the tower would still be structurally sound, he said, and the links would not even have to be replaced immediately.

It's like what happens after a fender bender, he said. "Your car is perfectly drivable, and it's designed that way, with a bumper that can take the shock.

"So you basically stop, just to make sure," he went on. "You see everything's O.K., and you can come in anytime you want to repair your bumper."

—*February 8, 2012*

The Aftermath of Hurricane Sandy: Costs of Shoring Up Coastal Communities

By CORNELIA DEAN

For more than a century, for good or ill, New Jersey has led the nation in coastal development. Many of the barrier islands along its coast have long been lined by rock jetties, concrete sea walls or other protective armor. Most of its coastal communities have beaches only because engineers periodically replenish them with sand pumped from offshore.

Now much of that sand is gone. Though reports are still preliminary, coastal researchers say that when Hurricane Sandy came ashore, it washed enormous quantities of sand off beaches and into the streets—or even all the way across barrier islands into the bays behind them.

But even as these towns clamor for sand, scientists are warning that rising seas will make maintaining artificial beaches prohibitively expensive or simply impossible. Even some advocates of artificial beach nourishment now urge new approaches to the issue, especially in New Jersey.

The practice has long been controversial.

Opponents of beach nourishment argue that undeveloped beaches deal well with storms. Their sands shift; barrier islands may even migrate toward the mainland. But the beach itself survives, because buildings and roads do not pin it down.

By contrast, replenishment projects often wash away far sooner than expected. The critics say the best answer to coastal storms is to move people and buildings away from the water, a tactic some call strategic retreat.

Supporters of these projects counter that beaches are infrastructure—just like roads, bridges and sewer systems—that must be maintained. They say beaches attract tourists and summer residents, conferring immense economic benefits that more than outweigh the costs of the projects. Also, they argue, these beaches absorb storm energy, sparing buildings inland.

New Jersey has embraced this approach with gusto. Stewart C. Farrell, a professor of marine geology at Stockton College of New Jersey, said that since 1985, 80 million cubic yards of sand had been applied on 54 of the state's 97 miles of developed coastline: a truckload of sand for every foot of beach.

Dr. Farrell and his colleagues have calculated that the work cost more than $800 million—before adjusting for inflation.

Typically, the federal government pays 65 percent of the cost; the state and towns share the rest.

By now in New Jersey, most beaches "are engineered dikes," said Thomas Herrington, a professor of ocean engineering at the Stevens Institute of Technology in Hoboken, who has been working with the state to assess its coastal protection. About half of its coastal communities have projects still under way, so their beaches are already approved—at least in theory—for topping up with sand as needed.

But even if there is money for that work, engineers must find the sand. Around the nation, that is getting more and more difficult. The problem is particularly acute in New Jersey.

"We know from geological surveys—and New Jersey is a prime example—that offshore sand, high-quality sand, is a highly finite resource," said S. Jeffress Williams, a coastal scientist with the United States Geological Survey and the University of Hawaii.

Underwater ridges of sand lie offshore, but engineers must go farther and farther (and spend more and more) to find them, Mr. Williams said, adding that eventually "it is not going to be there."

And while it is theoretically possible to replenish a beach with material mined inland, that approach would create other problems, said Robert Young, a coastal geologist who directs the program for the study of developed shorelines at Western Carolina University. "Trucks full of sand weigh a lot," he added. "There is a tremendous toll on highway infrastructure." And excavating inland sand "would create holes that would be miles in diameter."

Howard Marlowe, a prominent advocate of replenishment projects, agrees that the nation needs "a better way of managing short supplies of sediment."

Mr. Marlowe is president of Marlowe & Company, a lobbying concern that represents many Atlantic and Pacific coastal communities. He said in an interview that sand supplies should be managed "holistically"—and on a regional basis, not town by town, as is now largely the case.

Managers should look at inlets, ports and the Intracoastal Waterway, as well as offshore sand sources, Mr. Marlowe said, adding, "You have to have an interstate approach."

That is particularly true in New Jersey, he said, because towns in New York and Delaware also often find themselves on the hunt for sand. Elsewhere in the country, towns feud over who is entitled to offshore sand. Towns in Florida have gone to court over the issue.

Avalon, N.J., about 20 miles north of Cape May, looks to Townsends Inlet north of town for its sand, according to Harry deButts, who retired in 2008 as the town's director of public works and now works part time on its emergency management efforts. He said Avalon shares that sand with Sea Isle City, the town across the inlet.

Mr. deButts said the Army Corps of Engineers dredged about 450,000 cubic yards of sediment from the inlet in 2003, and applied "a couple hundred thousand cubic yards" more about five years later. The town is scheduled to receive more sand in December, to repair damage from Hurricane Irene last year.

He said "the beaches did their job" during Hurricane Sandy, saving buildings in Avalon from flood damage. The town had built a dune more than 20 feet high, adopting what he called "an education program" to explain the advantages of second-floor living rooms to residents whose views were blocked.

That can be a hard sell. In Harvey Cedars, on Long Beach Island, home-owners have sued for compensation over loss of ocean views because of a proposed project. The case is before the New Jersey Supreme Court.

Where artificial beaches failed to protect their communities, it was probably because "this storm just exceeded the design conditions," said Dr. Herrington, of the Stevens Institute, who has been working with the state to assess coastal protection. Typically, he said, projects in New Jersey are engineered to withstand the kind of storm that on average occurs only once every 75 years.

But as the climate warms, sea levels are rising and bad storms may come more frequently. And New Jersey is particularly vulnerable because of tectonic forces and changes in ocean currents.

When the glaciers retreated about 15,000 years ago, land in the region bounced up; now it is sinking again. Meanwhile, ocean circulation patterns are changing in ways that push water up against the mid-Atlantic coast.

"We cannot sustain the shoreline in the future as we have in the past," said Mr. Williams, of the Geological Survey. "Particularly from a beach nourishment standpoint."

—November 6, 2012

CHAPTER 5

The Environment: Challenges to Life

THE NATURAL WORLD

A Tree's Life Through Thirteen Centuries

There are few of us who have not been introduced by geographers, newspapers, and books to the Big Trees of California. But the chances are that our acquaintance is rather cursory, extending little beyond the knowledge that among trees their dimensions are unequaled, and that they are the oldest living thing in either the animal or vegetable kingdom. Some of us with good memories may even be able to quote figures: There are the "Two Sentinels," for instance, famous giants of the Calaveras Grove, over 300 feet high, the larger 23 feet in diameter, and both many hundred years of age.

Plain figures, however, are very unsatisfactory material out of which to reconstruct anything in the imagination. A tree 300 feet high! It means little to us until we get something for comparison a man standing at its base, no bigger relatively than a lady bug on a rose bush—then we can gasp an appreciative, "Gee whizz! But that's a whopper!"

But what does it mean to have lived hundreds and hundreds of years? How is the imagination to reconstruct that stretch of time?

The Life Story of "Mark Twain"

George H. Sherwood, one of the assistant curators in the American Museum of Natural History on Central Park West, has devised a graphic solution of this problem. He has told the history of "Mark Twain," one of the California forest giants, a 16-foot cross section of which is on exhibition in the museum, by erecting on the circles which tell the tree's age little flags or labels showing contemporaneous events in the world's history.

One of "Mark Twain's" first contemporaries was Justinian, emperor of the Roman Empire in the East; one of its last, Benjamin Harrison, president of the United States. It began life as a seedling while the Eastern emperors in the year that followed the fail of Rome were struggling hard and doubtfully to withstand the waves of barbarian inundation which constantly threatened to overwhelm Constantinople with the same awful calamities that had befallen the imperial city of the West; it came to an untimely death by violence in the days of the telegraph and telephone. With these historic contrasts before, we can begin to picture in our imagination the span of life that has been enjoyed by this hardy forest Methuselah.

The Big Tree has inherited its longevity. It belongs to a family, the Sequoia, which is not only noted for the long lives enjoyed by its individual members, but which can also put to blush all human genealogies—the Chinese not excepted. The Sequoia family can trace its sap, unmixed with the sap of any other tree family, back to preglacial days. Before the glacial period the Sequoia family flourished widely in the temperate zones of three continents. There are many species, and Europe, Asia, and America had each its share.

This is the recommendation which Gifford Pinchot, United States Forester, gives the Big Tree branch of the Sequoia family:

"The Big Trees are unique in the world—the grandest, the largest, the oldest, the most majestically graceful of trees—and if it were not enough to be all this, they are among the scarcest of known tree species, and have the extreme scientific value of being the living representative of a former tropic age. It is a tree which has come down to us through the vicissitudes of many centuries solely because of its superb qualifications. Its bark is often two feet thick and almost noncombustible. The oldest specimen's fungus is an unknown enemy to it. Yet with all these means of maintenance the Big Trees have apparently not increased their range since the glacial epoch. They have only just managed to hold their own on this little strip of country where the climate is locally favorable.

Selecting the Specimen

For the purpose of procuring a specimen of this remarkable tree for the American Museum of Natural History, S. D. Dill was sent to California in the summer of 1891. Through the courtesy and liberty of A. D. Moore, owner of one of the largest groves of Big Trees, and his son (manager of the King's River Lumber Company), Mr. Dill was permitted to select any tree that he desired. After diligent search he found a fine specimen growing at an altitude of 7,000 or 8,000 feet and bearing the name "Mark Twain." Nearly all the large trees have been christened by hunters or tourists, and several are marked with marble tablets. Such names as "Bay State," "Sir Joseph Hooker," "Pride of the Forest," and "Grizzly Giant" are familiar.

"Mark Twain" was a tree of magnificent proportions, one of the most perfect trees in the grove, symmetrical, fully 300 feet tall, and entirely free of limbs for nearly 200 feet. Eight feet from the ground the trunk was 62 feet in circumference, while at the ground it measured 90 feet. Mr. Moore took the contract of felling the tree and shipping to the museum a section suitable for exhibition.

We all know these "ring" or concentric circles, which are shown in a cross section from any exogenous tree in which botanical division is included nearly all trees growing in temperate climate; how each ring of wood corresponds to a period of vegetable growth during the spring and summer; and how the lines of separation represent periods of interrupted growth during the winter and fall. Each ring marks of a year in a tree's growth, and every tree therefore carries within itself its autobiography.

With the patient courage of the scientist, Mr. Sherwood deciphered the 16-foot 2-inch autobiography of "Mark Twain." In many places the records of neighboring years were less than a sixteenth of an inch apart.

In the outermost ring, which marks the last year of the tree's life, he inserted a pin upon which is mounted a black card bearing the date "1891." Patiently counting in toward the center of the tree ninety-one rings, he inserted another pin bearing the date "1800." Then as he made the next century of the rings he set up another card, "1700," and so on back through the centuries. It was as though he were reversing the order of nature, putting a giant tree through a shrinking process which gradually reduced it to its sapling days. In the very innermost ring, the one that records its first year of life, he inserted a pin which bears the date "550 AD."

At the Time of Its Birth

And so it was just a century, less one year, after Attila, the "Scourge of God," and his host of Huns had been forced to retreat across the Rhine after the battle of Châlons, a conflict which decided that the Christian Germanic races, and not the pagan Scythic Huns, should inherit the dominions of the dying Roman Empire and control the destinies of Europe, that a little seed, scarcely one-fourth inch in length, forced its tiny sprout above the soil of the Sierras and sought its inheritance as a member of the Big Tree family.

While still a mere sapling, fighting with the surrounding forest undergrowth for light and air, Europe was overrun by the Goths, Vandals, and Franks and a state of almost universal war prevailed.

When our tree had reached its fifteenth year and attained a circumference of twelve inches, one of those big events happened which have had an influence on the progress of civilization. This is the way that Mr. Sherwood tells of the event on a pin-card inserted in the fifteenth ring from the center of the tree: "Silk worms introduced from China into Europe."

At the Age of Fifty

When the tree completed in fiftieth year it had attained a diameter of twenty-one inches, and during the next century it added twenty-eight inches more. And so our tree is already a huge forester when we come to the next pin-card—"732, Battle of Tours." This was just five years after the conquest of Spain. The Saracens, crossing the Pyrenees, established themselves upon the plains of Gaul. . . . As Draper pictures it, the Crescent, lying in a vast semicircle upon the northern shore of Africa and the curving coast of Asia, with one horn touching the Bosphorus and the other the Straits of Gibraltar, seemed about to round to the full and overspread all Europe. The Franks and their allies met the Moslems upon the plains of Tours, dealt them a crushing blow, and forced them to retreat behind the Pyrenees.

The climatic conditions in California during the year 800 and the year preceding must have been very favorable for the growth of our tree, which had already attained the size of a large and venerable elm. Its growth during these two years, indicated by the large rings, was phenomenal.

This year, 800, was also notable for the crowning of Charlemagne, a monarch who, in addition to being a great warrior, established a school at his court, inviting thither the few learned men of his time, and laying "the foundation of all that is noble and beautiful and useful in the art of Middle Age."

During this century occurred also the effort of King Alfred to establish schools in England. The hardy Norsemen began their bold voyage in quest of treasure and adventure, colonized Iceland in 874, discovered Greenland (981), and, pushing farther westward, probably sailed down along the eastern shore of America.

The year that saw our tree getting on toward the shady end of its sixth century also saw the beginning of the Crusades. These great military expeditions against the Moslems, begun in 1096 and continuing for almost 200 years, brought the various European peoples into intercourse, which resulted in exchange of ideas and helped prepare the popular mind for the discoveries which were soon to follow.

The first half of the thirteenth century saw the founding of the universities. First, the University of Paris (1200), which became the center of theology; a few years later were founded the University of Bologna, famous for law, and the University of Padua, which attracted the greatest students in medicine. In England, Oxford University was founded in 1249.

Nearing the Thousand Mark

The fifteenth century bought those marvelous discoveries which were of so much importance in the advancement of civilization and which contributed to the growth of science. Printing with wooden block type was introduced by John Gutenberg in 1438, and his invention was followed in 1450 with the use of metal type, making the general dissemination of knowledge possible.

Our tree was getting well on toward its thousandth birthday when Columbus discovered America. And in this grizzled old age it was contemporary with much that sounds ancient to us moderns. The pin-cards thus tell the story: "1543, Copernicus publishes his system of astronomy"; "1609, Kepler announces his laws of planetary motion"; "1685, Newton discovers laws of gravitation"; "1775, American Revolution"; "1815, Battle of Waterloo"; and the last and most recent pin-card in the historical series, "1866, First successful Atlantic cable laid."

Had man left our big tree undisturbed on its Sierra hillside, what great men and events in the world's history might it be a contemporary of?

"I never saw a Big Tree," writes John Muir, "that had died a natural death; barring accidents they seem to be immortal, being exempt from all the diseases that afflict and kill other trees. Unless destroyed by man, they live on indefinitely until burned, smashed by lightning, or cast down by storms, or by the giving way of the ground on which they stand. A colossal scarred monument in the King's River forest is burned half through, and I spent a day in making an estimate of its age, clearing away the charred surface with an axe and carefully counting the annual rings with the aid of a pocket lens. The wood rings in the section I laid bare were so involved and contorted in some places that I was not able to determine its age exactly, but I counted over 4,000 rings, which showed that this tree was in its prime, swaying in the Sierra winds, when Christ walked the earth."

—January 12, 1908

There's Poison All Around Us Now

A Review of Rachel Carson's *Silent Spring*

By LORUS and MARGERY MILNE

Poisoning people is wrong. Yet, for the sake of "controlling" all kinds of insects, fungi and weed plants, people today are being poisoned on a scale that the infamous Borgias never dreamed of. Cancer-inducing chemicals remain as residues in virtually everything we eat or drink. A continuation of present programs that use poisonous chemicals will soon exterminate much of our wild life and man as well. So claims Rachel Carson in her provocative new book, *Silent Spring*.

Silent Spring is similar in only one regard to Miss Carson's earlier books (*Under the Sea Wind, The Sea Around Us, The Edge of the Sea*): in it she deals once more, in an accurate, yet popularly written narrative, with the relation of life to environment Her book is a cry to the reading public to help curb private and public programs which by use of poisons will end by destroying life on earth.

Know the facts and do something about the situation, she urges. To make sure that the facts are known, she recounts them and documents them with 55 pages of references. She intends to shock and hopes for action. She fears the insidious poisons, spread as sprays and dust or put in foods, far more than the radioactive debris from a nuclear war. Miss Carson, with the fervor of an Ezekiel, is trying to save nature and mankind from chemical biocides that John H. Baker (then president of the National Audubon Society) identified in 1958 as "the greatest threat to life on earth."

Her account of the present is dismal. It is not hopeless—at least not yet. But she demands a quick change in "our distorted sense of proportion." How can intelligent beings seek to control a few unwanted species by a method that contaminates the entire environment and brings the threat of disease and death even to our own kind? "For the first time in the history of the world," she writes, "every human being is now subjected to contact with dangerous chemicals from the moment of conception until death. . . . These chemicals are now stored in the bodies of the vast majority of human beings, regardless of age. They occur in the mother's milk, and probably in the tissues of the unborn child."

Albert Schweitzer has said, "Man can hardly even recognize the devils of his own creation." Yet *Silent Spring* will remind some people that a few years ago they went without cranberry sauce at Thanksgiving rather than risk eating berries contaminated with a cancer-inducing chemical used improperly by some growers as a weed-killer in the cranberry bogs. A few others may recall that tax money was paid not only to growers of cranberries, but also (a year or so earlier) to poultry raisers whose chickens retained dangerous amounts of a chemical included in poultry feed upon government recommendation and had to be condemned.

Miss Carson adds many other instances to the list, and points to programs that cost many millions of tax dollars, yet were doomed at the outset to failure. She gives details about the gypsy-moth campaigns that killed fish, crabs and birds as well as some gypsy moths; about the fire-ant program that killed cows, wiped out pheasants, but not fire ants; and dozens of others that led to more of the pest (or of new pests) by destroying the natural means of control.

Miss Carson gives most of her attention to insecticides, herbicides and fungicides, since these are the most dangerous poisons. She shows the futility of relying on them or any new substitutes offered to counteract the swift evolution of immunity to chemical control shown by more and more insects and fungus diseases. She quotes an authority on cancer, Dr. W. C. Hueper of the National Cancer Institute, who has given "DDT the definite rating of a 'chemical carcinogen'"—a cancer inducer. She notes that "storage of DDT [in the body] begins with the smallest conceivable intake of the chemical (which is present as residues on most foodstuffs) and continues until quite high levels are reached. The fatty storage depots act as biological magnifiers, so that an intake of as little as one-tenth of 1 part per million in the diet results in storage of about 10 to 15 parts per million, an increase of one hundredfold or more. . . . In animal experiments, 3 parts per million has been found to inhibit an essential enzyme in heart muscle; only 5 parts per million has brought about necrosis or disintegration of liver cells; only 2.5 parts per million of the closely related chemicals dieldrin and chlordane did the same." Other modern insecticides are still more deadly. Nor did the discovery of their poisonous character "come by chance: insects were widely used [during World War II] to test chemicals as agents of death for man."

In some of the chapters, Miss Carson does approve of alternatives to the widespread use of poisonous chemicals. She points to the successful controlling of scale insects with ladybird beetles, and Japanese beetles with the "milky disease." So often, harmful species, new to a given area, have ceased to be much of a problem as soon as their natural enemies or their equivalents appear or are introduced. The natural struggle for survival can then keep the numbers of the pests at a fairly low level. This approach, as Miss Carson emphasizes, rarely creates new pests, whereas extermination campaigns often do so.

Those who grow and store food and other products that can be hurt by pests will surely accuse *Silent Spring* of telling only part of the story. They will claim that today efficiency in raising and distributing food and wood depend upon the use of poisons. The traces of chemical compounds left in and on these materials are the price we must pay for such efficiency. If biological control methods were relied upon or hand labor required, the yield would be smaller and the market price higher. They might ask, "Do you want wormy apples and buggy flour, or traces of pesticides that by themselves have not yet been proved harmful?" Miss Carson can also count on vociferous rebuttal from many pesticide makers and users. Government agencies that have encouraged poisoning campaigns are more likely to remain silent, unwilling to take blame for extensive programs that seem senseless and heedless in retrospect. *Silent Spring* is so one-sided that it encourages argument, although little can be done to refute Miss Carson's carefully documented statements. Valiant attempts will certainly be made to defend the motives and methods of biocide users. These arguments will concede that these chemicals could be dangerous and that the substances might be helpful if used correctly. They are unlikely to cite the calamities that have followed application of the poisons—as *Silent Spring* does.

The book mentions that in 1960 private citizens of America invested more than $750,000,000 in poisons to kill insects, rats, unwanted fish, crabgrass and other pests. Federal, state and local governments spent an even larger amount to put poison on public lands (including national forests, parks and roadsides) and on private property (many of whose owners objected vehemently to such treatment). Understandably, the manufacturers, distributors and appliers of all these tons of chemicals hope that the demand for pesticides will increase. To expand their businesses, they are willing to invest a great many dollars in research and promotion. With so large a financial stake in the continuation of present programs, they can be counted upon to spend even more money to tell the public the other side of the story.

No amendment to the Constitution protects us from this new danger. "If the Bill of Rights contains no guarantee that a citizen shall be secure against lethal poisons distributed either by private individuals or by public officials, it is surely only because our forefathers, despite their considerable wisdom and foresight, could conceive of no such problem," she says.

Nor is Congress likely, unless urged by enough people, to vote appropriations to let the Food and Drug Administration monitor more adequately the poisonous residues in foods. The two criteria that legislators understand are votes and taxes. Few votes and few taxes come from outdoor groups, such as the National Audubon Society. These organizations and their small-circulation magazines have little money to spend on educating and influencing legislators.

About one-third of *Silent Spring* has appeared in the *New Yorker*; the book itself rounds out and documents the account. In answer to the charge that the balance of nature has been upset, it has been pointed out by some members of the chemical industry that modem medicine is equally upsetting. This sort of defense merely invites a pox on both the biocide and the drug industries. *Silent Spring* offers warnings in this direction too: trivial amounts of one poison often make trivial amounts of another suddenly disastrous; and poisons stored in the body may be tolerated during health, but take effect dramatically as soon as any sickness decreases the body's resistance. It is high time for people to know about these rapid changes in their environment, and to take an effective part in the battle that may shape the future of all life on earth.

—September 23, 1962

Tests Show Aerosol Gases May Pose Threat to Earth

By WALTER SULLIVAN

Two scientists have calculated that gases released by aerosol cans have already accumulated sufficiently in the upper air to begin depleting the ozone that protects the earth from lethal ultraviolet radiation.

The calculations, by scientists at Harvard University, follow the recent discovery that these gases, used as aerosol propellants for hair sprays, insecticides and the like, while inert chemically, are highly efficient in promoting ozone breakdown.

The finding has posed a new and ominous threat to stability of the ozone layer that lies primarily between 10 and 30 miles aloft. There has also been concern that the layer would be depleted by exhaust gases from a large fleet of supersonic transport planes or by extensive explosions of nuclear weapons.

On Sept. 5, Dr. Fred C. Iklé, director of the Arms Control and Disarmament Agency, said that nitric oxides injected into the stratosphere by a nuclear war could wipe out the ozone layer entirely.

Because certain wavelengths of ultraviolet light from the sun break down molecules essential to life, it is believed that land life did not emerge until the development of the ozone layer late in the earth's history. The lethal wavelengths cannot penetrate water.

The most prevalent concern, however, is not for total loss of the ozone, which is broken down and restored in a complex sequence of day and night chemical reactions. Rather, it is a fear of sufficient depletion to cause widespread skin cancer and other effects.

Furthermore, because ultraviolet absorption by ozone contributes substantially to upper air heating, radical reduction of such heating could alter climates.

Ozone is a gas whose molecules are formed of three oxygen atoms instead of the two that are paired in ordinary oxygen, providing individual atoms that can merge to convert two-atom molecules to the three-atom ozone molecule.

The Harvard calculations were made by Dr. Michael McElroy, professor of atmospheric science, and Dr. Steven C. Wofsy, an atmospheric physicist. They found that, even if dispersal of aerosol propellants and other such gases, widely known under the trade name Freon, is halted as soon as practicable, depletion of the ozone layer by 1990 could reach 5 percent.

They consider a halt by the end of this decade to be the earliest plausible time, in view of political and commercial considerations. As others have pointed out in a number of recent discussions of the danger, the effect will continue for some time after a cutoff because the gas at sea level must work its way up into the stratosphere.

If according to the Harvard scientists, the cutoff is delayed until its effect on the ozone layer, having reached 10 percent, becomes indisputable, the consequences could be more severe.

Basing their calculations on relatively conservative estimate of an annual increase of 10 percent in release of the gases, they predict that the depletion will not level off until the year 2000. By then, they believe the ozone layer will have been reduced 14 to 15 percent.

Slow Recovery Foreseen

If releases of Freon continue to increase 21 percent a year, as has recently been the case for the aerosol propellants, the ozone level will be down 7 percent by 1984 and 30 percent by 1994. A cutoff in 1987 would modify the effect to a maximum depletion, in 1995, of 21 per cent.

In all cases recovery would be slow since there are no chemical reactions that remove such gases from the air.

In an independent analysis, three University of Michigan scientists have concluded that, by 1985 or 1990, chlorine derived from the atmosphere's Freon content will have become the dominant factor in ozone breakdown.

This report, by Drs. Ralph J. Cicerone, Richard S. Stolarski and Stacy Walters, appears in tomorrow's issue of the journal *Science*.

The Freon gases in question are chlorofluoromethanes, one of which, marketed as Freon 11, is used as the propellant in aerosol cans since it is inert chemically and does not react with the substance being ejected. The other, Freon 12, is used as a refrigerant.

As recently as last year the release of these gases into the atmosphere was considered a boon rather than a hazard. In the British journal *Nature* on Jan. 19, 1973, it was noted that these gases, being released into the air at a rate growing rapidly, could be used as harmless traces of air movements.

As of 1971, it was estimated, one million tons of each kind of Freon was being released yearly into the atmosphere. Furthermore, almost all such gas produced to date by world industry is still in residence within the air, the report said.

However, wrote the authors, "the presence of these compounds constitutes no conceivable hazard."

But last June, Drs. Mario J. Molina and F. S. Rowland of the University of California at Irvine reported in *Nature* that the Freons, far from being innocuous constituents, are six times more effective in breaking down ozone than the oxides of nitrogen that had been the chief focus of attention.

Supersonic transports and nuclear explosions release oxides of nitrogen into the upper air, and it is known that they act as catalysts both in promoting the breakdown of ozone and in "stealing" free oxygen atoms that might otherwise mate with oxygen gas to form new ozone.

Hence, there has been concern regarding the effect of such transports and such explosions on the ozone layer. The Harvard group has calculated that the exhaust from 400 Concorde-type supersonic transports operating seven hours a day would deplete the ozone layer by about 1 percent.

Now it appears that sunlight breaks down the Freon, releasing chlorine, which has an even more powerful catalytic effect. The Harvard scientists had begun a year ago to look into the possible role of chlorine introduced into the stratosphere by passage of the projected space shuttle, whose exhaust would contain hydrogen chloride.

This enabled them to apply the same calculations to the Freon problem.

That the accumulation of Freon in the world's atmosphere is increasing rapidly was confirmed earlier this month by Dr. John W. Swinnerton of the Naval Research Laboratory at a meeting of the American Chemical Society.

In 1972, measurements on a cruise from Los Angeles to Antarctica, he said, showed an average level of 61 parts per trillion. A year later, over the Atlantic, it was 85 parts per trillion, and in January of this year measurements in the Arctic showed it to have reached 120 parts per trillion.

On June 14 of this year A. B. Pittock, a government atmospheric physicist in Australia, reported that balloon measurements of ozone over Australia showed a general decline in the amount of that gas over the eight years prior to 1973. Whether this was a global effect was uncertain.

As noted yesterday by Dr. McElroy of Harvard, there is a variation in ozone content of about 1 percent, apparently related to the 11-year sunspot cycle, and a suggestion of a slight, 20-year cycle of unknown origin.

While Freon is the trade name of du Pont de Nemours & Co., similar gases are manufactured in this country by Allied Chemical, Union Carbide, Pennwalt,

Kaiser Chemical and Racon. They are extensively used in air-conditioning as well as in refrigeration systems.

In response to recent reports of possible ozone effects, the Manufacturing Chemists Association in Washington, on behalf of producers of the gases in various parts of the world, has initiated its own program of laboratory studies.

Last night Charles S. Booz, spokesman for du Pont's Freon products division in Wilmington, Del., termed the assessments of ozone effects "largely hypothesis." He added that "very little is known" of chemical reactions in the special environment of the upper atmosphere.

Dr. McElroy, in a telephone interview, himself emphasized that the analysis, which is being submitted to the journal *Science,* is obviously theoretical. He expressed the hope that there would now be extensive measurements of Freon levels and of upper air contamination by hydrogen chloride.

He said he assumed that others would take a hard look at the Harvard calculations. The surest way to assess the predictions would be to wait and see what happens, he added, but that is hardly acceptable.

"It is," he said, "a very unusual situation for science."

—*September 26, 1974*

The Endangered West

A sample of recent bulletins from the Old West: Montana rewrites some of the country's strongest water pollution laws as a favor to the mining industry. Idaho lawmakers award potential polluters a major voice in setting clean water standards. Utah's governor rebuffs the stated wishes of Utah's citizens to set aside 5.7 million acres of state land as protected wilderness. Washington State's legislature passes the nation's most far-reaching "takings" law, weakening essential land-use controls. Wyoming's legislature authorizes a bounty on wolves—recently re-introduced into Yellowstone National Park and protected under the federal Endangered Species Act.

Clearly, the United States Congress is not the only place where laws protecting the environment are under siege. Throughout the West, particularly in the Rocky Mountains, state legislators and governors, egged on by commercial interests and by small but noisy groups of property-rights advocates, are engaged in full-scale mutiny against federal and state regulations meant to protect what is left of America's natural resources.

What we are seeing is an updated but more ominous version of the Sagebrush Rebellion of the early Reagan years. That revolt was dominated by ranching interests protesting federal regulation of public lands. The present explosion embraces not only those familiar despoilers but mining companies, timber barons, developers, big commercial farmers and virtually anyone else who stands to profit from relaxation of environmental controls.

The war in the West and the war in Congress on basic environmental protections have much in common. First, both are being driven and in some cases underwritten by big business. Second, both are being waged to save the "little guy" from federal tyranny. Third, this alleged little guy is nowhere to be found when the time comes to draft crippling legislation. Indeed, his wishes have been largely ignored. Poll after poll suggests that what ordinary citizens want is more environmental protection if it means a cleaner environment and a healthier society. But that is not what this Congress and its Western allies want to give them.

Montana and Idaho are particularly sad cases. Despite citizen complaints, and nearly unanimous editorial opposition, two bills whistled through the Montana legislature that would in effect permit higher levels of toxic wastes to reach the state's streams and lakes. They were signed, with some reluctance, by the governor. Mining lobbyists were conspicuous during the parliamentary maneuvering—including representatives from Crown Butte and its Canadian parent, Noranda Inc. These companies are working relentlessly for permission to build in geologically precarious terrain a gold mine that would leave a permanent reservoir of pollutants in the watershed of one of Montana's most important wilderness streams.

Idaho's people—not to mention its endangered Snake River salmon—face a double threat. Under a new statute, acceptable water quality levels will be set by watershed advisory groups. These groups will be well stocked with large landowners and representatives from timber, mining, and agribusiness companies who are almost certain to write new and more permissive regulations. Meanwhile, back in Washington, an Idaho Republican, Dirk Kempthorne, is leading the Senate charge to cripple the Endangered Species Act, which provides what little protection the salmon have. If Senator Kempthorne succeeds in transferring protection of endangered species from Washington to Boise, it will be goodbye salmon, with grizzlies and wolves to follow.

There are, of course, honorable exceptions. In Colorado, for example, ranchers, environmentalists and state officials were able to agree on less destructive grazing practices—although it took a half-dozen or so exhausting visits from Interior Secretary Bruce Babbitt to get the agreement. But nearly everywhere one turns, the anti-Washington ideologues seem to have the upper hand.

The most conspicuous example is Nevada, where officials in Nye County passed a series of ordinances claiming ownership of federal lands and then set about physically intimidating employees from the Forest Service and the Bureau of Land Management. The Justice Department has now sued to reaffirm federal jurisdiction, but Nye County's rebels have inspired imitators: More than 70 rural Western counties have passed or proposed laws to "take back" the public lands.

Lost in all the rhetoric about individualism and states' rights is one basic legal fact: At no time have the Western public lands belonged to the states. They were acquired by treaty, conquest or purchase by the federal government acting on behalf of all the citizens of the United States. Lost, too, is a colossal irony. Western ranchers have traditionally fed well at the trough of federal beneficence. In their war against Washington, they are biting the hand that has fed them lavish subsidies and protected them against the disasters of nature and the vagaries of the marketplace.

But all of this escapes the Sons-of-Sagebrushers. The fact that there might be an overriding national interest in preserving the public lands and forests from exploitation is not something that quickly pops to their minds. Nor does this fact seem to register with the newer breed of rebels in the statehouses and state legislatures who would nullify more than two decades of struggle to clean America's waterways, preserve its wetlands and otherwise protect its dwindling natural heritage.

There can be no satisfaction in any of this—except perhaps to the enemies of the environment in a Congress that is well on its way to abandoning any pretense to national stewardship.

—June 18, 1995

A Wild, Fearsome World Under Each Fallen Leaf

By JAMES GORMAN

Dr. Edward O. Wilson, Pellegrino university research professor and honorary curator in entomology of the Museum of Comparative Zoology at Harvard University, winner of two Pulitzer prizes and scientific honors too numerous to recount, is on his hands and knees, pawing in the leaf litter near Walden Pond.

He eases into a half-sitting, half-reclining position and holds out a handful of humus and dirt. "This," he says, "is wilderness."

Just a dozen yards from the site of Thoreau's cabin, Dr. Wilson is delving into the ground with a sense of purpose and pleasure that would instantly make any 10-year-old join him. His smile suggests that at age 73, with a troublesome right knee, he still finds the forest floor as much to his liking as a professor's desk.

These woods are not wild; indeed they were not wild in Thoreau's day. Today, the beach and trails of Walden Pond State Reservation draw about 500,000 visitors a year. Few of them hunt ants, however. Underfoot and under the leaf litter there is a world as wild as it was before human beings came to this part of North America.

Dr. Wilson is playing guide to this micro-wilderness—full of ants, mites, millipedes and springtails in a miniature forest of fungal threads and plant detritus in order to make a point about the value of little creatures and small spaces. If he wrote bumper stickers, rather than books, his next might be "Save the Microfauna" or "Sweat the Small Stuff."

He begins his most recent book, *The Future of Life*, with a "Dear Henry" letter, talking to Thoreau about the state of the world and the Walden Pond woods.

"Untrammeled nature exists in the dirt and rotting vegetation beneath our shoes," he writes. "The wilderness of ordinary vision may have vanished—wolf, puma and wolverine no longer exist in the tamed forests of Massachusetts. But another, even more ancient wilderness lives on."

In their world, centipedes are predators as fearsome as saber-toothed tigers, ants more numerous than the ungulates of African plains. And, in contrast to the vast preserves required by the world's most revered megafauna—grizzlies and elephants, jaguars and condors—maintaining biodiversity among the little creatures, shockingly rich in unexplored behavior and biochemistry, can

174

be done on the cheap, in relatively tiny patches, as small as a few acres, around the world.

Dr. Wilson is by no means turning from the grand plans for conservation. Indeed, he has suggested that 50 percent of the globe ought to be reserved for nonhuman nature. But he is a realist, and, as he describes himself, "a lover of little things."

He has been turning over logs and rocks looking into the world of insects and other tiny creatures since he was a boy in Alabama and Florida. And he has not stopped. During the walk, he talks enthusiastically about a coming field trip to the Dominican Republic to investigate ants there, and about the publication this fall of a book-length monograph on the genus *Pheidole* describing all 625 species of ants, including 341 new to science.

Researchers tend to share a kind of acquisitive passion to see, touch, grasp the world. Nothing passes without comment. As he strolled along the shore of Walden Pond, on the way to the woods, Dr. Wilson spotted a butterfly and interrupted his discussion of the sizes of reserves needed for mammals, reptiles and amphibians, complete with references and citations to scientific studies.

When the butterfly landed on the beach, Dr. Wilson stopped, leaning forward like a heron on the hunt, and peered at it. "It's hard to identify," he said. "It's a very beat-up little butterfly," probably the variety called a question mark, because of the design on its wings.

Having reached the woods and having begun to talk about what lay under the surface of the forest floor, he held the crumbled leaf litter and humus in his hand as if he were savoring what lovers of certain wines call the *goût de terroir,* or taste of the soil, a certain earthy specificity that the wine owes to the ground, not the grape.

"When I go on a field trip," he said, "providing you can get me up to the edge of a natural environment, I usually don't go more than a hundred yards or so in, because when I settle down immediately I start finding interesting stuff."

"This ground," he said, "we see it as two-dimensional because we're gigantic, like Godzilla. When you just go a few centimeters down, then you're in a three-dimensional world where the conditions change dramatically almost millimeter by millimeter. In one square foot of this litter you're looking at into the tens of thousands of small creatures that you can still spot with your naked eye."

The ground was drier than usual, and Dr. Wilson speculated that the drought might have affected insects. Some he had hoped to see were not there. "If we looked long enough," he said, breaking open several rotting acorns, "we would find entire colonies of very small ants living in an acorn."

As he moved on, from log to log, he uncovered relatives of cave crickets, predatory rove beetles, termites, several varieties of ants, spiderlings, beetle larvae—not quite in the abundance he had hoped for. Still, the small wilderness was teeming.

"The exact perception of wilderness is a matter of scale," he writes in *The Future of Life*, going on to say that "microaesthetics" is "an unexplored wilderness to the creative mind." But he also notes that while microreserves are "infinitely better than nothing at all, they are no substitute for macro- and megareserves." He continues, "People can acquire an appreciation for savage carnivorous nematodes and shape-shifting rotifers in a drop of pond water, but they need life on the larger scale to which the human intellect and emotion most naturally respond."

Dr. Wilson is no sentimentalist, about nematodes or people. His proposals for microreserves are practical and hardheaded. The idea is fairly simple. While areas of nearly 25,000 acres are needed to have a good chance at preserving most large forms of life, plants and insects can sometimes be preserved in plots of 25 or even 2.5 acres.

In the Amazon, for instance, Dr. Wilson said, where the land is being savaged, "You'll see hanging on the side of a ravine somewhere a patch a farmer hasn't farmed, one hectare, to maybe 10." Such a small area may not catch the eye of most conservationists, he said, but, he added, "The entomologist and the botanist is likely to say, hold on a minute."

A researcher, he said, may find species not found anywhere else, and such plots can grow, with care and reseeding of the surrounding area. "You can do this in most parts of the world, in most developing countries, where a farmer or village elders would happily take a thousand bucks for you to set aside 10 or 100 hectares and even hire them to help with the reseeding," he said.

But it is not just the developing world where biodiversity can be preserved, bit by bit. City parks may hold small wonders. Even at Walden Pond, in the midst of the Massachusetts suburbs, he said: "Many of the species you find here are new to science. The basic biology of most of these things is poorly known or not known at all."

In the Walden woods live two relatively unknown ant species in the genus *Myrmica*. "They've been noticed, but not named or described," he said. As to the nematodes and mites, he said, lifetimes can be spent and careers can be made studying them.

It is not, of course, entomologists, or even weekend naturalists, who need convincing about the richness of the forest floor. And there are many reasons to try to preserve biological diversity at the near-microscopic level and below. There is always potential economic value in new biochemical discoveries. There is real and present economic value in clean air and water, to which the plants and insects and microbes contribute. In fact, Dr. Wilson points out, the life of the planet is built on a foundation of tiny creatures. In ecological systems it is the giants that stand on the shoulders of mites.

Finally, there is the simplest argument of all, that life itself, in all its variations, is astonishing and mysterious, and that humans have a responsibility to preserve it.

Under one log turned over on our walk, the environment was moist enough to provide a widely varied selection of insects. Dr. Wilson picked up a gooey white worm without a trace of discomfort. "That is a fly larva, a maggot actually," he said. "I don't know the kind of fly. It's very slimy."

Asked if he might be demonstrating the very reason many people do not like to turn over logs and dig under leaves, he laughed. "I'll admit it's an acquired taste," he said. "Don't mistake me, I don't expect legions of people, particularly Americans, going out and seeing how many different kinds of oribatid mites or fly larvae they can find.

"But," he continued, "they can get a feel, one way or the other, that what's at their feet is not dead leaves and dirt, but a living world with a diversity of creatures, some of which are so strange to the average experience that they beat most of the things you see in *Star Wars*."

It is not so hard to imagine. Butterfly fanciers are legion. Dragonflies are now attracting watchers with binoculars. "But mainly," he said, "there's got to be some sense of the beauty and integrity and the extreme age of these areas."

At the next and last log, he pointed out a predatory rove beetle, a millipede (a detritovore), a spiderling, a nematode he caught only a glimpse of ("like a very tiny silvery strand"), more ants and a wood cockroach. As he began to describe the thousands of inoffensive species of cockroach throughout the world, a siren went off on a nearby street, a reminder of where he was.

Dr. Wilson looked up from the timeless environment under the log, "That's sure a sound that Thoreau never heard."

—September 24, 2002

Afloat in the Ocean,
Expanding Islands of Trash

By LINDSEY HOSHAW

Aboard the *Alguita*, 1,000 miles northeast of Hawaii: In this remote patch of the Pacific Ocean, hundreds of miles from any national boundary, the detritus of human life is collecting in a swirling current so large that it defies precise measurement.

Light bulbs, bottle caps, toothbrushes, Popsicle sticks and tiny pieces of plastic, each the size of a grain of rice, inhabit the Pacific garbage patch, an area of widely dispersed trash that doubles in size every decade and is now believed to be roughly twice the size of Texas. But one research organization estimates that the garbage now actually pervades the Pacific, though most of it is caught in what oceanographers call a gyre like this one—an area of heavy currents and slack winds that keep the trash swirling in a giant whirlpool.

Scientists say the garbage patch is just one of five that may be caught in giant gyres scattered around the world's oceans. Abandoned fishing gear like buoys, fishing line and nets account for some of the waste, but other items come from land after washing into storm drains and out to sea.

Plastic is the most common refuse in the patch because it is lightweight, durable and an omnipresent, disposable product in both advanced and developing societies. It can float along for hundreds of miles before being caught in a gyre and then, over time, breaking down.

But once it does split into pieces, the fragments look like confetti in the water. Millions, billions, trillions and more of these particles are floating in the world's trash-filled gyres.

PCBs, DDT and other toxic chemicals cannot dissolve in water, but the plastic absorbs them like a sponge. Fish that feed on plankton ingest the tiny plastic particles. Scientists from the Algalita Marine Research Foundation say that fish tissues contain some of the same chemicals as the plastic. The scientists speculate that toxic chemicals are leaching into fish tissue from the plastic they eat.

The researchers say that when a predator—a larger fish or a person—eats the fish that eats the plastic, that predator may be transferring toxins to its own tissues, and in greater concentrations since toxins from multiple food sources can accumulate in the body.

Charles Moore found the Pacific garbage patch by accident 12 years ago, when he came upon it on his way back from a sailing race in Hawaii. As captain, Mr. Moore ferried three researchers, his first mate and a journalist here this summer in his 10th scientific trip to the site. He is convinced that several similar garbage patches remain to be discovered.

"Anywhere you really look for it, you're going to see it," he said.

Many scientists believe there is a garbage patch off the coast of Japan and another in the Sargasso Sea, in the middle of the Atlantic Ocean.

Bonnie Monteleone, a University of North Carolina, Wilmington, graduate student researching a master's thesis on plastic accumulation in the ocean, visited the Sargasso Sea in late spring and the Pacific garbage patch with Mr. Moore this summer.

"I saw much higher concentrations of trash in the Pacific garbage patch than in the Sargasso," Ms. Monteleone said, while acknowledging that she might not have found the zone with the highest concentration of trash.

Ms. Monteleone, a volunteer crewmember on Mr. Moore's ship, kept hoping she would see at least one sample taken from the Pacific garbage patch without any trash in it. "Just one area—just one," she said. "That's all I wanted to see. But everywhere had plastic."

The Pacific garbage patch gained prominence after three independent marine research organizations visited it this summer. One of them, Project Kaisei, based in San Francisco, is trying to devise ways to clean up the patch by turning plastic into diesel fuel.

Environmentalists and celebrities are using the patch to promote their own causes. The actor Ted Danson's nonprofit group Oceana designated Mr. Moore a hero for his work on the patch. Another Hollywood figure, Edward Norton, narrated a public-service announcement about plastic bags, which make their way out to the patch.

Mr. Moore, however, is the first person to have pursued serious scientific research by sampling the garbage patch. In 1999, he dedicated the Algalita foundation to studying it. Now the foundation examines plastic debris and takes samples of polluted water off the California coast and across the Pacific Ocean. By dragging a fine mesh net behind his research vessel *Alguita*, a 50-foot aluminum catamaran, Mr. Moore is able to collect small plastic fragments.

Researchers measure the amount of plastic in each sample and calculate the weight of each fragment. They also test the tissues of any fish caught in the nets to measure for toxic chemicals. One rainbow runner from a previous voyage had 84 pieces of plastic in its stomach.

The research team has not tested the most recent catch for toxic chemicals, but the water samples show that the amount of plastic in the gyre and the larger Pacific is increasing. Water samples from February contained twice as much plastic as samples from a decade ago.

"This is not the garbage patch I knew in 1999," Mr. Moore said. "This is a totally different animal."

For the captain's first mate, Jeffery Ernst, the patch was "just a reminder that there's nowhere that isn't affected by humanity."

—November 10, 2009

The Changing Climate

Our Melting North

By WALDEMAR KAEMPFFERT

Two pieces of evidence were recently presented to substantiate the views held by most geologists that someday there will be no frozen North and that vessels will sail in Arctic seas now imperiled by ice floes. One piece of evidence comes from Greenland, the other from Alaska.

A party of scientists from the University of Michigan, headed by Professor Ralph I. Belknap, found a pile of stones with a very artificial look. At the bottom of the pile was a piece of paper which had been torn from a notebook and which bore a few words signed by the late Professor R. S. Tarr of Cornell and dated 1896. Tarr had gone to Greenland to explore the great Cornell glacier. His note had been deposited at what was then the edge of the glacier, 400 miles north of the Arctic Circle. But the pile found by the Michigan expedition was about three-fifths of a mile further front. There was but one conclusion to be drawn. The ice had receded.

Turn now to Alaska and look at it through the eyes of Professor Robert F. Griggs of the University of Washington. The forest line is advancing in the tundra or treeless flat country at the rate of one mile in a century. Evidently Alaska's climate is growing warmer. Her trees are the first that have grown in 20,000 years.

Greenland Growing Warmer

On one point Professor Griggs seemingly disagrees with the scientists from Michigan. He finds that Greenland's climate has been growing increasingly colder for several centuries. Bodies of Norse colonists who settled in Greenland four centuries before Columbus discovered America were buried in ground now icy. The roots of trees long ago frozen to death have pierced many a grave. Who can doubt that Greenland must have had a much milder climate 1,000 years ago than it has now?

The findings of Professors Griggs and Belknap are not actually in conflict. Both scientists are dealing with comparatively short periods. The ice may again creep down upon the spot where Professor Tarr left his note, or it may continue its apparent recession and convert Greenland into a country as hospitable as the Norsemen once found it.

Still in the Ice Age

Geologists are convinced that we are living in the Quaternary Ice Age, which began about 600,000 or 700,000 years ago and which will continue for thousands more. Although there can be no doubt that the ice will eventually melt at both Poles, it is possible that it will increase in thickness and area before it finally disappears. And what a mass there is to melt! Five million square miles in the Antarctic and 1,000,000 in the North.

The truth is that the earth has passed through several Ice Ages. Those of the past were probably much like that in which the earth is still wrapped. In other words, there were relatively cold and mild periods. Although an ice age may last for 700,000 or 1,000,000 years, it is but a passing phase in the history of the earth. The mild intervals between Ice Ages are measured by tens of millions and even hundreds of millions.

What will happen if all the ice should melt? According to Dr. W. J. Humphreys of the United States Weather Bureau, the ocean levels would be raised 151 feet, which means that Manhattan Island would be completely inundated and some of the world's most fertile regions would be destroyed.

—January 28, 1934

Warming Arctic Climate Melting Glaciers Faster, Raising Ocean Level, Scientist Says

By GLADWIN HILL

A mysterious warming of the climate is slowly manifesting itself in the Arctic, engendering a "serious international problem," Dr. Hans Ahlmann, noted Swedish geophysicist, said today.

Dr. Ahlmann, Professor of Geography at the University of Stockholm and director of the Swedish Geographical Institute, discussed the phenomenon, on the basis of personal research over two decades, at a seminar of the Geophysical Institute at the University of California here.

Since 1900, Dr. Ahlmann said, Arctic air temperatures have increased 10 degrees Fahrenheit, an "enormous" rise from a scientific standpoint.

In the same period, ocean waters in the militarily strategic Spitsbergen area have risen 3 to 5 degrees in temperature, and, apparently because of the accelerated melting of glaciers, one to one and one-half millimeters yearly in level, he said.

"We do not even know the reason behind this climatic change in recent years," Dr. Ahlmann added.

If, however, the cause were of global nature, and "if the Antarctic ice regions and the major Greenland ice cap should be reduced at the same rate as the present melting, oceanic surfaces would rise to catastrophic proportions," he said. "Peoples living in lowlands along the shores would be inundated."

The climatic change was not implausible, Dr. Ahlmann suggested, in view of the fact that "we know that the tropics have felt a marked climatic change in the last fifteen or twenty years, especially in the vicinity of West Africa. Many smaller lakes have actually disappeared and larger ones are drying up. Even huge Lake Victoria has dropped seven inches in the past decade."

The Arctic change, the scientist asserted, "is so serious that I hope an international agency can be formed to study conditions on a global basis. That is most urgent."

One effect of the change, he said, has been to improve navigation conditions along the northern rim of Europe, a development of chief interest to Russia.

"In 1910 the navigable season along western Spitsbergen lasted only three months," he said. "Now it lasts eight months. This is of world strategic importance."

—May 30, 1947

The Weather Is <u>Really</u> Changing

By LEONARD ENGEL

Weathermen have a saying that weather is never unusual. It's just different from day to day and place to place, so the "unusual" is bound to occur.

Nevertheless, this year the weather has been somewhat unusual. New York had its third wettest spring since the keeping of local weather records began in 1826. Nearly twenty-six inches of rain—nine inches above normal—fell in the city between Jan. 1 and May 31. Other sections of the East had as much or more. Meanwhile, rainfall was well below normal throughout the West. Texas had a spring drought that was early and severe even for Texas.

It has also been a year for tornadoes, though the record number reported may have been due in part to recent expansion of the Weather Bureau's corps of volunteer storm spotters. By June 19, 266 twisters had been counted—about fifty more than have ever been reported for a similar period before and three times as many as is usual. Tornadoes occurred this spring, for the first time in many years, in Utah, up near the Canadian border and in Massachusetts.

Unusual weather inevitably stirs up speculation as to cause, in part, no doubt, because we like to talk about the weather anyway. Any prominent event coincident with the exceptional weather is apt to be blamed. Heavy rains during World War I were popularly attributed to artillery bombardments in France. During the twenties and thirties it was fashionable to lay abnormal weather (along with other odd occurrences) to changes in the sunspot cycle. Today the popular villains of freak weather are atom-bomb tests and the activities of rainmakers.

Meteorologists put no stock in any of these as sources of aberrant weather. Rains come on battlefields whether or not guns are firing. Similarly no correlation has ever been found, though many researchers have sought one, between earthly weather and the periodic march of "storm clouds" across the face of the sun. Sunspots do generate disturbances in the earth's magnetic field which blot out radio communications and one type of telegraph line, but only communications men are aware of these "magnetic storms."

As for atom bombs, by man's standards atomic explosions release vast amounts of energy and stir up enormous clouds of dust. Even a moderate rainstorm, however, releases energy at a rate equivalent to several hundred atom bombs a minute. And the great dust storms of the thirties raised clouds with a billion times the dust concentration of an atomic cloud without perceptible effect on rainfall anywhere in the country. Rain has sometimes occurred in large cities set ablaze by mass "fire raids"; it may also follow very large forest fires. But here, too, the release of energy is enormously greater than in an atom bomb, and it is continued over a period of days or weeks. An atomic explosion is awesomely powerful, but not powerful enough to open the heavens. Close study of test blasts shows that they do not even produce local rainstorms.

What about rainmakers? Weathermen have still to agree on the efficacy of seeding clouds with particles of dry ice or silver iodide as a means of bringing down rain. The great majority feel, though, that cloud-seeding has, at most, local effects. Three years ago Dr. Irving Langmuir, Nobel Prize physicist and pioneer rainmaker, turned on a cloud-seeding smoke generator at Socorro, N.M., once a week for eighty-four weeks. He believes this was responsible for weekly rains that fell in many parts of the East for some months. A similar experiment the next year proved inconclusive. In any event, no long-distance rainmaking effect of this sort could have happened this year, for the downpour in the East this spring finally let up late in May, though the rainmakers have continued as busy as ever out West.

If the experts thus dispose of popular theories on the origin of the cockeyed weather this year, they themselves have none to offer. They simply set down unusual weather to the daily vagaries of weather. They are more interested in the basic forces that shape weather day in and day out and in long-term trends.

There is abundant evidence that the world's weather is currently undergoing a long-term change. Since the middle of the last century, mean annual temperatures have risen by 1 to 4 degrees throughout the middle and high latitudes of the Northern Hemisphere. In Philadelphia, the rise has been 4 degrees; in Montreal, nearly 2½ degrees in the 70 years records have been kept; and in Britain and Scandinavia, 1 to 2 degrees since 1850.

Summers have become warmer, but the most striking changes have been in winters. Fifty years ago, a horse and carriage could be driven for several weeks each winter across the Hudson River between Nyack and Tarrytown; iceboating was a favorite sport. The last of the iceboating clubs there is long since gone. In the Alps, in Greenland, in Alaska, winters are no longer long or severe enough to make up for summer melting of the glaciers, and nearly all the great ice sheets are in retreat. In Spitsbergen, the mean winter temperature has risen 18 degrees since 1910 and the harbor is now open to shipping 200 days a year.

This is by no means the first major climatic change experienced by earth. Before the Ice Ages, tropical palms grew in northern Europe. At the height of the ice advance, lands near the Tropic of Cancer had the climate of New York today. Within historical times, grapes could be cultivated in England, and Greenland and Iceland supported a flourishing Viking culture whose accomplishments we are only beginning to realize.

Specialists in climate have many suggestions as to possible causes of these climatic swings, from changes in the tilt of the earth's axis to increased or decreased output of energy by the sun. Careful measurements over many years, however, reveal neither any significant tendency on the part of the earth's axis to wobble, nor important changes in the sun itself. For an explanation of such climatic changes as occurred in medieval England and are occurring now, at least most meteorologists prefer to look to the forces at work within the great weather engine of the world—the earth's atmosphere.

Like the ocean sea, the sea of air in whose depths we deal is in ceaseless motion. Warm humid masses of air move, in the Northern Hemisphere, up from the Equator. Cool, dry masses come down from the Arctic. Winds blow, in a clockwise direction, around high-pressure areas above the Equator, and the opposite way around high-pressure areas below it.

As yet, the mechanisms involved in the motion of the air are imperfectly understood. There are all sorts of confusing local movements caused by the varied surface topography of the earth, such as winds from offshore along coasts, and from glaciers to the warmer areas around them. But certain basic atmospheric movements can be distinguished. There are movements of the air due to the rotation of the earth about its axis. There are movements due to the fact that the upper atmosphere is very cold and many parts of the earth's surface are very warm; just as with water in a kettle on a stove, warm air tends to rise and cold air to sink. And there are movements connected with the temperature differences between polar areas (which receive much less energy from the sun in the course of a year) and equatorial areas.

These movements of the atmosphere constitute a gigantic engine for transferring heat from one part of the earth to another. What we know as "weather"—storms, heat waves, cold waves, fair skies and foul—is merely a byproduct of the working of this gigantic atmospheric heat-transfer engine.

The British meteorologist C. E. P. Brooks has pointed out that the atmospheric engine, though very powerful, is also delicately balanced. Suppose, for example, something were to occur to cut down the amount of sunlight reaching even a single continental area. This would reduce the temperature difference between that area and adjacent sections of the Arctic. Air circulation and heat transfer between the two would, therefore, be slowed down. According to computations by Dr. Brooks, this state of affairs need not obtain for more than a few years to result in appreciable growth of Arctic ice. In turn, cold winds from the growing ice sheet would cool the "warm" continental area, further diminishing temperature differences and heat transport. In not too long, the ice sheet would be very much larger and the continental area substantially cooler than before. Conversely, if something were to raise the temperature of the continental area by a few degrees even for a comparatively short time, heat transfer would be speeded up enough to put the ice in full retreat.

Dr. Brooks himself makes no definite suggestion as to the event or events that could have started the warm cycle we find ourselves in. But Dr. Harry Wexler, chief of the Science Services Division of the Weather Bureau, has suggested one development that could very easily have started it—the decline in volcanic activity, especially in the Northern Hemisphere, in recent years.

Dr. Wexler's volcano story begins on August 27, 1883. On that day, the island of Krakatoa in the East Indies blew up with the mightiest explosion in the memory of man. The explosion, which was heard several thousand miles away, threw thirteen cubic miles of rock, dust and ash into the air. Huge clouds of fine ash rose twenty miles or more and spread across the Northern Hemisphere. Wherever the cloud drifted it turned the sun and moon purple, blue or green, and produced extravagantly colored sunsets such as had never been seen. Even more remarkable, when the cloud finally arrived over Europe, astronomers at the Montpellier Observatory in France noted a sudden, sharp drop in solar radiation reaching the ground. The sunfall at Montpellier remained 10 percent below normal for three years.

Astronomers thereafter systematically measured sunfall following volcanic explosions. In the next three decades, there were a number of eruptions that threw sizable quantities of ash into the air. Following each, there was a pronounced decline in solar energy striking the ground in many parts of the world. For instance, Dr. Wexler relates, the explosion of Mount Katmai in the Aleutians in 1912 reduced the sunfall in Algeria, more than 7,000 miles away, by 20 percent.

Swiss and American meteorologists suggested that these changes in the amount of solar energy reaching the ground must play a part in shaping the world's weather. Their theory was premature; there were then too many gaps in the world weather map to permit conclusions. Moreover, too little was known of the workings of the atmospheric weather engine to show how changes in sunfall in a part of the world could affect the weather of the whole.

Dr. Wexler points out that we now know enough about winds and air circulation to assert that a decline in sunfall as a result of volcanic explosions would reduce atmospheric circulation and heat transfer enough to bring about colder weather throughout a hemisphere. And, of course, a rise in sunfall following a decline in volcanic activity would have the opposite effect, making many parts of the world warmer.

It happens that there has been a decline in volcanic activity. The explosion of Mount Katmai was the last eruption in the Northern Hemisphere to throw any amount of ash into the air. The only significant eruptions since 1912 have occurred in the Southern Hemisphere, where the climate has warmed up very little, if at all. North of the Equator, on the other hand, the air seems to have been getting clearer (despite urban air pollution), the sunfall greater and the atmospheric circulation faster. Consequently, northern ice is shrinking, our summers are warmer, and our winters shorter and milder.

Another theory, advanced by some meteorologists, attributes at least part of the rise in temperatures to a small but definite increase in the past century in the percentage of carbon dioxide in the atmosphere. The air's content of this product of combustion is important because carbon dioxide has heat-conserving properties similar to greenhouse glass.

In 1850 the air contained somewhat less than thirty parts of carbon dioxide per 1,000 parts of air. In the hundred years since, industrialized, urbanized man has poured unprecedented quantities of carbon dioxide out of home and factory chimneys. (The heat that also went up the chimneys has had little effect, except immediately above cities.) At the same time, deforestation has greatly reduced the number of trees absorbing carbon dioxide for the manufacture of carbohydrates by photosynthesis. As a result, there are now thirty-three parts of the gas per 1,000 in the atmosphere instead of thirty. Calculations by physicists show that this is enough of an increase to make a detectable difference in the temperature at the surface of the earth.

Meteorologists are not at all certain that the change in our climate is due solely or principally to changes in atmospheric circulation as a result of the decline in volcanic activity and to the rise in the carbon dioxide content of the air. Other influences may also be at work. For there are still many gaps in our knowledge of weather and this is the first major climatic change they have had an opportunity to examine at first hand. The only certainties are that our climate is changing, and that the changes will affect us deeply.

The first signs of the change are already evident. In 1919, cod appeared for the first time off Godthaab, at 64 degrees North Latitude on the west coast of Greenland. By 1948 they were at 73 degrees North, and cod had become a staple of Eskimo diet, and Greenland a major source of the fish. Haddock, halibut, herring and other commercial food fish have likewise migrated north. In Canada and Siberia, the area of permanently frozen ground is retreating poleward at a rate of dozens of yards to several miles a year. In Finland and Scandinavia, farmers are plowing upland fields that had lain under the ice for 600 years. Meanwhile the tropics seem to be cooling off very slightly, though observations are too few for anyone to be sure.

Thus, the current swing of climate seems to be opening up new opportunities for man. The warming-up process, however, also poses problems. For one, it seems to have been accompanied to date by a marked decline in rainfall in many parts of the world. For another, if the warm-up continues for another several decades, shrinkage of the Arctic ice cap could cause a troublesome rise in ocean levels. The rise would not, as alarmists predict, wipe out all our port cities. But it could be troublesome enough to demonstrate anew that, for all his central heating and air conditioners, climate still makes man more than man makes climate.

—July 12, 1953

Warmer Climate on the Earth May Be Due to More Carbon Dioxide in the Air

By WALDEMAR KAEMPFFERT

The general warming of the climate that has occurred in the last sixty years has been variously explained. Among the explanations are fluctuations in the amount of energy received from the sun, changes in the amount of volcanic dust in the atmosphere and variations in the average elevation of the continents.

According to a theory which was held half a century ago, variation in the atmosphere's carbon dioxide can account for climatic change. The theory was generally dismissed as inadequate. Dr. Gilbert Plass re-examines it in a paper which he publishes in the *American Scientist* and in which he summarizes conclusions that he reached after a study made with the support of the Office of Naval Research. To him the carbon dioxide theory stands up, though it may take another century of observation and measurement of temperature to confirm it.

Abundant Gases

In considering the theory, Dr. Plass reminds us that the most abundant gases in the atmosphere are nitrogen and oxygen. There is also a little argon. These cannot absorb much of the heat radiated by the earth after it has been warmed by the sun. If they could, the climate would be far colder than it is today, because the passage of heat to outer space would not be stopped.

Three other gases could check the radiation of heat. These are carbon dioxide (the gas that fizzes in ginger ale), water vapor and ozone. All these are relatively rare.

To explain what happens, Dr. Plass resorts to the familiar greenhouse analogy. The rays of the sun pass through the transparent glass, but the outgoing energy (heat) from the plants in the greenhouse cannot pass through. Heat is trapped in the greenhouse, with the result that it is warmer inside than outside.

The atmosphere acts like the glass of a greenhouse. Solar radiation passes through to the earth readily enough, but the heat radiated by the earth is at least partly held back. That is why the earth's surface is relatively warm. Carbon dioxide, water vapor and ozone all check radiation of heat.

Of the three gases that check radiation, carbon dioxide is especially important even though the atmosphere contains only 0.03 percent of it by volume. As the amount of carbon dioxide increases, the earth's heat is more effectively trapped, so that the temperature rises.

All this was first brought to the attention of scientists by Tyndall in 1861. In his day the facilities for studying the atmosphere and measuring its temperature were crude. Today they are highly refined. According to Dr. Plass, the latest calculations indicate that if the carbon dioxide content of the earth were doubled the surface temperature would rise 3.6° C. and that if the amount were reduced by half the surface temperature would fall 3.8° C.

Striking Changes

Such a comparatively small fluctuation seems of no importance. Nevertheless it can bring about striking changes in climate. If the average temperature should fall only a few degrees centigrade, glaciers would cover a large part of the earth's surface. Similarly a rise in the average temperature of only 4° C. would convert the polar regions into tropical deserts and jungles, with tigers roaming about and gaudy parrots squawking in the trees.

Dr. Plass examines the various factors that enter into what is called the "carbon dioxide balance," including the exchange of carbon dioxide between the oceans and the atmosphere. That balance must be preserved. Photosynthesis (the process whereby plants with the aid of sunlight assimilate carbon dioxide to produce sugars and starches) causes a large loss of carbon dioxide, but the balance is restored by processes of respiration and decay of plants and animals.

Despite nature's way of mainlining the balance of gases the amount of carbon dioxide in the atmosphere is being artificially increased as we burn coal, oil and wood for industrial purposes. This was first pointed out by Dr. G. S. Callendar about seven years ago. Dr. Plass develops the implications.

Generated by Man

Today more carbon dioxide is being generated by man's technological processes than by volcanoes, geysers and hot springs. Every century man is increasing the carbon dioxide content of the atmosphere by 30 percent—that is, at the rate of 1.1° C. in a century. It may be a chance coincidence that the average temperature of the world since 1900 has risen by about this rate. But the possibility that man had a hand in the rise cannot be ignored.

Whatever the cause of the warming of the earth may be, there is no doubt in Dr. Plass' mind that we must reckon with more and more industrially generated carbon dioxide. "In a few centuries," he warns, "the amount of carbon dioxide released into the atmosphere will be so large that it will have a profound effect on our climate."

Even if our coal and oil reserves will be used up in 1,000 years, seventeen times the present amount of carbon dioxide in the atmosphere must be reckoned with. The introduction of nuclear energy will not make much difference. Coal and oil are still plentiful and cheap in many parts of the world, and there is every reason to believe that both will be consumed by industry so long as it pays to do so.

—October 28, 1956

Meeting Reaches Accord to Reduce Greenhouse Gases

By WILLIAM K. STEVENS

Negotiators from around the world agreed today [in Kyoto, Japan] on a package of measures that for the first time would legally obligate industrial countries to cut emissions of waste industrial gases that scientists say are warming the earth's atmosphere.

But details on one contentious issue—the possible trade or sale of emission permits between countries—remain unsettled, and may remain unsettled for months, and the United States has said it wants this issue resolved before it signs the treaty. Any treaty is subject to approval by the United States Senate.

The agreement reached by delegates from more than 150 nations creates a landmark environmental policy to deal with global warming, and innovative new mechanisms to carry it out.

The nations would have a year to ratify the treaty, starting next March. Talks on "trading" of emissions are expected to take place next November.

Despite the uncertainties, some environmentalists hailed the agreement as a remarkable political and economic innovation, in that it would establish a global system for dealing with what many scientists believe is the overarching environmental concern.

Opponents of the treaty condemned it as economically ruinous.

The accord—known as the Kyoto Protocol—would require the industrial nations to reduce their emissions of carbon dioxide and five other heat-trapping greenhouse gases to 5.2 percent below those of 1990. The United States would be required to cut emissions 7 percent below 1990 levels, on average, in the years from 2008 through 2012.

The agreement appears to represent a significant concession by the United States, which previously had insisted on the less stringent course of reducing the emissions to 1990 levels, not below them. It is also a measure of the flexibility in the American bargaining position that Vice President Al Gore said he had ordered negotiators to adopt on a visit here Monday.

It remains to be seen, however, whether the White House can persuade the Senate to accept the agreement, or even whether the United States will sign it.

Many senators had already expressed reservations about the less stringent Administration proposal. By 2000, emissions of greenhouse gases by the United States are expected to be about 13 percent higher than they were in 1990. They are expected to be perhaps 30 percent higher in 2010, if trends in energy use continue and no other action is taken.

Senator Chuck Hagel, a Nebraska Republican observing the talks, said here today, "Any way you measure this, this is a very bad deal for America." He predicted the Senate would not approve it.

For his part Stuart Eizenstat, the senior American negotiator, said the United States was still waiting for more signs of flexibility from developing nations on the question of emission trading.

"We do not yet have that meaningful commitment coming out of Kyoto," he said. But, he added, the possibilities for such a commitment "are certainly pregnant."

President Clinton, visiting New York, said in a statement that he was "very pleased" by the agreement, which he said was "environmentally strong and economically sound." But he said developing nations had to do more.

And in Washington, Vice President Gore called the agreement a "vital turning point." He added, "Clearly, more work is needed. In particular we will continue to press for meaningful participation by key developing nations. We are confident that can be achieved. "

Under the accord, different countries would be assigned different targets, depending on their national circumstances and economic profiles. The European Union's target was set at a reduction in emissions of 8 percent below 1990 levels, and that of Japan at 6 percent. Other targets, albeit within a narrow range, also applied to other developed countries. Some countries may be allowed to increase emissions, but globally, emissions are to be reduced by 30 percent from the levels currently projected for 2010.

After an all-night session that ran into an unscheduled 11th day of discussion, the protocol was approved by representatives of more than 150 countries.

The countries were modifying an agreement, negotiated in Rio de Janeiro in 1992, that called for voluntary efforts to limit emission of greenhouse gases. The burning of fossil fuels like coal and oil is responsible for most emissions of carbon dioxide, the most important greenhouse gas.

As a means of promoting reductions in developing countries, the delegates established a special mechanism for transferring energy-efficient technologies and nonpolluting forms of energy production from richer nations to poorer ones. Greater efficiency means fewer emissions, and alternative energy sources like solar and wind power mean none.

The Clean Development Mechanism, as the new arrangement was named, is intended to encourage companies in industrial countries to invest in emissions-reduction projects in developing countries—modern, fuel-efficient power plants, for example—and get some credit for reducing their own emissions.

But the agreement was held up for hours last night and this morning by the resistance of some developing countries, including China, India and Saudi Arabia, to the inclusion of a provision enabling the industrialized nations to trade or purchase emissions rights. The United States considers this arrangement the cheapest and most efficient way of cutting emissions. Without it, American negotiators said, it would be impossible for the country to meet its emission targets.

In emissions trading, a country or industrial company will be able to meet its emissions reduction target by cutting some of its emissions itself, while at the same time purchasing part of its required reduction from another country or company that achieved excess cuts.

Based on its success with trading emissions of sulfur dioxide, a chemical implicated in acid rain, the United States argues that larger and cheaper reductions can be achieved with this mechanism.

The objectors said that the mechanism could lead to shifting the burden to less developed countries, and that countries and companies might be able to buy their way out of their obligations.

After hours of argument, delegates agreed to allow the parties to the 1992 treaty to postpone negotiations on the principles, rules, guidelines and operations of the trading system until November.

Experts advising the negotiators here say that if emissions are not reduced, the average surface temperature of the globe will rise by 2 to 6 degrees Fahrenheit over the next century, causing widespread climatic, environmental and economic disruption.

Philip E. Clapp, president of the National Environmental Trust in Washington, who has been observing the talks here, called the agreement "a historic landmark in environmental protection." He said it would be remembered "as a central achievement of the Clinton-Gore Administration."

But representatives of the American fossil fuel and heavy manufacturing industries saw disaster in the agreement.

"It is a terrible deal and the president should not sign it," said William K. O'Keefe, chairman of the Global Climate Coalition, an industry group. Mr. O'Keefe said that "business, labor and agriculture will campaign hard and will defeat" the treaty if it is submitted to the Senate for ratification.

The protocol is generally viewed only as an early step in a continuing attempt to deal with the question of climate change. It is generally acknowledged that any attempt to deal with the problem requires a global solution involving all countries.

—December 11, 1997

CHAPTER 6

Exploration:
New Worlds, Down Here
and Out There

OUR OWN PLANET

The Discoverer of the Source of the Nile

Neither the American nor the English press appears to have noticed that Mr. Stanley has won the great prize of African exploration—the right to be recognized as the discoverer of the true source of the Nile. The river flows out of the Albert N'yanza, but no one regards that lake as its proper source, since it has a more elevated and remote source in the great Victoria Lake. Neither can the latter lake be adjudged to be the final source of the Nile, since the waters of the River Shimeeyn, which fall into the Victoria Lake on the side opposite to that from which the Victoria Nile issues, are carried by the Nile to the Mediterranean. The Shimeeyn has a course estimated by Mr. Stanley to be at least 350 miles in length. As it is the largest tributary of the Victoria N'yanza, it must be regarded as the extreme upper course of the Nile, and its fountain-head as the fountain of the great river of Africa. Now, to Mr. Stanley alone belongs the honor of having discovered the Shimeeyn, and hence he deserves to outrank Baker and Speke in the field of Nile exploration, and to be hailed as the fortunate man who has finally solved the most famous problem of geographical science.

Of course, if Dr. Livingstone's theory that the Lualaba, which has its source near the 10th parallel of south latitude, belongs to the Nile system, he—or rather his predecessor the Portuguese Lacerda—is the discoverer of the furthest source of the Nile. But it is not only improbable, but actually impossible, that the Lualaba should have any connection with the Nile. Livingstone assumed that the Lualaba falls into the Albert Lake. At the most northerly point of his last journey, he was at least four hundred miles from the Albert Lake, and at that point he found the Lualaba running with a very rapid current. Of course, if it is asserted that the Albert Lake extends in a southwesterly direction below the equator for a vast and indefinite distance, it may easily be claimed that it extends nearly to the most northerly point of Livingstone's route. This, however, would compel us to ignore Schweinfurt's explorations, since his discovery of the River Welle, flowing from east to west and having its origin in the mountains west of the Albert N'yanza and in the neighborhood of the equator, shows that he was on another watershed than that of the Nile system, and renders it extremely improbable that there can be any very great extension of the lake in a southwesterly direction. Moreover, it has been ascertained that Lake Tanganyika is drained by a stream falling into the Lualaba. This stream, on which Cameron embarked months ago, flows with a hardly perceptible current, showing that the Lualaba, at the point where this sluggish stream joins it, is but slightly below the level of Tanganyika. Now this lake is but ten feet higher than the Albert Lake. If the Lualaba flows into the latter, it has a fall of only ten feet in a course of at least five hundred miles. And yet Livingstone found it flowing with a current so swift that it was difficult and dangerous to cross it. This swift current necessarily means a rapid descent, not only from the junction of the Lualaba and the outlet of Tanganyika, which junction is distant fully fifty, and perhaps a hundred, miles south of Livingstone's last sight of the river, but also below, or, in other words, further north of Livingstone's furthest. There can be hardly any doubt that where Livingstone last saw the Lualaba it was many feet lower than the Albert N'yanza. At any rate, its rapid current is wholly inconsistent with the theory that the river has only a fall of ten feet in a course of five or six hundred miles. Its rapid current, taken in connection with the fact that it drains the Tanganyika, and that the latter is but ten feet higher than the Albert N'yanza, demonstrates that it does not fall into the latter lake, and hence that it has no possible connection with the Nile.

Stanley's fame is safe. He has found the true fountain of the Nile in the River Shimeeyn. Livingstone never saw a drop of water that belonged to the Nile, and the discoveries of Speke and of Baker, important as they were, have been interpreted and completed by the bold American who has finally grasped the prize which has cost so many precious lives, and which has eluded so many gallant and persevering efforts.

—November 11, 1875

Peary Discovers the North Pole
After Eight Trials in 23 Years

Commander Robert E. Peary, U.S.N., has discovered the North Pole. Following the report of Dr. F. A. Cook that he had reached the top of the world comes the certain announcement from Mr. Peary, the hero of eight polar expeditions, covering a period of twenty-three years, that at last his ambition has been realized and from all over the world comes full acknowledgment of Peary's feat and congratulations on his success.

The first announcement of Peary's exploit was received in the following message to *The New York Times*:

> Indian Harbor, Labrador, via Cape Bay, N.F., Sept. 6.
>
> The New York Times, New York:
>
> "I have the pole, April sixth. Expect arrive Chateau Bay, September seventh. Secure control wire for me there and arrange expedite transmission big story. PEARY.

Following the receipt of Commander Peary's message to *The New York Times*, several other messages were received in this city from the explorer to the same effect.

Soon afterward The Associated Press received the following:

> INDIAN HARBOR, Via Cape Ray, N.F., Sept. 6.—To Associated Press, New York:
>
> Stars and Stripes nailed to the pole.
>
> PEARY.

To Herbert L. Bridgman, Secretary of the Peary Arctic Club, he telegraphed as follows:

> Herbert L. Bridgman, Brooklyn, N.Y.: Pole reached. *Roosevelt* safe.
>
> PEARY.

This message was received at the New York Yacht Club in West Forty-fourth Street:

INDIAN HARBOR, Via Cape Ray, N.F., Sept. 6.—

George A. Carmack, Secretary New York Yacht Club:

Steam yacht *Roosevelt*, flying club burgee, has enabled me to add north pole to club's other trophies.

(Signed) PEARY.

Cipher Shows Authenticity

The telegram to Mr. Bridgman was sent in cipher. The cipher used was a private one and indicated clearly that the dispatch was undoubtedly from Commander Peary.

Commander Peary also sent a message to his wife at South Harpswell, Me., where she has been spending the summer.

"Have made good at last," said the explorer to his wife. "I have the old pole. Am well. Love. Will wire again from Chateau."

The message was signed simply: "Bert," an abbreviation of Robert, Commander Peary's first name. Mrs. Peary sent a wife's characteristic reply, with love and a blessing and a request for him to "hurry home."

By a strange coincidence, Mrs. Frederick A. Cook, too, was in South Harpswell, Me., when she received the first news from her husband.

Peary's Companion Reports

Two messages were received in this country also from Donald B. McMillan, who accompanied Peary. Mr. McMillan was an instructor in mathematics and physical training at the academy in Worcester, Mass., until the close of school last year, when he obtained a leave of absence of two years to go on the Peary expedition.

In addition to his message to Dr. D. L. Abercrombie, Principal of the academy, Mr. McMillan sent the following to Mrs. W. C. Fogg, his sister, who is Postmistress at Freeport, Me.:

Indian Harbor, Sept. 6, 1909.

Mrs. W. C. Fogg, Freeport, Me.:

Arrived safe. Pole on board. Best year of my life. BEN.

Follows Cook's Report Quickly

These messages, flashed from the coast of Labrador to New York and thence to the four corners of the globe while Dr. Frederick A. Cook is being acclaimed by the crowned heads of Europe and the world at large as the discoverer of the North Pole, added a remarkable chapter to the story of an achievement that has held the civilized world up to the highest pitch of interest, since Sept. 1, when Dr. Cook's claim to having reached the "top of the world" was first telegraphed from the Shetland Islands.

The two explorers, Dr. Frederick A. Cook and Commander Robert E. Peary, both Americans, had been in the arctic seeking the goal of centuries, the impossible north pole, whose attainment has at times seemed beyond the reach of man. Both were determined and courageous, and both had started expressing the belief that their efforts would be crowned with success.

Peary the Better Known

Peary was well known to both scientists and the general public as a persistent striver for the honor of reaching the "farthest north." Dr. Cook, on the other hand, had held the public attention to a lesser degree. He made his departure quietly and his purpose was hardly known except to those keenly interested in polar research.

Then suddenly, and with no word of warning, a steamer touched at Lerwick, in the Shetland Islands, and Dr. Cook's claim to having succeeded where expedition after expedition of the hardiest explorers of the world had failed was made known. Dr. Cook's announcement was that he had reached the pole on April 21, 1908.

Three days later Dr. Cook arrived at Copenhagen and received a welcome such as no explorer had ever received before.

Peary Announces Success

Five days after the receipt of the Lerwick message, almost to the hour, came the sensational statement from Indian Harbor, Labrador, that Commander Peary also had been successful on his third expedition to the coveted goal, the date being April 6, 1909.

He filed his brief messages and continued on his way to the south, leaving the world to marvel at a dramatic situation such as has seldom been recorded—

the double achievement of a purpose that for almost ten centuries had baffled the endeavor of man and had taken many an explorer to his death in the frozen north.

It is almost certain that Commander Peary did not know of Dr. Cook's announcement when he sent his messages from Indian Harbor.

Under ordinary circumstances Commander Peary's announcement would have evoked worldwide interest, but the existing conditions conspired to add many times to the importance of his communication.

According to Dr. Cook's account of his expedition, he buried the American flag at the pole in a metal tube; Peary's words would indicate that the Stars and Stripes were raised by him and left standing.

How the News Came

The message from Commander Peary to *The New York Times* was received in New York at 12:39 yesterday through the Postal Telegraph Company. It was handed in at Indian Harbor; Labrador, and was sent from there by wireless telegraph to Cape Ray, Newfoundland, and from Cape Ray to Port aux Basques by the Newfoundland Government land lines; thence to Canso, Nova Scotia, by cable, and to New York from there over the lines of the Commercial Cable Company.

—September 7, 1909

Cook Answers *Times* Queries

D r. Frederick A. Cook sat in a parlor of the Waldorf-Astoria for two hours yesterday afternoon and answered questions put to him by reporters representing all the newspapers of this city concerning his trip to the arctic and his assertion that he reached the North Pole. He answered every question put to him excepting those which had a bearing upon the definite observations he made when, as he says, he had reached the pole. In refraining from giving information upon this point he, however, explained, as he had before, that his proofs are to go to the Danish University at Copenhagen, and that before the results are announced by the Danish scientists he does not feel that he should give the information sought.

The most important and definite information elicited from Dr. Cook was in answer to a series of twenty-six questions presented to him by *The New York Times,* which Dr. Cook had had in his possession since Tuesday. These questions and the answers to them are printed in full on Page 2 of this issue in conjunction with comments by Prof. Harold Jacoby, head of the Department of Astronomy at Columbia University. Prof. Jacoby finds that Dr. Cook's answers, especially his refusal to supply details of any single observation at the pole, make impossible at present a decision upon the question whether he did or did not actually reach the pole.

Adhering to his announced intention when he arrived on Tuesday on the *Oscar II,* Dr. Cook in the interview would not permit himself to make any criticism of Commander Peary. He refused to comment upon Peary's denunciation of him as a false claimant to the discovery of the pole and reiterated his statement of Tuesday that when Peary comes to the city he will face him and stand his own ground. Dr. Cook insists that no cause exists for ill feeling between him and Peary so far as he knows.

Concerning the obligation of secrecy imposed by the explorer on Harry Whitney, the New Haven sportsman; Pritchard, the cabin boy of the *Roosevelt,* and the two Eskimos who accompanied him in his dash for the pole, Dr. Cook declared that he wanted to be the first to announce to the world that he had accomplished the object of his quest. He further explained that he did not wish Commander Peary by any chance to learn that he had been to the pole before he himself had an opportunity to convey to civilization the news of the discovery.

Asked for the reason for this attitude toward Peary, the explorer replied that he felt that if Commander Peary heard of the discovery he (Peary) might try to "beat him" with an announcement Peary had made the discovery. Dr. Cook made it plain, however, that he did not mean to insinuate that Peary would have put up a fictitious claim.

He Meets the Reporters

Forty reporters gathered in the Waldorf parlor to meet Dr. Cook. He had set the time for the interview at 4 p.m., but it was half an hour after that time when the explorer walked into the room. With him was his daughter Rose. After greeting the reporters with a graceful courtesy, she disappeared upstairs to the suite occupied by the explorer and his family.

Dr. Cook greeted his interviewers cheerily. He stood for a moment surveying the group. Then he asked:

"Now, what can I do for you?"

"Dr. Cook, there are many questions we would like to ask you," replied one of the reporters.

"Well, fire away," said Dr. Cook, drawing a chair to the center of the room and proceeding to make himself comfortable.

Throughout the interview he sat with one leg resting on a knee. He met his interrogators with a frank gaze. At times when some question amused him a merry twinkle would appear in his eye. The explorer gave his answers in a first tone. He has a voice of pleasant conversational quality.

His replies were, as a rule, given slowly, the explorer evidently carefully weighing each word before giving it utterance. Yet there seemed to be no reluctance on his part to meet the questions fairly and fully.

The first question put to Dr. Cook related to the proofs he said on Tuesday that he had brought back with him as evidence of the discovery of the North Pole.

"Are you ready to produce the proofs you have spoken of?" the explorer was asked.

"No, I am not," replied Dr. Cook. "I would like to explain exactly why this is although I have given my reasons before. I am preparing to send the proofs to the University of Copenhagen, whose scientists will examine them. They will pass upon them, and I have no reason to doubt that they will pronounce them to be correct." . . .

Kept the News from Peary

Dr. Cook was asked if he knew that Whitney had written a letter, shortly after meeting him at Etah, to Whitney's mother, telling her of the discovery of the pole.

"No, I hadn't any information of that sort," he replied. Mr. Whitney had the privilege of writing letters, so long as he did not allow the secret I had entrusted to him to reach civilization before I did. My idea was to keep the news from Peary."

"Why did you not want Peary to know?"

"Why should I have let him know it? Wasn't it the natural thing for me to desire that I should be the one to make the announcement?"

"Did you think Peary would try to make any improper use of the information, that is that he might make an endeavor to proclaim himself the discoverer although he had not been to the pole?"

"No, I would not say that," replied Dr. Cook. "I merely thought it a wise policy to use every means to keep him from knowing of my success."

"Had anything ever occurred between you and Peary to cause bitterness or enmity?"

"Nothing that I know of."

"There never was any trouble between you?"

"None at all."

"Do you now consider him a friend or an enemy?"

"I don't know. I have always treated him as a friend, and I shall do so until I know more about the situation."

"Did you ever say anything to indicate you feared violence at the hands of Peary or that you feared staying at Etah?"

"No, nothing of the kind."

"Would you meet Peary in debate?"

Dr. Cook smiled wearily.

"I am willing to meet Peary when he comes here," he replied. "But as far as I am concerned this controversy is a closed incident. Peary is not the dictator of my affairs, nor I the dictator of his."

"Peary says he can disprove your claim to the discovery of the pole."

Dr. Cook leaned forward, and his eyes snapped as he replied:

"Now, I am through with that discussion. I shall have nothing more to say of Peary." . . .

At this juncture the list of questions prepared by *The Times* was read to Dr. Cook, and he answered them as they were read. The questions brought out many others, which kept the explorer busy for an hour. He was questioned as to the supplies he had with him during the last 460 miles to the pole, when he left his last supporting party.

"I had provisions for eighty days for myself, two Eskimos, and twenty-six dogs." he replied. "We counted on dog to eat dog for a part of the eighty days. The sledges we had weighed about 600 pounds. There were two of them."

"Did you establish any caches of food?"

"You can't put them in ice. There was no land."

"What was it that impelled you to try the dash for the pole when it had been understood when you left Gloucester, Mass., that you were merely going on a hunting expedition?"

"We had equipped ourselves on leaving Gloucester with everything we needed to make a start for the pole, although no announcement was made of our intention. We did not have any government backing or any private funds back of us. We had no reason for crying out to the world what we were going to do."

Equipped for a Polar Trip

"When we sailed we were prepared for the polar trip. Arriving at Etah we found that everything was favorable for the attempt. The Eskimos had had their most successful season of hunting in years. They were prepared to go with us and hunt for bears to add to our stores. Never were conditions so favorable for the working out of our plans. The opportunity was seized eagerly upon. We made our start and found that we had calculated correctly. Never was a trip toward the pole made under more favorable circumstances."

Dr. Cook, speaking of the search for food on the way down from the pole, remarked that he did not trust to providence so much as to himself and the Eskimos with him. They made their supplies last as long as they could, and when they all ran out the little party was in fear that they would perish.

The explorer smiled when told that Henson, the negro who went to the pole with Peary, had started in to interview at Sydney that Dr. Cook's two Eskimos broke down when he talked with them at Etah after Peary came from the pole and admitted that Cook did not go beyond the land.

"Those two Eskimos will come here and they will answer all questions," replied the explorer. "They will say that I was at the pole and nothing else. I know that they can be depended upon. They were with me when I reached the pole and will not lie."

"Henson says the Eskimos told him you had ordered them to say at first that you had been to the pole."

"Is that so?" exclaimed Dr. Cook. "Well, just question Henson when he comes back. He may be able to tell something else."

"What do you mean?"

"Just wait—that's my advice," replied the explorer, with a smile. . . .

"Will your records prove beyond all doubt that you were at the pole?" was asked as the interview drew to a close.

"I am sure they will," replied Dr. Cook. "I have said that all my observations were carefully made. They will show I found the pole."

—September 23, 1909

A Clash of Polar Frauds and
Those Who Believe

By JOHN TIERNEY

In September 1909, Dr. Frederick A. Cook and Robert E. Peary each returned from the Arctic with a tale of having reached the North Pole. Neither provided any solid proof or corroborating testimony; both told vague stories with large gaps. They couldn't even convincingly explain how they had plotted their routes across the polar ice.

Yet each explorer's claim immediately attracted its supporters, and no amount of contradictory evidence in the ensuing years would be enough to dissuade the faithful.

A century later, the "discovery" of the North Pole may qualify as the most successful fraud in modern science, as well as the longest-running case study of a psychological phenomenon called "motivated reasoning."

The believers who have kept writing books and mounting expeditions to vindicate Cook or Peary resemble the political partisans recently studied by psychologists and sociologists. When the facts get in the way of our beliefs, our brains are marvelously adept at dispensing with the facts.

The first people to believe Cook and Peary had obvious motivations: scooping rival newspapers and increasing circulation.

When Cook cabled his tale to *The New York Herald* (the newspaper promptly devoted its entire front page to the news: "Fighting Famine and Ice, the Courageous Explorer Reaches the Great Goal").

Several days later Peary cabled his claim to *The Times*, which had helped sponsor his expedition. *The Times* hailed his triumph, reporting that "the world accepts his word without a shadow of hesitation" and quoting Peary's denunciation of Cook as a fraud who "has simply handed the public a gold brick."

Each explorer promised to provide proof, but neither had taken along a trained navigator to corroborate the feat with independent celestial observations. Cook wasn't even competent himself to make the observations.

Peary was an expert navigator and traveled with companions who could also use a sextant, but he left them behind for the final week's push. Then, with no other trained navigator present, his daily rate of progress suddenly doubled.

Most puzzling of all, his expedition traveled for hundreds of miles across the ice without making any celestial observations to determine their longitude and to make sure they hadn't veered off course to the east or west. Then, after five weeks, Peary made an observation and refused to reveal the results to his companions. He was reported to look disappointed, and he left his diary pages blank that day. But he would later tell the rest of the world that his observation had confirmed his arrival at the pole.

How, in moving across jumbled pack ice continuously drifting in the wind and ocean currents, did Peary unerringly travel right to the North Pole? How did he achieve a nearly 500-mile "pole-in-one," as the historian Dennis Rawlins would later dub it?

In 1909, such questions didn't trouble *The Times*, the National Geographic Society and Peary's other supporters. They were so busy denigrating Cook's claim—"the most astonishing imposture since the human race came on earth," according to *The Times*—that they overlooked flaws in their own hero. This is not surprising, really, at least not to researchers who have studied both Democrat and Republican partisans using brain scans and other techniques.

When we contemplate contradictions in the rhetoric of the opposition party's candidate, the rational centers of our brains are active, but contradictions from our own party's candidate set off a different reaction: the emotional centers light up and levels of feel-good dopamine surge.

With our rational faculties muted, sometimes the unwelcome evidence doesn't even register, and sometimes we use marvelous logic to get around the facts.

In one study, Republicans who blamed Saddam Hussein for the attacks of Sept. 11, 2001, were presented with strong counterevidence, including a statement from President George W. Bush absolving Hussein. But most of the people in the study went on blaming Hussein anyway, as the researchers report in the current issue of *Sociological Inquiry.*

Some of the people ignored or rejected the counterevidence; some "counter-argued" that Hussein was evil enough to do it; some flatly said they were entitled to counterfactual opinions. And some came up with an especially creative form of motivated reasoning that the psychologists labeled "inferred justification": because the United States went to war against Hussein, the reasoning went, it must therefore have been provoked by his attack on Sept. 11.

This is the sort of backward logic employed by Peary's supporters in recent decades. As scholars and explorers with much more Arctic experience than Peary have rejected his claim, the supporters have tried furnishing the missing proofs and explanations: if Peary said he made it to the pole, there must have been a way to do it.

They have dreamed up ways for him to navigate precisely north by studying wind patterns in the snow, looking at the sun or observing shadows. They have suggested he navigated by compass (even though it is notoriously difficult to use near the magnetic pole). They've tried to match his speeds near the pole (but have failed even when guided by GPS).

They have analyzed Peary's photographs and concluded that the shadows offer the long-sought proof he was at the pole, according to a report for the National Geographic Society in 1989. The society hailed the report as "unimpeachable" and today stands by it and by Peary's claim to the pole.

But the report was criticized by outside experts, who concluded that the photos could have been taken more than 100 miles from the pole. Another of the report's assertions, that Peary's accurate steering was plausible because Roald Amundsen had reached the South Pole in a similar manner, was directly contradicted by evidence that Amundsen had relied on regular observations to determine longitude.

Among polar experts today, the consensus is that Peary got much closer than Cook, but not to the pole. Some suggest Peary gave up the day he took that solitary observation because he realized how far off course he had gone; some suspect he had earlier avoided taking longitude observations so as not to leave a paper trail of his route. (For more on the continuing debate—and for who really reached the pole first—go to nytimes.com/tierneylab.)

Mr. Rawlins and another prominent polar scholar, Robert M. Bryce, doubt that Peary got much closer than 100 miles to the pole. Mr. Bryce, who recently discovered the draft of the Cook telegram that started the controversy, figures that Cook stopped more than 400 miles short.

Mr. Bryce is the author of *Cook & Peary* (1997), a 1,100-page book subtitled *The Polar Controversy, Resolved*, but Mr. Bryce knows it's not resolved in all minds. Although some of the loyalists have lost faith (*The Times* ran a formal correction in 1988, citing Peary's "unreliable" records and his "incredible" speeds), both explorers still have their supporters at the Frederick A. Cook Society, the National Geographic Society and elsewhere.

Mr. Rawlins, who is the editor of *Dio*, a science history journal, says he cannot think of any modern scientific fraud that has been so profitable and popular and endured a century.

The only longer-lived example that comes to mind, he says, are the second-century astronomical "observations" of Ptolemy that were apparently derived not from the sky but from his theories.

Ptolemy's tables were used for more than 14 centuries, which seems like a hard record to beat. But with sufficiently motivated reasoning, who knows? In 1909, after Cook's loyalists ignored the evidence of fraud provided by Cook's own traveling companions, the *Independent* magazine wearily predicted, "There will be a 'Cook party' to the end of time, no matter how strong the evidence brought against him in the future." A century later, there is still a Peary party, too.

—September 8, 2009

Amundsen Describes
His Polar Dash

Capt. Roald Amundsen received me today, and not only materially added to the information already cabled by him respecting his journey to and from the South Pole, but also discussed the question of whether it is likely that he was preceded by his English rival, Capt. Scott. He said:

"I saw no traces whatever showing that Capt. Scott had been at the pole, but it is possible that he had been there and had left some unsubstantial memorial which had afterward been destroyed by the storms.

"The chances, however, are heavily against this theory, for during the three days that I was there the weather was calm and still, and I think that is the prevalent condition. There was nothing but the vast level plains of snow, and hence there was no possibility of erecting a permanent cairn of stones.

"The season was very favorable, and therefore it is exceedingly likely that Capt. Scott did reach the pole later, if not sooner, than myself. I most sincerely hope he did arrive there, for he well deserves success.

How the Dash Was Made

"On my sledging journey to the pole a new plan was tried. At first the expedition daily traveled fifteen miles in five hours, then spent two hours eating and in feeding the dogs, and then we attempted to spend the other seventeen hours in sleep. This period of rest was found to be too long both for the men and dogs, and a new plan was tried. This was to march fifteen miles in about six hours, spend two hours eating and attending to the dogs, sleep six hours, and then breakfast and march again. This accounts for the remarkable speed of over twenty miles a day attained on the return journey.

Had Difficulty in Breathing

"The greatest difficulties of the expedition were caused by the heights encountered. During the latter part of the journey to the pole we spent nearly six weeks at great elevations, which sometimes reached 16,750 feet. [This may be an error in cabling, as Capt. Amundsen, in an earlier report, mentioned 10,750 feet as the greatest reached.] The pole itself is at an elevation of 10,500 feet. When we were working hard, great difficulty was experienced in breathing at these heights, and we panted and struggled for breath.

"With regard to food, we had full rations all the way, but in that climate full rations are a very different thing to having as much as a man can eat. There seems little limit to one's eating powers when doing a hard sledging journey. However, on the return journey we had not merely full rations, but as much as we could eat from the depots after passing 86 degrees.

Eating Dog Meat No Hardship

"The first dogs were eaten on the journey to the pole in 85½ degrees, when twenty-four were killed. In spite of the fact that they had not always been able to obtain full meals, the dogs were fat and proved most delicious eating. It is anything but a real hardship to eat dog flesh.

"Two skua gulls were seen at 84½ degrees. A small cairn had previously been erected as a mark to guide us on our return, and just when he had left this the gulls came flying past and alighted on the cairn.

Three Dogs Deserted

"Three of the best dogs deserted the party at 83 degrees. We had killed a female dog at 82½ degrees, and the dogs went back, searching for her. This caused us great anxiety, for it was feared that the dogs would pillage the depots on which the party depended. When we returned to 83 degrees, after being at the pole, we saw fresh dog tracks around the large snow cairn used as a depot there. Curiously enough, the pemmican in the depot was untouched.

"Traces of the dog were followed to 82½ degrees, where the female dog had been killed. They had found the body, which was placed on the top of a heap of snow as a food reserve for the party, and, having eaten it, the dogs had gone to the depot at 82 degrees, where a large number of cases were piled up. They had got at one of the cases of pemmican and had not only eaten that, but had also eaten the leather straps and other indigestible articles. They had also eaten two dogs which we had killed and left for food at this depot.

"Eleven dogs survived the whole journey and safely reached the *Fram*.

Christmas Dinner Near the Pole

"I and my four companions on the pole party kept the Christmas festival in the high mountains, not a great distance from the pole. The feast consisted of an extra allowance of biscuits cooked in a porridge, a poor substitute for the abundance of the Norwegian Christmas, but enjoyed it heartily.

"On the return journey we had not a single day's rest. We did not even rest on Christmas Day, but passed on, day after day, through all weathers. There was little that was adventurous about the trip, but it was very hard work.

"I attribute my success to my splendid comrades and to the magnificent work of the dogs, and next to them to our skis.

"The splendid condition of the dogs on landing in the Antarctic was due mainly to the precautions taken on the *Fram*. In order to keep the dogs in good health while crossing the tropics a special double-deck planking had been fixed above the deck of the *Fram* with a space several inches deep left for the circulation of fresh air. This device was constructed before leaving Norway, and in the hot weather sails stretched above kept the dogs always in the shade. As a result, little trouble was experienced with the dogs, which landed fit for work.

Kept Fat on Polar Dash

"Of real hardships in the way of food on the pole journey there were none. Rather the reverse, for when my companions reached the ship they were almost fat, and could not eat as much as when they started. The dogs, too, were fat, and that they had lived well during the last part of the journey was shown by the fact that they would hardly touch the seal meat which was lying in large quantities about the base of the camp.

"Washing was a luxury never indulged in on the journey, nor was there any shaving, but as the beard has to be kept short to prevent ice accumulating from one's breath, a beard-cutting machine which we had taken along proved invaluable.

"Another article taken was a tooth extractor, and this also proved valuable, for one man had a tooth which became so bad that it was absolutely essential that it should be pulled out, and this could hardly have been done without a proper instrument.

"The party which explored King Edward VII land reports seeing a bird of a new species. They are certain of that, as they got close to it and saw it distinctly.

Protecting His Narrative

"I am gratified to learn that the steps taken to prevent the results of the expedition prematurely leaking out proved successful, and that *The New York Times* was the first to publish the news in America.

"The method adopted here was simple. Nobody was allowed on board the ship, and no member of the expedition other than myself was allowed to go ashore. All necessary business and other business for the crew was transacted by myself.

"I arrived Thursday and sent you a brief cablegram announcing my success, and on Friday dispatched a long account of my journey to the pole and the discoveries we had made.

"From all over the world telegrams had been pouring into Hobart, chiefly from newspapers in England and America, begging for messages from me. I had arranged that *The London Chronicle*, *The New York Times*, and their associates should have the story exclusively, and that promise I have been glad to fulfill.

Will Now Seek the Arctic

"I have been most gratified today to receive hearty congratulations from King George.

"Before sailing to Buenos Aires I have arranged to go on a month's lecturing tour in Australia. From Buenos Aires I intend to make my way to the arctic via Bering Strait, thus carrying out my original programme."

More than 100 telegrams of congratulation were received at Hobart yesterday by Capt. Amundsen, fifty being delivered to him in one bundle.

—March 11, 1912

Down into Davy Jones's Locker

By WILLIAM BEEBE

After years of peering down through blue water from the deck of a steamer, of lying flat in a glass-bottomed boat, of climbing down through ten fathoms in a helmet and then looking over a submarine precipice with longing at the unattainable blue-blackness still further beneath, I have at last been able to go to, and see, and return from, depths in mid-ocean which I never hoped to reach. Three times I touched 800 feet beneath the surface and once 1,428 feet—well over a quarter of a mile.

First of all, full credit must be given to Otis Barton. For a year he has devoted his spare time and large sums of money to the devising and making of his great steel sphere with its windows of fused quartz and the heavy bolted door. In the frequent consultations I have had with him he has never for a moment relaxed his faith and optimism in the ultimate success of the adventure.

We joined forces in Bermuda, he with the sphere and 3,500 feet of the finest, non-twisting steel cable and electric wire hose, and I with the great winch and sheaves of the *Arcturus* equipment and my trawling tug *Gladisfen* for towing. When the hoisting machinery was installed on the barge *Ready* she was towed to an anchorage near Nonsuch and we prayed for fair weather. After a young gale had blown itself out this came and with six members of the Zoological Society's expedition, we put out to sea.

We were towed through Castle Roads and found the ocean calm, almost without a ripple, with a long, low, heaving swell. Five miles off shore, due south of Nonsuch, with a mile of water beneath our keel, we headed across the swells, stopped and sent down the sphere for a test dive, empty, to 1,500 feet. It reached the deck again two hours later having taken in only about a quart of water.

After putting in the oxygen tanks, the telephone, electric lights, fans, thermometer and the chemicals for drying the air and absorbing the carbon dioxide, I crawled in and Otis Barton followed and we arranged ourselves on the curved sides of the hard steel. John Tee-Van took charge of the deck crew and Gloria Hollister put on the earphone. The 400-pound door was hoisted up and clanged into place and the interminable screwing down and hammering of the nuts and the big central bolt began. The boats had by this time drifted seaward until we were at Lat. 32 degrees 11 minutes N. and Long. 64 degrees 39 minutes W., ten miles from land.

Inside the sphere, Barton had the headphone on and took charge of our line of communication. He also regulated the turning on and off of the electric light, the amount of flow of oxygen and watched the floor for possible accumulation of water. I did the observing through the window and intermittently illumined the door, the hose-inlet and the thermometer and carried on what experimenting I could with colors.

Peering out through my six-inch porthole I could see Tee-Van on the deck waiting for a moment of quiet. Old Captain Millet on the distant poop signaled that a spot of calm was coming, and soon we were gently lifted, and exactly at 1 o'clock, with the momentary feeling of an airplane takeoff, we swung over the bulwarks and out above the water. After a delay for rearranging the tackle, we splashed gently into the surface and sank to a level with the keel of the *Ready*. This was a landscape familiar to me from the diving helmet—a solid, moving reef with banners of seaweed, long sponges and masses of rough-spined pearl shells. Here we hung for a time and then the keel passed slowly from view and we could sense, rather than feel, our descent.

A few seconds after we lost sight of the hull, word came down the hose that we were at fifty feet, and I looked out at the brilliant blue and could not realize that this was almost the limit of my diving helmet range. Then "100 feet" was called out and still the only change was a slight twilighting and chilling of the blue. In a few seconds I realized we were at 132 feet, the depth where Commander Ellsberg labored so gallantly to free the men in the submarine *S-51*. "Two hundred feet" came and we stopped with the slightest jerk and hung suspended while a clamp was attached, binding the hose and cable together. Then word came down that all was clear and again I knew we were sinking, although only by the upward passing of the small motes of life in the water.

Here we were far from any touch of Mother Earth: ten miles from the shore of Bermuda and one and one-half miles from the sea bottom far beneath us. Barton gave an exclamation at 800 feet and I turned the flash on the door and saw a slow trickle of water beneath it. About a pint had already collected in the bottom of the bell. I wiped away the meandering trickle and still it came. I remember that there flashed across my mind the memory of gentle rain falling on a windowpane and the first drops finding their way with such difficulty over the dry surface of the glass. Then I looked out through the crystal clear quartz at the deep blue and the contrast closed in on my mind like the slowly deepening twilight.

We watched the trickle. I knew the door was solid enough—a mass of 400 pounds of steel—and I knew the inward pressure would increase with every foot of depth. So we gave the signal to descend. But after that, the flashlight was turned on the doorsill a dozen times during our descent, but the stream did not increase.

Another few seconds and "400 feet" was called out; 500 feet came and passed, then 600 feet, where we remained for five minutes. The electric searchlight now became useful, shooting a shaft out through the dark blue, faint, but serving to illuminate anything which crossed it.

Now when I cupped my face in my hands and stared and stared out, I began to see what a strange illumination this was. In fact, the most amazing thing about our whole descent was the color. My companion kept exclaiming at its brilliance and saying he could read by it, and while the excitation of my optic nerves agreed with him, my mind told me it was not true. I brought all my logic to bear, I put out of mind the excitement of our position in the watery space, and tried to think color, and I failed utterly. I was dealing with something too different to be classified.

We sent up an order, put out our light and sat quietly, sensing that we were sinking. The twilight (the word had become absurd, but I could coin no other) deepened, but we still spoke of its brilliance. It seemed to me it must be like the last terrific upflare of a flame before it is quenched, but sunlight and flames meant nothing to us at this moment. As long as the electric light was off I could not imagine red or yellow or orange. And yet no blue or violet I had ever seen was anything like this.

We spoke very little now. Barton examined the dripping door, took the temperature, watched and adjusted the oxygen tank, and now and then asked, "What depth now?""Yeah, we're all right.""No, the leak's not increasing.""It's as brilliant as ever."

And we both knew it was not as brilliant as ever, but our eyes kept telling us to say so. It actually seemed to me to have a brilliance and intensity which the sunshine lacked; sunshine, that is, as I remembered it in what seemed ages ago.

"Eight hundred feet" now came down the wire and I called a halt. There seemed no reason why we should not go on to 1,000; the leak was no worse, our palm leaf fan kept the oxygen circulating so that we had no sense of stuffiness and yet some hunch, some mental warning which I have had at half a dozen critical times in my life spelled bottom for this trip. This settled, I concentrated on the window for five minutes.

The three exciting internal events which marked this first trip were, first, the discovery of the slight leak through the door at 300 feet, which lessened as we went down; next, the sudden short-circuiting of the electric light switch, with attendant splutterings and sparks, which was soon remedied. The third was absurd for it was only Barton pulling out his palm leaf fan from between the wall of the sphere and the wire lining of the chemical rack. I was wholly absorbed at the time in watching some small fish, when the sudden shrieking rasp in the confines or our tiny cell gave me all the reactions which we might imagine from the simultaneous caving in of both windows and door! After that, out of regard for each other's nerves, we squirmed about and carried on our various duties silently.

What proved to be our record dive for extreme depth came four days later. The preceding night a fire broke out on the *Ready* near her boiler and was put out only after a three hours' fight. The water line was untouched, so we were taken in tow, and again began to drift cross-swell five miles from shore.

We took a chance on everything being in order, and having fastened on the flags of the Explorers' Club and that of the tropical research department of the Zoological Society, we were screwed and bolted in at 10 o'clock. Twenty-five minutes later we had again reached our record floor—800 feet. This time I had no hunch—reasonable or unreasonable—and within three minutes we were passing through a mist of small shrimp at 900 feet. We could distinguish illustrations in a book from a page of type, but fish swam around the loop of our hose. White shrimps like ghosts drifted past, and luminous sparks burned and were quenched upon unseen bodies. Our searchlight cut a sharp swath through the blue murk, and at 10:44 we heard "1,400 feet" called down the hose. For some reason a few more meters were reeled out and when we returned we found that our exact, extreme depth had been 1,426 feet—over a quarter of a mile.

I pressed my face close against the glass and peered down, and again felt the old longing to go further, although it looked like the black pit-mouth of hell itself—yet still showed blue. I thought I saw a new fish flapping close to the sphere, but it proved to be the waving edge of the Explorers' Club flag—black as jet at this depth.

My window was clear as crystal, in fact, clearer, for fused quartz is one of the most transparent of substances and transmits all wavelengths of sunlight. But the outside world I saw through it was a blue-black world. A tiny, semitransparent jelly throbbed slowly past. For seconds at a time there was little to tell that I was not inside my helmet diving on the reefs at night near the surface—little, that is, but the light, or absence of it which persisted in making us gabble "brilliant" when it was not. There seemed no reason why we should not swing open the door and swim out, and yet I knew that on the single square inch of glass which my nose pressed there was a dead weight of more than 650 pounds.

On the outside of our steel sphere weight—400 pounds—seemed to have unpleasant "last straw" significance!

For many months, since first we planned this descent, I had been rather pessimistic about the amount of life we might hope to see from the window. The constant swaying, due to the rolling of the barge overhead, the great white sphere itself looming up through the blue murk, the scarcity of organisms at best in the depths of the ocean and the small size of the aperture, seemed handicaps too great to be overcome.

Yet the hope of such observation was the sole object of the entire experiment. We had never thought of it as a stunt, as "The-first-white-man-who-had-ever," &c.

I think it was the greatest thrill of my life when I realized that I was actually seeing a generous number of deep-sea organisms, and that I was really observing luminous fish functioning in their own environment. For a year and a half I had been studying the fish of a tiny section of mid-ocean off Nonsuch, and now I was actually in that section myself. Once before I had somewhat of a similar sensation, when at last I was able to pull myself up to the top of a gigantic jungle tree, to find myself in an entirely new world of life. My present researches in this submarine area, during the present and perhaps next year, will take on a wholly new and more intimate meaning. It was as if an astronomer should have rocketed himself around Mars and back to earth.

I found that without the electric light for 600 feet, and with it from there down, I could distinctly see many forms of life through our little window. One hundred feet below the surface I saw two of the curious *Linuche* jellyfish which I had observed and photographed by the hundreds of thousands in Haiti. Large jellies appeared at 200 feet, and then small and larger spots began passing. At first they were just meaningless motes, but by following carefully with my eye as long as they were within sight I perceived that they were shrimps of various sizes. This was the dominant form of life at all depths below 800 feet, as indeed it is in our nets. A pilot fish suddenly appeared and swam close in front of the window, appearing dead white with many vertical black bands.

My next fish were below the 400-foot level, and I was delighted to be able to identify them as lantern fish, or myctophids. No hint of luminosity was visible—it was still too light for that—but their momentary hesitation near the glass rendered their identification certain. The amazing thing is that only once have we ever taken a member of the family in the daytime from so shallow a depth. If my four lantern fish were not rare strays, then in this upper stratum the light must enable them to see and avoid our nets. This fact solves a half dozen problems unanswered since the days of the *Arcturus*.

At 600 feet I saw a long string of salpa—colonial animals which drift slowly through the depths, living lives as unreal as their surroundings. More small fish passed, and once a dark, large-headed creature twisted up to the lower part of the window, turned and vanished. Always the gentle rain of shrimps continued.

At 700 feet I began to see more luminescence. Fish would glisten like sliver as they crossed the narrow shaft of electric light, and then, entering the blue-blackness, they would become instantly disembodied and colorless, with a flash of fiery light to mark their path.

Sparks as large as a penny would flash in the distance and be quenched almost in the same instant, leaving no slightest hint of their origin. I saw no yellow or reddish gleam among the magic pyrotechnics; the momentary glow was always pale greenish or silvery. In the case of the shrimps I could detect no separate lights, only an occasional general luminescence. Once a little school of almost a dozen fish crossed my diminutive comet's tail, all astonishingly swimming upright on their tails. I have no idea what they were.

Another puzzle was the passing now and then of slender gray wraiths—almost string-like—usually drifting quietly, but occasionally darting ahead with a sudden wriggle. I thought of everything which had come up in our nets, but without avail, until the image of the most abundant of deep-sea illuminated fish occurred to me—*Cyclothones* (they have no common name). These are small, with many lights, faceted like cut diamonds, and all from the upper strata are pale gray.

At 900 feet I saw nothing startling, except a pair of four-inch slender, black fish, one of which came head on for a moment, then darted through the light and joined his companion. More than once I thought I saw large moving forms in the distance, but my eyes and senses were too alien to this illumination to be trusted for anything except what was in perfect focus.

When in a position of great interest, or at some critical moment at the very climax of a long period of preparation, we often realize its full significance only when it is all over. Fortunately this was not the case in our sphere at this great depth. I was curled up in a ball on the cold steel, Barton's voice was droning out questions and assurances, the ticking of my watch came as a strange memory sound of another world and the fan swished through the air.

I sat crouched with my eyes close against the cold surface of the tiny circle of quartz—that transparent bit of Mother Earth which held so sturdily its burden of nine tons of water back from my face—and there came over me a tremendous wave of emotion, a real appreciation which was momentarily almost superhuman, cosmic, of the whole situation: our barge tossing high overhead in the blazing sunlight like the merest chip in the midst of the ocean, the long cobweb of a cable leading down through the spectrum to our tiny sphere, where, sealed tight, we dangled in mid-water, lonely as a lost planet in space.

Here, under a pressure which, if loosened, in a fraction of a second world make tissue paper of a human being, breathing our own homemade atmosphere, sending a few comforting words chasing up and down a wire thread; here I was privileged, by the ingenuity and generosity of Otis Barton and, from the Zoological Society side, of Harrison Williams and Mortimer Schiff, to peer out and actually see the creatures which had evolved (shall we say for 100,000,000 years?) in the blackness of a blue midnight which, since the ocean was born, had known no following day.

Small wonder that at the time I could not stop to write futile words; my sole contribution to science and literature I afterward found was: "Am writing at a depth of a quarter of a mile. A luminous fish is outside the window." My own particular sensation was Herbert Spencer's lifelong dread of being "an infinitesimal atom floating in illimitable space."

We hope, with Mr. Barton's help, to carry on with this splendid deep-sea chamber—this bathysphere—next year. We shall probably, with fresh hose and new windows, try to see something of the life of the sea a half mile down. But there is sufficient work for a lifetime well within the zone of new territory in the open ocean which we have added to human activities. During the past two seasons, like an astronomer, I have had to reach out blindly into an inaccessible region, and, like a student of fossils, I have had to work for the most part on dead organisms. But this new means of exploration changes all this. It is as if the astronomer could conquer space, or the paleontologist annihilate time. I can see, and photograph, and paint—only I cannot touch them—the creatures of the deep abysses. I can be certain at first hand of the temperature and light in which they live and whether they are solitary wanderers or gathered into schools.

It seems to be the habit with modern explorers to measure the new land they have added to human ken in terms of States or countries. I, alas, have nothing but additional water to offer, but if the water of the oceans which we have made our own for a quarter of a mile depth was equally distributed over the whole of the dry land the average depth of our contribution over America, Europe, Asia and all the rest would be well over a half mile. Or, if you like this sort of thing, I may add that we have provided the possibility of exploring at first hand 35,000,000 cubic miles of water—count them!—and these if laid end to end—but the reader can carry on for himself!

The importance of the whole adventure may be summed up in a single sentence: The margin of safety, as we have demonstrated, makes future research in this direction possible and reasonable; and the scientific results have proved to be greater, both in sheer number and accuracy of facts, and in philosophical values, than our utmost hopes had led us to believe possible.

—July 13, 1930

A Voyage into the Abyss:
Gloom, Gold and Godzilla

By WILLIAM J. BROAD

Flashes of light from luminescent creatures swirled in the darkness past our windows as we descended a mile and a half to a fiery gash on the ocean floor, to what scientists increasingly see as a colossal, hidden engine of creation.

After falling for more than an hour, we switched on the lights of our tiny, three-person submarine. There, barely illuminated in the gloom, some 250 miles off the Oregon coast, lay endless fields of gnarled lava, frozen in the midst of an eruptive frenzy by the icy seawater. A string of scientific discoveries have revealed that over the eons such rocks have played a major role in shaping the planet. Volcanic fissures in the ocean floor gird the earth like seams on a baseball. All told, they measure some 40,000 miles. Though hidden from sight, zigging this way and that through the deep, often veiled by undersea mountains, they make up one of the planet's dominant and most dynamic features.

Spewing molten lava and blistering hot water, these rifts of fire are the place where planetary crust is formed, where oceans and continents are created, where the sea gets much of its chemical identity, where precious metals like gold are concentrated and where bizarre creatures that never see sunlight feast on chemicals made by the earth's inner heat. Perhaps most important, they are the place where life itself may have begun billions of years ago.

Our tiny submarine, *Alvin*, was just ending a 16-day voyage of discovery to one corner of this alien world, the Juan de Fuca ridge off the West Coast of the United States.

The effort was prompted by a recent undersea eruption here, which, in an oceanographic first, was monitored by scientists as the seafloor shook and then studied by six expeditions from the United States and Canada. The last and largest of these featured *Alvin*, whose mission was to probe deeper than ever before into the heart of undersea creation.

Among other things, we found the seafloor alive with fresh flows of lava and blizzards of white bacteria surging upward on hot-water plumes. After days of frustration caused by bad weather and enigmatic clues of deep activity in one area, we discovered a field of towering chimneys spewing super-hot water and surrounded by strange forms of life, some apparently never before seen.

There were thickets of tiny tube worms, odd growths of beard-like bacteria, legions of tiny white crab-like creatures and, on rocks facing some mineral towers, delicate networks of colonial organisms whose tentacles reached out toward the monoliths like worshipers at a temple. Scientists said these creatures appeared to be corals, which if true would be a first.

"It's exciting as hell!" said Dr. John R. Delaney, a geologist from the University of Washington who led the expedition. "I'm blown away that they pay me to do this stuff."

The History from Dead Abyss to Cradle of Life

The ocean abysses used to be thought of as geologically dead and having only a thin population of bizarre fish, mainly scavengers living off a drizzle of detritus from above. It was considered a graveyard, basically inert and irrelevant.

Discoveries over the last 40 years have shown the truth to be the opposite. The abysses are the birthplace of the fissures where the ocean plates are moving apart, so that the line of fissures along the mid-Atlantic, say, mirrors the coastlines of the once united continents between which the Atlantic emerged.

The abysses also teem with special forms of life that cluster around vents of very hot water and are adapted to the heat and chemistry of the volcanic eruptions. The first community of strange animals was discovered in 1977 off the Galapagos Islands. These animals thrived on substances that to all others were toxic.

As similar communities were found on other mid-ocean ridges, scientists theorized that life on Earth may have begun at undersea geysers, nurtured by a steady diet of warm chemicals and protected from hurtling rocks and blasts of radiation from outer space that pounded the planet's surface in its early days.

Chimney Named Godzilla

Among the strangest discoveries at the hot vents were giant towers of rock, dubbed chimneys and smokers, that discharged very hot water, often gray or jet black. The towers built as suspended minerals in the super-hot fluid were precipitated by icy seawater. One such structure recently discovered on the Juan de Fuca ridge is more than 15 stories high. It is nicknamed Godzilla. Nearby vents gush water as hot as 750 degrees, which is kept from boiling by the crushing pressures of the abyss. Mineral-rich plumes from such vents have been found to spread across thousands of miles of ocean.

By some estimates, up to 70 percent of the earth's internal heat flows through such zones, which every eight million years filter a volume of water equal to all the world's oceans. Such processes play major roles in determining the planet's heat and chemical balances.

Gold was found in some vents during the 1980s, igniting a rush of precious-metal analysis. Many billions of dollars' worth of rich ores, including zinc, silver, gold and copper, were soon estimated to lie hidden among the mid-ocean rifts. In March 1983, President Ronald Reagan nearly doubled the size of the United States by declaring sovereignty over nearby ocean floors, inspired in part by the allure of seabed riches.

To date, private companies have found mining the deep ocean too expensive. But they have raced to exploit the new findings for clues to mineral wealth on land, hunting for slices of ancient seafloor.

"The work on the ridges is really in its infancy," said Dr. Randolph A. Koski, an economic geologist on the voyage from the United States Geological Survey. Among other things, his agency is assessing the likelihood of an undersea gold rush sometime in the future. To date, Dr. Koski said, less than 1 percent of the volcanic ridges have been studied close up.

The Breakthrough Researchers Rush to Site of Eruption

The mysteries of this world are many. For instance, no one knows for sure how water flows through the mid-ocean ridges, whether from the sides or along the axes. In general, deep hot rock is thought to cause the seafloor to crack. Water then trickles down, encounters magma and is heated. The hot fluid leaches minerals out of the ocean crust as it flows upward to the surface. But this is just an educated guess, scientists stress.

"The dynamics are basically unknown," said Dr. H. Paul Johnson, a physicist at the University of Washington who was on the voyage to study geomagnetism as a clue to vent action.

No detailed studies have been done of active volcanism at sea, especially the subtleties of how and when hot springs form. Opportunities for observation are few. Expeditions to sites of interest usually last only days and are often cut short by bad weather. Scientists have at times stumbled on volcanic outbursts, but have never witnessed the birth of hot-water vents and their associated colonies.

All that began to change on June 22 when a small laboratory of the National Oceanic and Atmospheric Administration in Newport, Ore., was electronically linked to a formerly secret Navy network for listening to ocean noises and enemy submarines. Only four days later, the lab recorded a seafloor eruption along the Juan de Fuca ridge. For two days, moving south to north, swarms of seaquakes shook an area some 25 miles in length, implying that lava was surging out of the earth to form new seafloor.

In response, an ad hoc alliance of universities, federal agencies and the Canadian government launched six forays (most rerouted from already planned voyages) to examine the site with undersea sensors, robots and a mini-submarine. The *Alvin* expedition was financed by the National Science Foundation and the National Oceanic and Atmospheric Administration, both of which have formal programs to study undersea ridges and vents.

The Expedition Probing for Clues in Inky Darkness

Alvin is a white, 25-foot-long research submersible that can carry three people to a depth of 2.5 miles. Inside the sub is a seven-foot titanium sphere that carries not only the pilot and two observers but all the vessel's electronics and life-support gear. *Alvin* and its support ship, *Atlantis 2,* which is 210 feet long, were built more than a quarter century ago but are still on the cutting edge of deep-sea exploration. They are operated by the Woods Hole Oceanographic Institution on Cape Cod.

The sub has two mechanical arms, and its exterior racks can hold cameras, sonars, temperature probes, animal traps, water-sample bottles and magnetometers. Its variable buoyancy allows it to carry scientific payloads of vastly different weights. The sub dives only in daylight, when launch and recovery are safest.

The *Atlantis 2* left Astoria, Ore., on Oct. 9 with 52 people on board. A day later, it reached the quake zone and lowered a network of electronic buoys to the seafloor. During dives, *Alvin* would chirp and the distant buoys would beep in response, letting the sub continuously fix its position even in the inky darkness.

Dr. Delaney, the expedition's chief scientist, told his team they had a "historic opportunity" to discover how the vents formed and evolved. The plan was for the submarine to work its way southward during the day, while at night unmanned sensors and cameras were towed on long cables from the ship in wide searches for likely observational targets.

The sky was gray and the sea was relatively calm during *Alvin's* first dive. One of two scientists who went down was Dr. Marvin D. Lilley of the University of Washington, a chemist. "Save the Tube Worms," read the logo on his T-shirt.

The scientists that day explored part of a 1.5-mile stretch of freshly erupted and still-glassy lava in the northern quake region. One volcanic vent spewed shimmering water whose temperature was measured at 97 degrees, up from the near-freezing 35 degrees in the surrounding waters. But there were no signs of new life or mineral chimneys.

Colonies of Bacteria

After three days, the team moved some 10 miles south to an area known from earlier work to have many particles in the water, mainly matted bacteria. Temperatures were measured at up to 65 degrees. Clumps of waving bacteria would occasionally break loose from the fissures, joining a swirl of biologic snow.

The scientists believed the bacterial storms were related to the recent outburst, though lava there was clearly old. Their thesis was that fresh magma lay just beneath the seafloor, heating the water and creating an ideal environment for bacterial growth.

At night, scientists from the Seattle-based Pacific Marine Environmental Laboratory of the National Oceanic and Atmospheric Administration would tow a sensor up and down through the water column as the ship moved forward. Over several nights, they found at the southern end of the quake region a strong signal indicating high heat, chemicals and particulate matter—signs of a large vent field.

That area was nicknamed the Source site, since it was where the seaquakes had begun as magma deep in the earth surged upward.

But *Alvin's* initial dives there were disappointing, and the expedition quickly moved back to the bacterial site. On one dive, a time-lapse video camera and a temperature probe, both battery powered, were left near an energetic vent to track its evolution for up to a year.

The Source a Tangle of Life about a Monolith

As the second week began, the scientists renewed their search for the Source. But dive after dive found little of interest, and even the unmanned instruments lost evidence of strong activity. The sky turned dark, the seas high. On the night of Oct. 22, just before all operations were shut down by bad weather, a towed sensor suddenly reported a series of exceptionally strong signals, helping pinpoint a likely target.

As *Atlantis 2* rode out the rough weather, *Alvin* could do no work during the last two scheduled dive days. With a discovery so close, Dr. Delaney asked Woods Hole if the expedition could be extended a day, gambling that the weather would break. Woods Hole said yes.

On Oct. 24, scattered clouds were shot with gold from the rising sun as Dr. Delaney and this reporter climbed into *Alvin* to descend more than a thousand fathoms into the darkness in search of the source. "May the force be with you," a dive controller aboard the ship said over the hydrophone as we began our descent.

The cramped sphere was surprisingly comfortable, lined with soft cushions, feeling like a college dorm packed with stereos and high-tech gear. Outside my observation port there flashed the lights of an endless procession of gelatinous organisms, some seemingly a foot or two in length. Ripples of light would pass along their bodies as we swept past them.

We reached bottom around 9:30 a.m. and proceeded to fly over endless fields of pillow-shaped lava. After an hour of fruitless searching, we came upon a Reebok shoe that had sunk into the abyss, a jarring discovery given the gravity of our pursuit.

Slowly our tiny sphere grew colder. Water vapor began to condense on exposed metal parts of the titanium hull, forming a glistening sheen. I put on a sweater.

Then, at 11:30 a.m., after what seemed like many hours of viewing interminable mounds of lava, we came upon a giant chimney looming up out of the dark, spewing water that shimmered with motion.

"It's hot," reported our pilot, Robert J. Grieve. "It's got tube worms all over it."

A riot of life flourished on and around the unworldly monolith, which loomed three or four stories high. The tube worms were four or five inches long, their red tips emerging from brown outer cases. The chimney's surface was spread with mats of white bacteria and iridescent, dark red palm worms about an inch long. Smaller still were orange worms covered with tiny bristles. The chimney's surface also crawled with swarms of miniature lobsters known as galatheids.

Remarkably, this tangle of life included at least two types of small corals that were heavy with spines, polyps and tentacles. "We've never seen corals among tube worms before," Dr. Cindy Van Dover, a biologist on the voyage from Woods Hole, said later. "It's going to be neat to figure out how the corals make a living."

Warm Bed of Creatures

Surrounding the tower, often lurking behind rocks, were other creatures. There were large sea anemones topped with explosions of tentacles, orangeish-brown spider crabs the size of dinner plates, and more corals, reaching with tentacles toward the chimney like religious devotees.

That day we had time to examine five large chimneys in all. Some small ones in the area were actively venting but naked of life and quickly crumbled in a cloud of debris when *Alvin*'s mechanical arm tried to grab them. The hottest water we measured coming out of vents that day was 543 degrees. At the base of one chimney we left a battery-powered temperature sensor, to be retrieved next year.

Suddenly, our pilot noticed that the temperature of the submarine's skin was starting to rise. By accident, we had positioned ourselves over a vent, a potentially dangerous thing to do since the sub's plastic windows can melt. We quickly began our ascent, exhausted and happy.

Later, back on the ship, Dr. Delaney said the field appeared to be a preexisting one that had been reactivated by the June eruption. The chimneys were generally big, he noted. But there were also small, fresh ones easily broken apart, as well as creatures that were unusually small, like the tiny lobsters, which were mostly youngsters.

Dr. Delaney held out the possibility that the field might be new, stressing how little was known about chimney growth rates. Ultimately, the field's age will be determined by future studies of the expedition's data and samples.

Over all, he said, the voyage would aid the worldwide effort to better comprehend deep volcanism.

"Catching an event at the beginning and watching its evolution is an essential part of the understanding process," Dr. Delaney said.

—November 2, 1993

THE QUEST FOR OUTER SPACE

Soviet Fires Earth Satellite into Space

By WILLIAM J. JORDEN

T he Soviet Union announced this morning that it successfully launched a man-made earth satellite into space yesterday.

The Russians calculated the satellite's orbit at a maximum of 560 miles above the earth and its speed at 18,000 miles an hour.

The official Soviet news agency Tass said the artificial moon, with a diameter of twenty-two inches and a weight of 184 pounds, was circling the earth once every hour and thirty-five minutes. This means more than fifteen times a day.

Two radio transmitters, Tass said, are sending signals continuously on frequencies of 20.005 and 40.002 megacycles. These signals were said to be strong enough to be picked up by amateur radio operators. The trajectory of the satellite is being tracked by numerous scientific stations.

Due Over Moscow Today

Tass said the satellite was moving at an angle of 65 degrees to the equatorial plane and would pass over the Moscow area twice today.

"Its flight," the announcement added, "will be observed in the rays of the rising and setting sun with the aid of the simplest optical instruments, such as binoculars and spyglasses."

The Soviet Union said the world's first satellite was "successfully launched" yesterday.

Thus it asserted that it had put a scientific instrument into space before the United States. Washington has disclosed plans to launch a satellite next spring.

The Moscow announcement said the Soviet Union planned to send up more and bigger and heavier artificial satellites during the current International Geophysical Year, an eighteen-month period of study of the earth, its crust and the space surrounding it.

Five Miles a Second

The rocket that carried the satellite into space left the earth at a rate of five miles a second, the Tass announcement said. Nothing was revealed, however, concerning the material of which the man-made moon was constructed or the site in the Soviet Union where the sphere was launched.

The Soviet Union said its sphere circling the earth had opened the way to interplanetary travel.

It did not pass up the opportunity to use the launching for propaganda purposes. It said in its announcement that people now could see how "the new socialist society" had turned the boldest dreams of mankind into reality.

Moscow said the satellite was the result of years of study and research on the part of Soviet scientists.

Several Years of Study

Tass said:

"For several years the research and experimental designing work has been under way in the Soviet Union to create artificial satellites of the earth. It has already been reported in the press that the launching of the earth satellites in the U.S.S.R. had been planned in accordance with the program of International Geophysical Year research.

"As a result of intensive work by the research institutes and design bureaus, the first artificial earth satellite in the world has now been created. This first satellite was successfully; launched in the U.S.S.R. October four."

The Soviet announcement said that as a result of the tremendous speed at which the satellite was moving it would burn up as soon as it reached the denser layers of the atmosphere. It gave no indication how soon that would be.

Military experts have said that the satellites would have no practicable military application in the foreseeable future. They said, however, that study of such satellites could provide valuable information that might be applied to flight studies for intercontinental ballistic missiles.

The satellites could not be used to drop atomic or hydrogen bombs or anything else on the earth, scientists have said. Nor could they be used in connection with the proposed plan for aerial inspection of military forces around the world.

An Aid to Scientists

Their real significance would be in providing scientists with important new information concerning the nature of the sun, cosmic radiation, solar radio interference and static-producing phenomena radiating from the north and south magnetic poles. All this information would be of inestimable value for those who are working on the problem of sending missiles and eventually men into the vast reaches of the solar system.

Publicly, Soviet scientists have approached the launching of the satellite with modesty and caution. On the advent of the International Geophysical Year last June they specifically disclaimed a desire to "race" the United States into the atmosphere with the little sphere.

The scientists spoke understandingly of "difficulties" they had heard described by their American counterparts. They refused several invitations to give any details about their own problems in designing the satellite and gave even less information than had been generally published about their work in the Soviet press.

Hinted of Launching

Concerning the launching of their first satellite, they said only that it would come "before the end of the geophysical year"—by the end of 1958.

Several weeks earlier, however, in a guarded interview given only to the Soviet press, Alexander N. Nesmeyanov, head of the Soviet Academy of Science, dropped a hint that the first launching would occur "within the next few months."

But generally Soviet scientists consistently refused to boast about their project or to give the public or other scientists much information about their progress. Key essentials concerning the design of their satellites, their planned altitude, speed and instruments to be carried in the small sphere, were carefully guarded secrets.

—October 5, 1957

U.S. Satellite Is
"Working Nicely"

By JOHN W. FINNEY

The United States' first scientific satellite, the bullet-shaped *Explorer*, whirled around the earth today, recapturing lost national prestige and gathering scientific information about space.

The long, thin satellite was orbiting a dozen times a day and reported to be "working nicely." Dr. Richard W. Porter, chairman of the earth satellite panel of the United States National Committee for the International Geophysical Year, gave the report.

The two miniature radio transmitters in the satellite were sending out a steady signal—a welcome sound to the free world after the "beep, beep" of the two Soviet satellites.

The satellite's small Geiger counter, designed to measure the intensity of cosmic radiation, reported thirty counts a second, which was in line with earlier scientific estimates. The satellite reported that its internal temperature was 30 degrees centigrade (86 degrees Fahrenheit).

Army Firing Succeeds

It was the Army, a late starter in the satellite race, which finally succeeded in launching this nation's first satellite—almost four months after the Soviet Union had scored the psychological and scientific triumph of being first into space.

At 10:48 last night an Army Jupiter-C four-stage rocket streaked aloft from its launching pad at Cape Canaveral, Fla., with the *Explorer* satellite in its nose.

About two hours later—after the *Explorer* had completed its first trip around the world—it was announced that the 30.8-pound satellite had been successfully placed in an earth-circling orbit. The announcement early this morning came jointly from President Eisenhower at his vacation retreat in Augusta, Ga., and scientists gathered in the high-domed Great Hall of the National Academy of Sciences here.

A Moment of Triumph

It was a belated moment of triumph for the United States, which first was bested in the race into space and then was frustrated in attempts to launch even a small test satellite.

In particular it was a moment of triumph for the Army missile team, which for more than two years had unsuccessfully been seeking a satellite-launching assignment. Army officials lost no time in declaring that its success on the first attempt demonstrated the Army's capabilities for conquest of space.

Army officials disclosed that they had received orders to launch a second satellite some time between now and April. The Navy's ill-fated Project Vanguard is scheduled to make another attempt to launch a small test satellite next week as a prelude to launching six full-scale scientific satellites during the remaining ten months of the International Geophysical Year.

In numbers and size the United States still cannot compare with the Soviet Union in the satellite race. The 30.8-pound satellite is no match for the 184-pound and 1,120-pound satellites launched by the Soviet on Oct. 4 and Nov. 3.

Dr. Wernher von Braun, the chief of the Army Ballistic Missile Agency's development operations and the man primarily responsible for the achievement, summed up the race this way today:

"We can compete only in spirit and not in hardware yet."

In the coming months, the United States will try to draw at least equal in numbers, if not in size of satellites.

By mid-morning today—after the *Explorer* had completed seven circuits around the world—Dr. Porter told a news conference that "we are getting lots of data back from the satellite and everything seems to be working nicely."

Elongated Elliptical Path

The *Explorer* was following an elongated elliptical path around the world. At its peak, or apogee, it was estimated to be 1,700 miles from the earth, and at its closest point, or perigee, around 200 miles.

This was closer to the earth than the preliminary estimates of last night when it was thought that the satellite might be as much as 2,000 miles from the earth.

The lower estimate reflected the fact that as the satellite was followed by tracking stations around the world and the information fed into a high-speed electronic computer, scientists were getting a more precise idea of its orbit.

The satellite was traveling at a speed of about 18,000 miles an hour, circling the earth every 114 minutes, or between twelve and thirteen times a day.

With this orbit, Dr. Porter said, the satellite, whose scientific name is 1958 Alpha, could be expected to have "a relatively long life." Army scientists said the life span should be at least several months.

The eleven pounds of instruments and battery-powered transmitters in the satellite nose were reported to be working efficiently, collecting information about conditions in space and radioing it back to earth.

Investigating Radiation

The satellite, part of this country's participation in the I.G.Y., an eighteen-month period that began last July, was designed primarily to conduct the investigation of the intensity of cosmic radiation in space. As secondary experiments, it also is collecting information on the density of micrometeorites, or cosmic dust, and temperatures inside and outside the satellite.

On the first pass, three or four of the satellite's twelve exterior erosion gauges were broken, which was more than had been expected. Scientists were uncertain, however, whether the gauges were broken by collision with cosmic particles or by the vibration of take-off. In addition, the satellite has a small microphone, which will detect the sound of small particles colliding with the satellite shell.

The information obtained by the scientific instruments is radioed back to earth in coded form by the two miniature transmitters, operating on a frequency of 108 and 108.03 megacycles. The latter is the more powerful transmitter, with an output of six-hundredths of a watt, and is expected to last for two to three weeks on its battery power. The other transmitter, with a power of one one-hundredth of a watt, will operate two to three months.

Messages Made Available

In keeping with the international spirit of the I.G.Y. year, the key to the satellite's coded messages has been made available in scientific publications so that all nations, including the Soviet Union, will be able to interpret the information transmitted by the satellite.

In contrast the Soviet Union has yet to disclose the secret of its satellites' codes or the detailed information they obtained. Four hours after the United States satellite was launched, however, the International Geophysical Year headquarters in Brussels was notified by the Soviet Union that preliminary information about the Soviet satellites was in the mail.

In its orbit, the *Explorer* is estimated to be traveling at an angle of 34 degrees to the equator, then reaching as far north and south on the globe as 34 degrees latitude.

In its travels, the satellite will pass over the southern tier of states in a line running from Santa Barbara County, Calif., to Cape Hatteras, N.C. The satellite will not pass over the Soviet Union, but its radio signals should be easily picked up in the southern part of the Soviet.

Because of its small size—80 inches long, 6 inches in diameter—it is unlikely that *Explorer* will be visible to the naked eye. Under the correct conditions, it may be possible to see the satellite with the aid of binoculars or telescopes.

At two news conferences held eight hours apart this morning, Dr. von Braun, the German-born scientist who helped design the Nazi V-2 missile, was obviously jubilant over the Army's successful first attempt at launching a satellite.

"So far I don't know of a single thing that went wrong," he exclaimed at the second news conference.

Dr. von Braun disclosed that a "rather exotic" high-energy fuel had been used in the modified Redstone missile that lifted the satellite-launcher off the ground and into the first part of its trajectory.

Normally the Redstone, a descendant of the V-2, uses alcohol for fuel. In this case, Dr. von Braun said, a special fuel called "hidyne" was used. The fuel, he said, furnished greater thrust and permitted the missile to launch a heavier satellite.

Heavier Satellite Seen

Using "a few more tricks within immediate reach," Dr. von Braun predicted that the Jupiter-C missile could launch a still heavier satellite weighing about 50 pounds.

Dr. von Braun, after revealing that the Jupiter-C was a four-stage rocket, gave this description of the launching of the satellite:

239

The first-stage Redstone, with its 78,000 pounds of thrust, lifted the 65,000-pound missile assembly to an altitude of 53 miles in 145 seconds.

With its fuel exhausted, the first stage dropped off, leaving the three other stages, consisting of clusters of small solid-propellant rockets, to coast on up to an altitude of about 212 miles. For guidance, the last three stages were set spinning on takeoff.

As they coasted upward, the final three stages gradually arched over until they were aimed parallel with the earth's surface. At this point, 405 seconds after takeoff, a remote control button was pushed on the ground, firing the three stages in rapid succession and accelerating the satellite to the orbital speed of 18,000 miles an hour.

The *Explorer* then went racing off into space, whirling at a rate of about 700 revolutions a minute.

Dr. von Braun, who has been dreaming of space ventures for 20 years, thought it unlikely that the 1958 Alpha would ever collide with the second Soviet satellite still in the air.

"Space is a pretty big place," he explained in his heavy German accent, "even bigger than Texas."

—*February 2, 1958*

Russian Orbited the Earth Once, Observing It Through Portholes

By OSGOOD CARUTHERS

Man's first flight into space took a 27-year-old Russian on a 108-minute single spin around earth today at a speed of more than 17,000 miles an hour.

The Soviet Air Force pilot, Maj. Yuri Alekseyevich Gagarin, returned safely to Russian soil and a triumphant hero's welcome at 10:55 a.m. Moscow time (2:55 a.m. Eastern Standard Time).

While on his swift flight around the earth, he told listeners on the ground by radio that he could see the earth clearly through portholes in the space vehicle.

"I can observe the earth," he said. "Visibility is good—one can see everything. Some areas are covered with piles of clouds."

Praised by Khrushchev

As additional details were learned here about the major's trip in the five-ton vehicle, praise poured in from all parts of the world. Tonight Premier Khrushchev hailed Major Gagarin with these words:

"You have made yourself immortal, because you are the first to penetrate into space."

The pilot, who traveled at a speed six times faster than man had ever moved before, was reported to be in good physical condition.

The vehicle was headed back to earth by the firing of retarding rockets. Then giant braking flaps slowed it as it hit the heavier atmosphere and, finally, it was eased to earth by parachute.

Suffers No Injuries

The Moscow radio said that on his return to earth Major Gagarin had said:

"I wish to report to the [Communist] party and the government and personally to Nikita Sergeyevich Khrushchev that the landing was normal. I feel fine and I have no injuries or bruises."

Soviet citizens in every walk of life throughout the land proudly hailed the Soviet victory in the race with the West to achieve the first manned space flight. Moscow's propaganda machine began at once to take the fullest advantage of the psychological impact the feat would have on the minds of men throughout the world.

There was a touch of gloating over the West in the celebration. In his message to the reserve Air Force pilot, Premier Khrushchev declared:

"You and I and all our people will solemnly celebrate this great feat in the conquering of space. Let the whole world see what our country is capable of, what our great people and our Soviet science can do."

To which the major replied:

"Now let the other countries try to catch us."

And this, in turn, brought from the Soviet premier:

"That's right. Let the capitalist countries try to catch up with our country, which has blazed a trail into space and which has launched the world's first cosmonaut."

The historic achievement of man's first flight into space culminated more than three years of efforts by Soviet scientists since the first Soviet *Sputnik* was launched into orbit around the earth Oct. 4, 1957.

With Major Gagarin strapped into his huge space vehicle, the multistage rocket started him on his flight from an undisclosed launching site at 9:07 a.m. Moscow time (1:07 a.m. Eastern Standard Time). The final stage of the rocket soon afterward put the cosmonaut into an elliptical orbit around the earth.

Fifteen minutes later he reported back to Soviet tracking stations over a two-way radio system that he was feeling well.

Fifty-three minutes after his first report, Major Gagarin said that he was over Africa.

Tape recordings of his voice as reportedly broadcast from the ship indicated that his comments had been laconic and that he had given little detailed description of what he had seen or felt during his flight. The recordings later were reproduced over the Moscow radio.

Calls Visibility Good

"I am observing the earth," the recording said. "Visibility is good. I hear you perfectly." His words seemed distinct and clearly received.

The major's broadcast continued:

"The flight is continuing well. I can observe the earth. Visibility is good—one can see everything. Some areas are covered with piles of clouds."

After a brief interval he resumed: "Flight continues. Everything is normal. Everything is working perfectly. Everything is working perfectly. I am going farther. I feel fine. I am in a cheerful mood. I am continuing my flight. Everything is going fine. The machine is working normally."

Earlier reports of his broadcasts back to Earth, which were not reproduced over the Moscow radio, quoted Gagarin as saying as he was flying over Africa: "The flight is progressing normally. I am withstanding well the state of weightlessness."

Three Radio Beams Used

His voice from space was being beamed by the ship's two-way radio on two shortwave bands and one ultra-short-wave band.

Ten minutes after he reported sighting Africa, the vehicle's retro-engines, rockets used for slowing down flight, were switched on at a command from the ground. Major Gagarin then began a half-hour descent from orbit to a predetermined landing area, apparently on dry land inside the Soviet Union.

The spot where he landed was not disclosed by official announcements but it was reportedly within the prearranged target area.

The whereabouts of Major Gagarin after his return was not made known. Even into the evening the Moscow television network had only a still photograph of him to show the waiting public.

It was assumed that the world's first space pilot had been whisked to a special laboratory for a thorough physical checkup and a rest.

The first announcement that a manned spaceship had been put into orbit was broadcast over the Moscow radio at 10:02 a.m. (2:02 a.m. Eastern Standard Time). It said that the satellite was circling the globe at an angle of 65 degrees 4 minutes to the equator.

It said that the vessel was traveling in an orbit that took it 302 kilometers (187¾ miles) away from the earth at the farthest point and 175 kilometers (109½ miles) at the nearest point. It said that a complete circuit would take 89.1 minutes.

This meant that Major Gagarin had made only one trip around the earth before he was brought back and that the space vehicle probably had been put into a long gliding descent before it had completed a full circuit.

It could not be determined whether the vessel had returned to the same area from which it had been launched.

News of the successful manned space flight had been preceded by rumors. Two days ago unofficial reports in Moscow said that the flight had already been made.

There was no hint in the official announcements that any previous manned space flight had been attempted.

All the stops were pulled out in hailing Major Gagarin as the first man to make flight into space.

Disarmament Appeal

The ruling Central Committee of the Communist party and the governing Council of Ministers took advantage of the occasion to issue a long appeal to peoples and governments throughout the world to strengthen their efforts for peaceful coexistence and disarmament.

Leading men in the Soviet Academy of Sciences, noting the East-West space race, declared that today's success showed that their side had been on the right track in concentrating on more powerful rocket engines and heavier vehicles.

News of the feat was broadcast over and over in slow dramatic tones to large crowds gathered in the streets of central Moscow. There was little doubt that the news was eagerly seized upon as a point of national pride and as something that brightened the day of the ordinary citizen.

The name of Yuri Gagarin quickly became a byword of Soviet heroism in every part of the land. He was hailed as a model for young Soviet citizens to emulate in work and study.

His 59-year-old father, Aleksei, a factory worker, and his 58-year-old mother, Anna, were congratulated. So were his young wife, Valentina, and their two daughters, Yelena, two years old, and Galia, one month old.

The spaceship in which Major Gagarin traveled was named *Vostok*, the word for "east" in Russian. The vehicle was considered a symbol of Soviet victory in the East-West space contest.

The success of today's flight will add a new theme and dimension to the Soviet Union's biggest spring holiday May 1. There was speculation here that Major Gagarin might be introduced at the big May Day parade through Red Square.

—*April 13, 1961*

Glenn Orbits Earth 3 Times Safely

By RICHARD WITKIN

John H. Glenn Jr. orbited three times around the earth today and landed safely to become the first American to make such a flight.

The 40-year-old Marine Corps lieutenant colonel traveled about 81,000 miles in 4 hours 56 minutes before splashing into the Atlantic at 2:43 p.m. Eastern Standard Time.

He had been launched from here [in Cape Canaveral, Fla.] at 9:47 a.m.

The astronaut's safe return was no less a relief than a thrill to the Project Mercury team, because there had been real concern that the *Friendship 7* capsule might disintegrate as it rammed back into the atmosphere.

There had also been a serious question whether Colonel Glenn could complete three orbits as planned. But despite persistent control problems, he managed to complete the entire flight plan.

Lands in Bahamas Area

The astronaut's landing place was near Grand Turk Island in the Bahamas, about 700 miles southeast of here.

Still in his capsule, he was plucked from the water at 3:01 p.m. with a boom and block and tackle by the destroyer *Noa*. The capsule was deposited on deck at 3:04.

Colonel Glenn's first words as he stepped out onto the *Noa*'s deck were: "It was hot in there."

He quickly obtained a glass of iced tea.

He was in fine condition except for two skinned knuckles hurt in the process of blowing out the side hatch of the capsule.

The colonel was transferred by helicopter to the carrier *Randolph*, whose recovery helicopters had raced the *Noa* for the honor of making the pick-up. After a meal and extensive "de-briefing" aboard the carrier, he was flown to Grand Turk by submarine patrol plane for two days of rest and interviews on technical, medical and other aspects of his flight.

The *Noa,* nearest ship to the capsule as it parachuted into the ocean, took just 21 minutes to close the six-mile gap, lift the capsule aboard with a bomb-block-and-tackle rig and place it gently on the deck.

Colonel Glenn first was set to wriggle out of the narrow top. But when difficulty was encountered in getting one of the bulkheads loose, the explosive side hatch was blown off, and the man from space stepped out on deck, apparently in excellent shape. He was soon afterward transferred to the carrier *Randolph.*

In the course of his three orbits, Colonel Glenn reported frequently to tracking stations at various points on earth and to the control center here. Invariably, he said that his condition was fine.

Shortly after Colonel Glenn was picked up by the *Noa,* he received congratulations on his feat from President Kennedy by radio telephone.

A situation that seemed at the moment to pose the greatest danger developed near the end of the flight.

A signal radioed from the capsule indicated that the heat shield—the blunt forward end made of ceramic-like material that dispels the friction heat of re-entry and chars in the process—might be torn away before it could do its job.

If it had, the flight would have had a tragic end.

Signal Is Received

The signal, received as the capsule was traveling between Hawaii and the West Coast, indicated that the heat shield had become unlatched from the main capsule body. This action was not intended to happen until the final stage of the parachute descent.

At that point, it would fall a few feet, and deploy, between it and the capsule base, a cloth landing bag to cushion the impact on the water.

Colonel Glenn was asked by radio to flip a switch to check whether the shield had, in fact, become unlatched. When the light did not go on, it appeared that the "unlatch" signal had been spurious.

But the Mercury team was taking no chances. It changed the sequence of re-entry events to try to insure that, even if unlatched, the heat shield would not fall away prematurely.

Colonel Glenn, apparently sensing possible serious trouble, asked: "What are the reasons for this? Do you have any reasons?"

"Not at this time," came the reply from the control center.

Normally, after the firing of the three braking rockets to bring the capsule out of orbit, the empty braking-rocket package is jettisoned.

Jettisoning was delayed today so that, in case the heat shield had become unlatched, the rocket-packet straps would hold the shield in place until this function was taken over by the force of re-entry into the atmosphere.

The package burned on re-entry. The heat shield did not drop away until it was supposed to. This indicated that the signal that had caused so much anxiety had, in fact, been a false one.

100,000 See Launching

The whole continent watched on television as Colonel Glenn's capsule was launched. The world listened by radio. And almost 100,000 persons had a direct view from here and the beaches around as the Atlas rocket booster bore the Project Mercury capsule upward with a thrust of 360,000 pounds.

The *Friendship 7* was lofted into a trajectory that varied between a low point, or perigee, of about 99 miles, and a high point, or apogee, of 162 miles.

It traveled at a speed of about 17,630 miles an hour and went from day to night three times before whirling east across the Pacific on the final leg of the flight.

Some 300 miles west of the California coast, three retro, or braking, rockets slowed the capsule enough to bring it out of orbit.

The elated astronaut on board radioed, "Boy, that was a real fireball of a ride!" as the capsule rammed back into the atmosphere.

Besides generating heat that gave him a spectacular moment of fireworks outside his capsule window, the re-entry ended Colonel Glenn's long hours of weightlessness and shoved him forcefully back against his contoured couch.

At 2:43 p.m., a 63-foot red-and-white parachute deposited the *Friendship 7* on gentle Caribbean waters.

After the capsule had been picked up by the *Noa* and safely placed on her deck, Colonel Glenn emerged triumphant in his gleaming silver space suit.

Sends Word of Trouble

It was on his first turn around the globe that Colonel Glenn sent word of erratic behavior by the attitude control system. This caused some concern almost to the end of the flight.

The system is designed to control the capsule's attitude in space.

This does not mean that it in any way alters the course of the capsule over the ground. The course is set once the Air Force Atlas booster has imparted to the capsule its speed and direction, and has been dropped away.

The astronaut exercises no control over the capsule attitude until after the Atlas booster rocket has finished burning and dropped away. During the climb to space, the Atlas provides the guidance and attitude control. Its engines swivel like a juggler's palm under a broomstick.

The attitude system, rather, controls the orientation of the capsule-whether the forward end tilts up or down; whether it yaws right or left; or whether the capsule rolls one way or the other.

If the capsule moves out of proper line on any of the three axes, it can be realigned by squirting hydrogen peroxide through tiny jets.

There are two completely independent systems for making these corrections. One is called automatic; the other manual. There are different ways to operate each system.

Used Automatic System

When the trouble developed, Colonel Glenn was flying by the completely automatic method. Gyroscopes were set to the desired attitudes. And when the capsule strayed too far, squirts of hydrogen peroxide were to be automatically ejected through the proper jets.

On this system, there are four jets for roll; four for pitch up and down; and four for yaw right and left. Two of the four jets in each set have a thrust of only one pound, while the other two have much larger thrusts.

Only the small jets are supposed to be brought into play during the main portion of the orbital journey. The large ones are mainly for more radical corrections necessary when attitude changes are likely to be most violent—coming back from orbit.

What happened to Colonel Glenn's capsule was that the small jets did not do their job. When the capsule drifted beyond the proper limits, and the small jets did not respond, the larger jets, with 24 pounds of thrust, automatically cut in.

Dangers Are Described

A similar malfunction occurred on the roll jets during the second orbit of the flight made by Enos the chimpanzee last year. Because there was no human aboard to analyze the trouble and make corrections, Enos's mission had to be ended one orbit ahead of schedule.

The danger today was that the large jets would consume the hydrogen peroxide too fast and that, when it came time to perform the important return-from-orbit maneuver, there would be none left to orient the capsule properly for re-entry.

Colonel Glenn initially met the problem by switching to a technique called fly-by-wire. He controlled the vehicle by manually moving the control stick.

This was not the regular "manual" system. The stick was electronically connected to the same jets used in the completely automatic system. But it had the result of making very finely calculated corrections that did not waste hydrogen peroxide.

At later stages of the flight, Colonel Glenn switched to the completely independent "manual" system that used six different jets, two for each axis. The amount of squirt ejected through these jets was also determined by how far the control stick moved, and it wasted no hydrogen peroxide.

The original trouble seemed to have disappeared by the time the colonel was ready to return from orbit.

He successfully used the automatic system for the difficult re-entry maneuver, while keeping himself ready to switch immediately to manual controls if the trouble recurred. It did not.

10 Previous Attempts

Today's orbital flight had been scheduled for just before Christmas. There had been 10 attempts to send Colonel Glenn on his trip, and 10 frustrating postponements, either because of weather or technical problems.

Last night, the weathermen talked about being "cautiously optimistic." But few observers agreed with them. It did not seem possible that the mess of weather bearing down on Florida could clear away in time, and that is the way it still looked when the swarm of official observers arrived here about 4 a.m.

Colonel Glenn had been awakened at 2:20 a.m. The countdown ritual was not much different from what had been witnessed on the suborbital 300-mile trips made last year by Comdr. Alan B. Shepard Jr., and on the attempt Jan. 27 to orbit Colonel Glenn.

A number of changes had been made in the mission plan since the short-range flights. The recovery system had been revised to minimize chances of another after-landing mishap that caused loss of Capt. Virgil I. Grissom's capsule and almost cost that astronaut his life.

Colonel Glenn was given a special camera with which to try to take various types of pictures of cloud cover and other phenomena.

He had a "bungee" cord—a "king-size rubber band"—on which he was to pull, like an oarsman pulling oars, to see how his blood pressure was affected by exercise when he was in a weightless state.

He had a medical kit of spring-loaded needles with which he could give himself various injections. One was to suppress nausea or other symptoms of motion sickness. (Colonel Glenn reported frequently that weightlessness bothered him not at all.)

He had also a painkiller, morphine; a stimulant, benzedrine; and a drug to counter shock.

Under his flying suit Colonel Glenn wore a plastic tube and container for bladder relief.

As the sun rose, the low-hanging clouds disappeared, and left conditions here as ideal as anyone ever had seen them. The weather in key recovery areas at sea was equally perfect.

While waiting for the count to proceed, Colonel Glenn had a chance to talk by phone with wife Anna, his 16-year-old son David, and his 15-year-old daughter Lynn. They watched the proceedings on TV from their home in Arlington, Va.

Visually, there was nothing particularly memorable about the takeoff, at 9:47. Emotionally, the atmosphere was charged, because a man was going into orbit.

There were the usual cries of "Go! Go!" at takeoff. Tears came to the eyes of some viewers in the blockhouse, at the observer's stand two miles from the launching pad, and on the beaches. But, generally, the emotions were held in. Everyone waited.

Colonel Glenn apparently had a fine, exhilarating time, right from the start. He experienced some vibration along with acceleration force, as he climbed through the atmosphere.

Then it smoothed out; the rocket burning stopped; the acceleration switched abruptly to weightlessness; and the capsule automatically turned its blunt end forward for the almost five hours he was to be in orbit.

"Capsule is turning around," he radioed. "Oh, that view is tremendous."

He was the professional test pilot, and at the same time a human being experiencing pure joy. The tone was full of enthusiasm.

On the first orbit, over the Canary Islands, Colonel Glenn reported that "the horizon is a brilliant blue."

One after another of the stations in the 18-station worldwide tracking net locked its radar on the *Friendship 7,* and most of them established voice communications with the astronaut on board.

Colonel Glenn received a special greeting from the citizens of Perth, Australia, who turned on the lights all over town.

"Thank everybody for turning them on," he radioed. About there, too, he tried the first of the special foods prepared for consumption in orbit, where there is no gravity to let liquids pour or meats stay on a dish. He ate tubes of food and meat, and malted milk tablets.

An odd phenomenon occurred when he was within range of Guaymas, Mexico. He reported "luminous particles around the capsule—just thousands of them—right at sunrise over the Pacific."

Maj. Gen. Leighton Davis, a Project Mercury officer here, suggested later that they might have been dust particles, or chips of paint from the capsule.

Moment of Decision

Then started the troubles with the attitude controls system—troubles that were to occupy the pilot the rest of the flight.

The moment of decision came near the end of the second orbit. Colonel Glenn was reporting continued erratic behavior of the controls, apparently even with the manual system—the alternative to the one that originally malfunctioned.

He also was reporting a warning light indicating that hydrogen peroxide fuel for at least one of the systems was getting low.

To many experts listening to these events unfold, there seemed no alternative to bringing Colonel Glenn back at the end of the second orbit rather than risking another circuit.

But the astronaut thought he could handle the situation without excessive trouble.

Between the technical talk, there was time for joking—the exuberance of a man who, no matter how experienced in combat and test flying, had never done anything quite like this.

Then he got down to the business of preparing himself for the critical firing of the braking rockets and the re-entry into the atmosphere.

Greatest Day in Space

Today's flight gave the United States, by any standards, its greatest day in space.

The achievement, however, could still not be considered quite up to what the Russians had done.

Colonel Glenn's flight was two orbits more than were flown by Maj. Yuri A. Gagarin, the Soviet space man, last April 12 but 14 less than another Russian, Maj. Gherman S. Titov, flew on Aug. 6.

In addition, there were some technical respects in which both Soviet orbital flights appeared to observers here to have an advantage: the size of the capsule orbited (five tons as against a ton and a half); the reliability of automatic controls; and the cabin atmosphere in which the pilot had to work.

But Colonel Glenn's trip was considered by most observers here to have gone a long way toward erasing this nation's "second-best" look in space.

—February 21, 1962

Men Walk on Moon

By JOHN NOBLE WILFORD

Men have landed and walked on the moon.

Two Americans, astronauts of *Apollo 11*, steered their fragile four-legged lunar module safely and smoothly to the historic landing yesterday at 4:17:40 p.m., Eastern Daylight Time.

Neil A. Armstrong, the 38-year-old civilian commander, radioed to earth and the mission control room here: "Houston, Tranquility Base here. The *Eagle* has landed."

The first men to reach the moon—Mr. Armstrong and his co-pilot, Colonel Edwin E. Aldrin Jr. of the Air Force—brought their ship to rest on a level rock-strewn plain near the southwestern shore of the arid Sea of Tranquility.

About six and a half hours later, Mr. Armstrong opened the landing craft's hatch, stepped slowly down the ladder and declared as he planted the first human footprint on the lunar crust: "That's one small step for man, one giant leap for mankind."

His first step on the moon came at 10:56:20 p.m. as a television camera outside the craft transmitted his every move to an awed and excited audience of hundreds of millions of people on earth.

Tentative Steps Test Soil

Mr. Armstrong's initial steps were tentative tests of the lunar soil's firmness and of his ability to move about easily in his bulky white spacesuit and backpacks and under the influence of the lunar gravity, which is one-sixth that of the earth.

"The surface is fine and powdery," the astronaut reported. "I can pick it up loosely with my toe. It does adhere in fine layers like powdered charcoal to the sole and sides of my boots. I only go in a small fraction of an inch, maybe an eighth of an inch. But I can see the footprints of my boots in the treads in the fine sandy particles."

After 19 minutes of Mr. Armstrong's testing, Colonel Aldrin joined him outside the craft.

The two men got busy setting up another television camera out from the lunar module, planting an American flag into the ground, scooping up soil and rock samples, deploying scientific experiments and hopping and loping about in a demonstration of their lunar agility.

They found walking and working on the moon less taxing than had been forecast. Mr. Armstrong once reported he was "very comfortable."

And people back on earth found the black-and-white television pictures of the bug-shaped lunar module and the men tramping about it so sharp and clear as to seem unreal, more like a toy and toy-like figures than human beings on the most daring and far-reaching expedition thus far undertaken.

Nixon Telephones Congratulations

During one break in the astronauts' work, President Nixon congratulated them from the White House in what, he said, "certainly has to be the most historic telephone call ever made."

"Because of what you have done," the president told the astronauts, "the heavens have become a part of man's world. And as you talk to us from the Sea of Tranquility it requires us to redouble our efforts to bring peace and tranquility to earth.

"For one priceless moment in the whole history of man all the people on this earth are truly one—one in their pride in what you have done and one in our prayers that you will return safely to earth."

Mr. Armstrong replied: "Thank you, Mr. President. It's great honor and privilege for us to be here representing not only the United States but men of peace of all nations, men with interests and a curiosity and men with a vision for future."

Mr. Armstrong and Colonel Aldrin returned to their landing craft and closed the hatch at 1:12 a.m., 2 hours 21 minutes after opening the hatch on the moon. While the third member of the crew, Lieut. Col. Michael Collins of the Air Force, kept his orbital vigil overhead in the command ship, the two moon explorers settled down to sleep.

Outside their vehicle the astronauts had found a bleak world. It was just before dawn, with the sun low over the eastern horizon behind them and the chill of the long lunar night still clinging to the boulders, small craters and hills before them.

Colonel Aldrin said that he could see "literally thousands of small craters" and a low hill out in the distance. But most of all he was impressed initially by the "variety of shapes, angularities, granularities" of the rocks and soil where the landing craft, code-named *Eagle*, had set down.

The landing was made four miles west of the aiming point, but well within the designated area. An apparent error in some data fed into the craft's guidance computer from the earth was said to have accounted for the discrepancy.

Suddenly the astronauts were startled to see that the computer was guiding them toward a possibly disastrous touchdown in a boulder-filled crater about the size of a football field.

Mr. Armstrong grabbed manual control of the vehicle and guided it safely over the crater to a smoother spot, the rocket engine stirring a cloud of moon dust during the final seconds of descent.

Soon after the landing, upon checking and finding the spacecraft in good condition, Mr. Armstrong and Colonel Aldrin made their decision to open the hatch and get out earlier than originally scheduled. The flight plan had called for the moon walk to begin at 2:12 a.m.

Flight controllers here said that the early moon walk would not mean that the astronauts would also leave the moon earlier. The lift-off is scheduled to come at about 1:55 p.m. today.

Their departure from the landing craft out onto the surface was delayed for a time when they had trouble depressurizing the cabin so that they could open the hatch. All the oxygen in the cabin had to be vented.

Once the pressure gauge finally dropped to zero, they opened the hatch and Mr. Armstrong stepped out on the small porch at the top of the nine-step ladder.

"O.K., Houston, I'm on the porch," he reported, as he descended.

On the second step from the top, he pulled a lanyard that released a fold-down equipment compartment on the side of the lunar module. This deployed the television camera that transmitted the dramatic pictures of man's first steps on the moon.

Ancient Dream Fulfilled

It was man's first landing on another world, the realization of centuries of dreams, the fulfillment of a decade of striving, a triumph of modem technology and personal courage, the most dramatic demonstration of what man can do if he applies his mind and resources with single-minded determination.

The moon, long the symbol of the impossible and the inaccessible, was now within man's reach, the first port of call in this new age of spacefaring.

Immediately after the landing, Dr. Thomas O. Paine, administrator of the National Aeronautics and Space Administration telephoned President Nixon in Washington to report: "Mr. President, it is my honor on behalf of the entire NASA team to report to you that the *Eagle* has landed on the Sea of Tranquility and our astronauts are safe and looking forward to starting the exploration of the moon."

The landing craft from the *Apollo 11* spaceship was scheduled to remain on the moon about 22 hours, while Colonel Collins of the Air Force, the third member of the *Apollo 11* crew, piloted the command ship, *Columbia*, in orbit overhead.

"You're looking good in every respect," Mission Control told the two men of *Eagle* after examining data indicating that the module should be able to remain on the moon the full 22 hours.

Mr. Armstrong and Colonel Aldrin planned to sleep after the moon walk and then make their preparations for the lift-off for the return to a rendezvous with Colonel Collins in the command ship.

Apollo 11's journey into history began last Wednesday from launching pad 39-A at Cape Kennedy, Fla. After an almost flawless three-day flight, the joined command ship and lunar module swept into an orbit of the moon yesterday afternoon.

The three men were awake for their big day at 7 a.m. when their spacecraft emerged from behind the moon on its 10th revolution, moving from east to west across the face of the moon along its equator.

Their orbit was 73.6 miles by 64 miles in altitude, their speed 3,860 miles an hour. At that altitude and speed, it took about two hours to complete a full orbit of the moon.

The sun was rising over their landing site on the Sea of Tranquility.

"We can pick out almost all of the features we've identified previously," Mr. Armstrong reported.

After breakfast, on their 11th revolution, Colonel Aldrin and then Mr. Armstrong, both dressed in their white pressurized suits, crawled through the connecting tunnel into the lunar module.

They turned on the electrical power, checked all the switch settings on the cockpit panel and checked communications with the command ship and the ground controllers. Everything was "nominal," as the spacemen say.

LM Ready for Descent

The lunar module was ready. Its four legs with yard-wide footpads were extended so that the height of the 16½-ton vehicle now measured 22 feet and 11 inches and its width 31 feet.

Mr. Armstrong stood at the left side of the cockpit, and Colonel Aldrin at the right. Both were loosely restrained by harnesses. They had closed the hatch to the connecting tunnel.

The walls of their craft were finely milled aluminum foil. If anything happened so that it could not return to the command ship, the lunar module would be too delicate to withstand a plunge through earth's atmosphere, even if it had the rocket power.

Nearly three-fourths of the vehicle's weight was in propellants for the descent and ascent rockets—Aerozine 50 and nitrogen oxide, which substituted for the oxygen, making combustion possible.

It was an ungainly craft that creaked and groaned in flight. But years of development and testing had determined that it was the lightest and most practical way to get two men to the moon's surface.

Before *Apollo 11* disappeared behind the moon near the end of its 12th orbit, mission control gave the astronauts their "go" for undocking—the separation of *Eagle* from *Columbia*.

Colonel Collins had already released 12 of the latches holding the two ships together at the connecting tunnel. He did this when he closed the hatch at the command ship's nose. While behind the moon, he was to flip a switch on the control panel to release the three remaining latches by a spring action.

At 1:50 p.m., when communications signals were reacquired, Mission Control asked: "How does it look?"

"*Eagle* has wings," Mr. Armstrong replied.

The two ships were then only a few feet apart. But at 2:12 p.m., Colonel Collins fired the command ship's maneuvering rockets to move about two miles away and in a slightly different orbit from the lunar module.

"It looks like you've got a fine-looking flying machine there, *Eagle*, despite the fact you're upside down," Colonel Collins commented watching the spidery lunar module receding in the distance.

"Somebody's upside down." Mr. Armstrong" replied.

What is "up" and what is "down" is never quite clear in the absence of landmarks and the sensation of gravity's pull.

Mr. Armstrong and Colonel Aldrin rode the lunar module around to the moon's far side; the rocket engine in the vehicle's lower stage was pointed toward the line of flight. The two pilots were leaning toward the cockpit controls, riding backwards and facing downward.

"Everything is 'go,'" they were assured by Mission Control.

Their on-board guidance and navigation computer was instructed to trigger a 29.8-second firing of the descent rocket, the 9,870-pound-thrust throttleable engine that would slow down the lunar module and send it toward the moon on a long, curving trajectory.

The firing was set to take place at 3:08 p.m. when the craft would be behind the moon and once again out of touch with the ground.

Suspense built up in the control room here. Flight controllers stood silently at their consoles. Among those waiting for word of the rocket firing were Dr. Thomas O. Paine, the space agency's administrator, most of the Apollo project officials and several astronauts.

At 3:46 p.m. contact was established with the command ship.

Colonel Collins reported, "Listen, baby, things are going just swimmingly, just beautiful."

There was still no word from the lunar module for two minutes. Then came a weak signal, some static and whistling and finally the calm voice of Mr. Armstrong.

"The burn was on time," the *Apollo 11* commander declared.

When he read out data on the beginning of descent, Mission Control concluded that it "looks great." The lunar module had already descended from an altitude of 65.5 miles to 21 miles and was coasting steadily downward.

Eugene F. Kranz, the flight director, turned to his associates and said: "We're off to a good start. Play it cool."

Colonel Aldrin reported some oscillations in the vehicle's antenna but nothing serious. Several times the astronauts were told to turn the vehicle slightly to move the antenna into a better position for communications over the 230,000 miles.

"You're 'go' for PDI," radioed Mission Control, referring to the powered descent initiation—the beginning of the nearly 13-minute final blast of the rocket to the soft touchdown.

When the two men reached an altitude of 50,000 feet, which was approximately the lowest point reached by *Apollo 10* in May, green lights on the computer display keyboard in the cockpit blinked the number 99.

This signaled Mr. Armstrong that he had five seconds to decide whether to go ahead for the landing or continue on its orbital path back to the command ship. He pressed the "proceed" button.

The throttleable engine built up thrust gradually, firing continuously as the lunar module descended along the steadily steepening trajectory to the landing site about 250 miles away.

"Looking good," Mission Control radioed the men.

Four minutes after the firing the lunar module was down to 40,000 feet. After five and a half minutes, it was 33,500 feet. At six minutes, 27,000 feet.

"Better than the simulator," said Colonel Aldrin, referring to their practice landings at the spacecraft center.

Seven minutes after the firing, the men were 21,000 feet above the surface and still moving forward toward the landing site. The guidance computer was driving the rocket engine.

The lunar module was slowing down. At an altitude of about 7,200 feet with the landing site still about five miles ahead, the computer commanded control jets to fire and tilt the bug-shaped craft almost upright so that its triangular windows pointed forward.

Mr. Armstrong and Colonel Aldrin then got their first close-up view of the plain they were aiming for. It was then about three and a half minutes to touchdown.

The brownish-gray panorama rushed below them—the myriad craters, hills and ridges deep cracks and ancient rubble on the moon, which Dr. Robert Jastrow, the space agency scientist, called the "Rosetta Stone of life."

"You're 'go' for landing," Mission Control informed the two men.

The *Eagle* closed in, dropping about 20 feet a second until it was hovering almost directly over the landing area at an altitude of 500 feet.

Its floor was littered with boulders.

It was when the craft reached an altitude of 300 feet that Mr. Armstrong took over semimanual control for the rest of the way. The computer continued to have control of the rocket firing but the astronaut could adjust the craft's hovering position.

He was expected to take over such control anyway, but the sight of a crater looming ahead at the touchdown point made it imperative.

As Mr. Armstrong said later, "The auto-targeting was taking us right into a football-field-sized crater, with a large number of big boulders and rocks."

For about 90 seconds, he peered through the window in search of a clear touchdown point. Using the lever at his right hand, he tilted the vehicle forward to redirect the firing of the maneuvering jets and thus shift its hovering position.

Finally, Mr. Armstrong found the spot he liked and the blue light on the cockpit flashed to indicate that five-foot-long probes, like curb feelers, on three of the four legs had touched the surface.

"Contact light," Mr. Armstrong radioed.

He pressed a button marked "Stop" and reported "Okay, engine stop."

There were a few more cryptic messages of functions performed.

Then Maj. Charles M. Duke, the capsule communicator in the control room, radioed to the two astronauts: "We copy you down, *Eagle.*"

"Houston, Tranquility Base here. The *Eagle* has landed."

"Roger, Tranquility," Major Duke replied. "We copy you on the ground. You got a bunch of guys about to turn blue. We are breathing again. Thanks a lot."

Colonel Aldrin assured Mission Control it was a "very smooth touchdown."

The *Eagle* came to rest at an angle of only about four and a half degrees. The angle could have been more than 30 degrees without threatening to tip the vehicle over.

The landing site is on the right side of the moon as seen from earth. The position: Lat. 0.779 degrees N., Long. 23.46 degrees E.

Although Mr. Armstrong is known as a man of few words, his heartbeats told of this excitement upon leading man's first landing on the moon.

At the first time of the descent rocket ignition, his heartbeat rate registered 110 a minute—77 is normal for him—and it shot up to 156 at touchdown.

At the time of the landing, Colonel Collins was riding the command ship *Columbia* about 65 miles overhead.

Mission Control informed the colonel, "*Eagle* is at Tranquility."

"Yea, I heard the whole thing," Colonel Collins, the man who went so far but not all the way, replied. "Fantastic."

When the *Apollo* astronauts landed on the Sea of Tranquility, the temperature at their touchdown site was about zero degrees Fahrenheit in the sunlight, even colder in the shade.

During a lunar night, which lasts 14 earth days, temperatures plunge as low as 280 degrees below zero. Unlike earth, the moon, having no atmosphere to act as a blanket, is unable to retain any of the day's warmth during the night.

During the equally long lunar day, temperatures rise as high as 280 degrees. By the time of *Eagle*'s departures from the moon, with the sun higher in the sky, the temperatures there will have risen to about 90 degrees.

This particular landing site was one of five selected by *Apollo* project officials after analysis of pictures returned by the five Lunar Orbiter unmanned spacecraft.

All five sites are situated across the lunar equator on the side of the moon always facing Earth. Being on the equator reduces the maneuvering for the astronauts to get there. Being on the near side of the moon, of course, makes it possible to communicate with the explorers.

—July 21, 1969

Postscript:
Standby Update Moon Poem

By A. M. ROSENTHAL

The moment *Apollo 11* went up, we knew the newspaper would need help four days later, when the astronauts landed on the moon, to tell the joy that was in us.

Even though we were farther from the action than journalists had ever been—about a quarter of a million miles away—it was the biggest story the editors of *The New York Times* had ever had a hand in, or would have.

But like every person who watched we felt we personally were part of the beauty and achievement, the great soaring. We loved those three men because we knew their adventure was born of the elegance of the human mind and desire.

They allowed us to feel part of that elegance. Humanity was loving itself, which does not happen often.

We made no great preparation for the possibility of disaster. So many big stories involve tragedies—wars, earthquakes, assassinations. But for once our journalistic minds were set for happiness and for once a big story filled a newsroom with joy.

But how would we express that, when the moment came when man set foot on the moon?

We decided on the simplest of pages—one story, pictures and whatever talk would be recorded between Houston and the moon. And we ordered a special headline type to be cast, one inch high. Bigger than ever used in the history of the paper!

Shouting is one way to express joy, but what else? We decided what the front page of *The Times* would need when the men landed was a poem.

What the poet wrote would count most, but we also wanted to say to our readers, look, this paper does not know how to express how it feels this day and perhaps you don't either, so here is a fellow, a poet, who will try for all of us.

We called one poet who just did not think much of moons or us, and then decided to reach higher for somebody with more zest in his soul—for Archibald MacLeish, winner of three Pulitzer Prizes. He turned in his poem on time and entitled it "Voyage to the Moon."

The poem was written on the assumption that the astronauts themselves had touched the moon. But the moon walk was taking place at about deadline time. Suppose it was delayed. We would need a poem rewrite, fast.

Henry R. Lieberman, then director of science news, was asked to call Mr. MacLeish and tell him to stand by to update the moon poem. After the moon walk, Mr. MacLeish was informed he could stand down; the poem was running in all editions.

The poem set other poets to work. A couple of weeks later, *The Times* ran a whole batch of moon poems. The MacLeish poem was reprinted in books and received everlasting recognition and distribution on reproductions of the front page on paperweights, coffee mugs and plastic shopping bags.

The shopping bag shows the headline, much of the lead story and the excerpts of the moon-Houston talk.

But I regret to say the shopping bag, a *Times* promotional item, ends before the bottom of the page, cutting off the poem at the byline. . . .

There was something of an intellectual struggle in the newsroom about the headline of the main news story. Some of us wanted it to say, "Man Walks on Moon," even though Neil A. Armstrong and Col. Edwin E. Aldrin Jr. both had stepped onto the moon's surface by the time we went to press.

But the purists said two walkers is plural, the record is the record. After all, only a few days earlier we had run a correction on a *Times* piece saying that scientifically, space travel was impossible—even though the article appeared on Jan. 13, 1920.

The purists won: "Men Walk on Moon." I still think we should have spoken for the race.

But maybe they were right, and poets should do that kind of thing, not editors.

Certainly our poet did, on the front page of *The Times* of July 21, 1969. "Our hands have touched you," he wrote, our hands.

Perhaps we will again, one day. I hope so. Then we, or our children, will again understand the elegance of the human mind and desire and rejoice in being part of it, which seems worth the money.

—July 18, 1989

Voyage to the Moon

By ARCHIBALD MacLEISH

Presence among us,

        wanderer in our skies,

dazzle of silver in our leaves and on our
waters silver,

      O
silver evasion in our farthest thought—
"the visiting moon" . . . "the glimpses of the moon" . . .

and we have touched you!

           From the first of time,
before the first of time, before the
first men tasted time, we thought of you.
You were a wonder to us, unattainable,
a longing past the reach of longing,
a light beyond our light, our lives—perhaps
a meaning to us . . .

     Now
our hands have touched you in your depth of night.

Three days and three nights we journeyed,
steered by farthest stars, climbed outward,
crossed the invisible tide-rip where the floating dust
falls one way or the other in the void between,
followed that other down, encountered
cold, faced death—unfathomable emptiness . . .

Then, the fourth day evening, we descended,
made fast, set foot at dawn upon your beaches,
sifted between our fingers your cold sand.

We stand here in the dusk, the cold, the silence . . .

and here, as at the first of time, we lift our heads.
Over us, more beautiful than the moon, a
moon, a wonder to us, unattainable,
a longing past the reach of longing,
a light beyond our light, our lives—perhaps
a meaning to us . . .

O, a meaning!

over us on these silent beaches the bright
earth,
        presence among us
                    —July 21, 1969

Shuttle Rockets into Orbit on First Flight

By JOHN NOBLE WILFORD

The space shuttle *Columbia*, its rockets spewing orange fire and a long trail of white vapor, blasted its way into Earth orbit today, carrying two American astronauts on a daring journey to test the world's first re-usable spaceship.

Soon after they settled into orbit, John W. Young, a civilian, and Capt. Robert L. Crippen of the Navy focused a television camera on the *Columbia's* tail section and discovered that more than a dozen heat-shielding tiles had ripped off, possibly because of the stresses of launching.

Project officials said that the tile loss should not shorten the flight or endanger the lives of the astronauts when the *Columbia* plunges back into the atmosphere, glowing red-hot from frictional heat, to attempt a runway landing Tuesday. The projected 36-orbit, 54½-hour flight is scheduled to end at Edwards Air Force Base in California.

"We've Got a Super Vehicle"

"I'm just not concerned about it," Neil B. Hutchinson, a flight director at Mission Control in Houston, said in discussing the tile problem at a news conference this afternoon. "We've got a super vehicle up there."

Mr. Hutchinson said that the small gaps in the fragile silica-fiber tile coating were in a "noncritical area," the two identical pods housing *Columbia's* orbital maneuvering rockets, and that an underlayer of insulation "appears to be intact."

"It's not going to bother us on the way home," Mr. Hutchinson said. The launching of the *Columbia* occurred on time at 7 a.m., after a smooth countdown that was in notable contrast to the many delays and technical problems that had plagued the project, which was begun in 1972 and has cost almost $10 billion. The liftoff had been rescheduled from Friday because of a breakdown in computer communications.

First of Four Orbital Tests

The mission is the first of four planned orbital tests of the space shuttle, a revolutionary complex of machinery designed to take off like a rocket, cruise in orbit

266

like a spacecraft and return to Earth like a giant glider. No other space vehicle has been reflown.

If the shuttle lives up to expectations, the *Columbia* and three sister ships now under construction should each be capable of making as many as 100 round trips into space, deploying and servicing satellites and also carrying scientific laboratories and planetary probes.

It was the first launching of American astronauts in nearly six years, and officials at the Kennedy Space Center were elated and visibly relieved. Controllers in the firing room waved small American flags as soon as they received assurances that Mr. Young and Captain Crippen were safely in orbit.

George F. Page, the director of shuttle operations, arrived at a post-liftoff news conference, smiling and waving his flag. "I've been on a lot of first launches," he said. "I've been in the business 20 years, and I never felt anything like today." Neither did Captain Crippen, the 43-year-old astronaut who made his first trip into space after waiting 15 years. His heart rate jumped to 130 beats a minute, from a normal 60, in the 12-minute ascent. The increased heart rate was not unusual for astronauts in a critical phase of a mission.

"That was one fantastic ride!" he exclaimed. "I highly recommend it." Though busy checking out the shuttle systems, Captain Crippen stole a few glances out the cockpit window at the earth and the airless space all around, remarking: "Oh, man, that is so pretty!"

"It sure hasn't changed any," added the laconic Mr. Young, 50 years old, who was on his fifth mission beyond Earth. Mr. Young's heart rate ran between 85 and 90, leading Mr. Hutchinson to joke, "John was kind of asleep at liftoff." In a scheduled interruption in the countdown nine minutes before liftoff, Mr. Page read a message from President Reagan to the two astronauts. It was a message that had been prepared and also released last week when the shuttle was scheduled for a Friday liftoff.

Message From Reagan

"Through you, today," the president said, "we all feel as giants once again. Once again we feel the surge of pride that comes from knowing we are the first and we are the best and we are so because we are free."

When the countdown resumed, computers commanded final checks of all the systems, exercised the elevons and rudder of *Columbia*, swiveled the rocket nozzles, monitored fuel pressures and finally issued the sequence of ignition orders.

Starting as a 4.5-million-pound, 184-foot-high complex of machinery, consisting of the orbiter *Columbia* attached to a huge external fuel tank and two solid-fuel rockets, the space shuttle struggled upward to achieve orbit.

First, while still rooted to launching pad 39A, the *Columbia*'s three main rocket engines ignited one by one in fractions of a second. Billowing clouds of vapor erupted from under the spaceship, first to the south, then to the north. The combustion of hydrogen and oxygen, the engine's propellants, produces steam. This had never been seen before in a launching since hydrogen-oxygen engines have previously always been used in upper-stage rockets, which were operated in the far reaches of the atmosphere or in orbit.

Sudden Burst of Flame

As soon as the three rockets reached a 90 percent thrust level in a little over three seconds, there was a sudden burst of flame from the ignition of the two booster rockets that burn a solid aluminum mixture. Eight hold-down bolts were severed explosively to release the spaceship.

The getaway was swifter than the slow, straining liftoffs of the Saturn 5 moon rockets, primarily because of the power of the solid fuel rockets. The shuttle cleared the 247-foot-high launching tower in six seconds. The explosions of energy sent blasts of sound and rumbling vibrations across the sandy flats, but they lacked the intensity of the thunderclaps of Saturn 5, the most powerful rocket ever built.

Just over two minutes into the ascent, 30 miles in the sky, the two solid rockets ceased firing and were cast away to float down by parachutes into the Atlantic Ocean, 150 miles offshore. The rockets were retrieved for refurbishment, refueling and re-use by two waiting ships, under the scrutiny of a Soviet trawler.

For the next six minutes, the three main engines continued to boost the shuttle to the brink of orbit before they, too, shut off. The external tank, which had been feeding liquid hydrogen and oxygen to the engines, was jettisoned and crashed back through the atmosphere, falling in pieces in the Indian Ocean far to the southeast of India.

Craft Attains Orbit

After a coasting period, the *Columbia*'s two smaller orbiting maneuvering rockets fired for a minute and a half to nudge the astronauts into a low earth orbit 12 minutes after liftoff. The engines were fired briefly three more times over the next seven hours to put *Columbia* in its final orbit, ranging from 169.8 miles to 171.8 miles.

On each 90-minute revolution of Earth the *Columbia* crosses the equator at angles of 40.2 degrees, which means that it sweeps north as far as 40.2 degrees latitude and as far south as 40.2 degrees latitude.

The same two maneuvering rockets are not to be operated again until they are called upon to start the *Columbia* on its return to Earth. About 8:30 a.m., toward the end of their first revolution of the earth, Mr. Young and Captain Crippen put the 122-foot-long *Columbia* through one of its most crucial tests. They sent commands to open the clamshell doors on the 60-foot-long cargo bay in the fuselage. This is where future shuttles will carry satellites and other payloads into space.

The latches opened and shut without flaw, the astronauts reported. If they had been unable to open the cargo doors, the astronauts would have had to cut short their mission since the opened doors expose radiators that dump into space the heat built up by the shuttle's electronic systems.

Inspection of Cargo Bay

In an inspection of the cargo bay with a television camera mounted there, but operated from inside the flight deck, the missing tiles were discovered. If this caused the astronauts any anxiety, their voices never betrayed it.

And before the *Columbia* could circumnavigate the earth again, engineers at the Johnson Space Center in Houston were also reported to be confident that the tile loss did not represent a serious problem. A spokesman at the center announced that no consideration was being given to sending one of the astronauts on a space walk to assess the damage. This is a contingency in the mission plan. But the astronaut would have no means for repairing or replacing the tiles, though such a capability had once been considered.

Instead, three hours into the flight, Mission Control told the crew: "You guys did so good, we're going to let you stay up there for a couple of days."

Late in the afternoon, the astronauts settled down for dinner and sleep. They had been up since 2 a.m., when they were awakened to prepare for their journey. In a televised status report from their orbiting craft, Mr. Young, the commander of the *Columbia*, told Mission Control:

"The vehicle is performing beautifully—much better than anyone expected on a first flight. The vehicle is really performing like a champ. It's beautiful."

—*April 13, 1981*

The Shuttle Explodes

By WILLIAM J. BROAD

The space shuttle *Challenger* exploded in a ball of fire shortly after it left the launching pad today, and all seven astronauts on board were lost.

The worst accident in the history of the American space program, it was witnessed by thousands of spectators who watched in wonder, then horror, as the ship blew apart high in the air.

Flaming debris rained down on the Atlantic Ocean for an hour after the explosion, which occurred just after 11:39 a.m. It kept rescue teams from reaching the area where the craft would have fallen into the sea, about 18 miles offshore.

It seemed impossible that anyone could have lived through the terrific explosion 10 miles in the sky, and officials said this afternoon that there was no evidence to indicate that the five men and two women aboard had survived.

No Ideas Yet as to Cause

There were no clues to the cause of the accident. The space agency offered no immediate explanations, and said it was suspending all shuttle flights indefinitely while it conducted an inquiry. Officials discounted speculation that cold weather at Cape Canaveral or an accident several days ago that slightly damaged insulation on the external fuel tank might have been a factor.

Americans who had grown used to the idea of men and women soaring into space reacted with shock to the disaster, the first time United States astronauts had died in flight. President Reagan canceled the State of the Union message that had been scheduled for tonight, expressing sympathy for the families of the crew but vowing that the nation's exploration of space would continue.

Killed in the explosion were the mission commander, Francis R. (Dick) Scobee; the pilot, Comdr. Michael J. Smith of the Navy; Dr. Judith A. Resnik; Dr. Ronald E. McNair; Lieut. Col. Ellison S. Onizuka of the Air Force; Gregory B. Jarvis, and Christa McAuliffe.

Mrs. McAuliffe, a high-school teacher from Concord, N.H., was to have been the first ordinary citizen in space.

After a Minute, Fire and Smoke

The *Challenger* lifted off flawlessly this morning, after three days of delays, for what was to have been the 25th mission of the reusable shuttle fleet that was intended to make space travel commonplace. The ship rose for about a minute on a column of smoke and fire from its five engines.

Suddenly, without warning, it erupted in a ball of flame.

The shuttle was about 10 miles above the earth, in the critical seconds when the two solid-fuel rocket boosters are firing as well as the shuttle's main engines. There was some discrepancy about the exact time of the blast: The National Aeronautics and Space Administration said they lost radio contact with the craft 74 seconds into the flight, plus or minus five seconds.

Two large white streamers raced away from the blast, followed by a rain of debris that etched white contrails in the cloudless sky and then slowly headed toward the cold waters of the nearby Atlantic.

The eerie beauty of the orange fireball and billowing white trails against the blue confused many onlookers, many of whom did not at first seem aware that the aerial display was a sign that something had gone terribly wrong.

There were few sobs, moans or shouts among the thousands of tourists, reporters and space agency officials gathered on an unusually cold Florida day to celebrate the liftoff, just a stunned silence as they began to realize that the *Challenger* had vanished.

Among the people watching were Mrs. McAuliffe's two children, her husband and her parents and hundreds of students, teachers and friends from Concord.

"Things started flying around and spinning around and I heard some oh's and ah's, and at that moment I knew something was wrong," said Brian Ballard, the editor of *The Crimson Review* at Concord High School. "I felt sick to my stomach. I still feel sick to my stomach."

Ships Searching the Area

At an outdoor news conference held here [in Cape Canaveral] this afternoon, Jesse W. Moore, the head of the shuttle program at NASA, said:

"I regret to report that, based on very preliminary searches of the ocean where the *Challenger* impacted this morning, these searches have not revealed any evidence that the crew of *Challenger* survived." Behind him, in the distance, the American flag waved at half-staff.

Coast Guard ships were in the area of impact tonight and planned to stay all night, with airplanes set to comb the area at first light for debris that could provide clues to the catastrophe. Some material from the shattered craft was reported to be washing ashore on Florida beaches tonight, mostly the small heat-shielding tiles that protect the shuttle as it passes through the earth's atmosphere.

Films of the explosion showed a parachute drifting toward the sea, apparently one that would have lowered one of the huge reusable booster rockets after its fuel was spent.

Pending an investigation, Mr. Moore said at the news conference this afternoon, hardware, photographs, computer tapes, ground support equipment and notes taken by members of the launching team would be impounded.

The three days of delays and a tight annual launching schedule did not force a premature launching, Mr. Moore said in answer to a reporter's question.

"Flight Safety a Top Priority"

"There was no pressure to get this particular launch up," he said. "We have always maintained that flight safety was a top priority in the program."

Several hours after the accident, Mr. Moore announced the appointment of an interim review team, assigned to preserve and identify flight data from the mission, pending the appointment of a formal investigating committee.

The members of the interim panel are Richard G. Smith, the director of the Kennedy Space Center; Arnold Aldridge, the manager of the National Space Transportation System, Johnson Space Center; William Lucas, director of the Marshall Space Flight Center; Walt Williams, a NASA consultant, and James C. Harrington, the director of Spacelab, who will serve as executive secretary.

A NASA spokesman said a formal panel could be appointed as soon as Wednesday by Dr. William R. Graham, the director of the space agency.

All American manned space launchings were stopped for more than a year and a half after the worst previous American space accident, in January 1967, when three astronauts were killed in a fire in an Apollo capsule on the launching pad.

"Hope We Go Today"

This year's schedule was to have been the most ambitious in the history of the shuttle program, with 15 flights planned. For the *Challenger*, the workhorse of the nation's shuttle fleet, this was to have been the 10th mission.

Today's launching had been delayed three times in three days by bad weather. The *Challenger* was to have launched two satellites and Mrs. McAuliffe was to have broadcast two lessons from space to millions of students around the country.

All day long, well after the explosion, the large mission clocks scattered about the Kennedy Space Center continued to run, ticking off the minutes and seconds of a flight that had long ago ended.

Long before liftoff this morning, skies over the Kennedy Space Center were clear and cold, reporters and tourists shivering in leather gloves, knit hats and down coats as temperatures hovered in the low 20s.

Icicles formed as ground equipment sprayed water on the launching pad, a precaution against fire.

At 9:07 a.m., after the astronauts were seated in the shuttle, wearing gloves because the interior was so cold, ground controllers broke into a round of applause as the shuttle's door, whose handle caused problems yesterday, was closed.

"Good morning, Christa, hope we go today," said ground control as the New Hampshire schoolteacher settled into the spaceplane.

"Good morning," she replied, "I hope so, too." Those are her last known words.

The liftoff, originally scheduled for 9:38 a.m., was delayed two hours by problems on the ground caused first by a failed fire-protection device and then by ice on the shuttle's ground support structure.

The launching was the first from pad 39-B, which had recently undergone a $150 million overhaul. It had last been used for a manned launching in the 1970s.

"Go With Throttle Up"

Just before liftoff, *Challenger*'s external fuel tank held 500,000 gallons of liquid hydrogen and oxygen, which are kept separate because they are highly volatile when mixed. The fuel is used in the shuttle's three main engines.

At 11:38 a.m. the shuttle rose gracefully off the launching pad, heading into the sky. The shuttle's main engines, after being cut back slightly just after liftoff, a normal procedure, were pushed ahead to full power as the shuttle approached maximum dynamic pressure when it broke through the sound barrier.

"*Challenger,* go with throttle up," said James D. Wetherbee of Mission Control in Houston at about 11:39 a.m.

"Roger," replied the commander, Mr. Scobee, "go with throttle up."

Those were the last words to be heard on the ground from the winged spaceplane and her crew of seven.

As the explosion occurred, Stephen A. Nesbitt of Mission Control in Houston, apparently looking at his notes and not the explosion on his television monitor, noted that the shuttle's velocity was "2,900 feet per second, altitude 9 nautical miles, downrange distance 7 nautical miles." That is a speed of about 1,977 miles an hour, a height of about 10 statute miles and a distance down range of about 8 miles.

The first official word of the disaster came from Mr. Nesbitt of Mission Control, who reported "a major malfunction." He added that communications with the ship had failed 1 minute 14 seconds into the flight.

"We have no downlink," he said, referring to communications from the *Challenger.* "We have a report from the flight dynamics officer that the vehicle has exploded."

His voice cracked. "The flight director confirms that," he continued. "We're checking with the recovery forces to see what can be done at this point."

Tapes Showed Small Fire

In the sky above the Kennedy Space Center, the shuttle's two solid-fuel rocket boosters sailed into the distance.

The explosion, later viewed in slow-motion televised replays taken by cameras equipped with telescopic lenses, showed what appeared to be the start of a small fire at the base of the huge external fuel tank, followed by the quick separation of the solid rockets. A huge fireball then engulfed the shuttle as the external tank exploded.

At the news conference, Mr. Moore would not speculate on the cause of the disaster.

The estimated point of impact for debris was 18 to 20 miles off the Florida coast, according to space agency officials.

"The search and rescue teams were delayed getting into the area because of debris continuing to fall from very high altitudes, for almost an hour after ascent," said Mr. Nesbitt of Mission Control in Houston.

Speaking at 1 p.m. in Florida, Lieut. Col. Robert W. Nicholson Jr., a spokesman for the rescue operation, which is run by the Defense Department, said range safety radars near the Kennedy Space Center detected debris falling for nearly an hour after the explosion. "Anything that went into the area would have been endangered," he said in an interview.

In addition, the explosion of the huge fuel supply would have created a cloud of toxic vapors. NASA officials said tonight that the hazardous gases presented no danger to land, but the Coast Guard was advising boats and ships to avoid the area.

"Not a Good Ditcher"

In an interview last year, Tommy Holloway, the chief of the flight director office at the Johnson Space Center in Houston, talked about the possibility of a shuttle crash at sea.

"This airplane is not a good ditcher," he said. "It will float O.K. if it doesn't break apart, and we have hatches we can blow off the top. But the orbiter lands fast, at 190 knots. You come in and stop in 100 yards or so. You decelerate like gangbusters, and anything in the payload bay comes forward. We don't expect a very good day if it comes to that."

On board *Challenger* was the world's largest privately owned communication satellite, the $100 million Tracking and Data Relay Satellite, which with its rocket boosters weighed 37,636 pounds.

This morning, water froze on the shuttle service structure, used for fire-fighting equipment and for emergency showers that technicians would use if they were exposed to fuel. The takeoff was delayed because space agency officials feared that during the first critical seconds of launching, icicles might fly off the service structure and damage the delicate heat-resistant tiles on the shuttle, which are crucial for the vehicle's re-entry though the earth's atmosphere.

—*January 29, 1986*

CHAPTER 7

Life on Earth:
Biology, Paleontology,
Zoology

How Life Evolved

The Origin of Species

A Review of *On the Origin of Species by Means of Natural Selection or the Preservation of Favored Races in the Struggle for Life,* by Charles Darwin

I.

It has been calculated on data that may be considered as tolerably satisfactory, that the number of specific types of organic life at present existing on the globe considerably exceeds half-a-million.

Enormous as this aggregation is, it is at the same time probable that each of the long suite of Geologic ages—all characterized by their own systems of animal and vegetable life—was at least as rich as at present in the number and variety of its organic structures.

Embracing in the imagination this vast ensemble of organisms—this ocean of forms amid which creative energy has sported—we realize the magnitude of the problem that seeks to account for the genesis of these millionfold incarnations of life. We touch, in fact, the grand and complex question of the Origin of Species—that mystery of mysteries, as one of our greatest philosophers has styled it.

With regard to this question the doctrine that has long prevailed is that each species has been independently created. The idea of the permanence of species is, in fact, embodied in one shape or another in every definition of the term which has been framed. All the most eminent paleontologists, Cuvier, Owen and Agassiz among them, and all our greatest geologists, Lyle, Murchison and the rest, have unanimously, often vehemently, maintained the immutability of species. No article of scientific faith is of more canonical authority, and though some bold speculator like LaMarck, or some ingenious theorist like the author of the *Vestiges*, has ventured to question the soundness of its basis, yet it has given no outward sign of instability, and is commonly regarded as one of those doctrines which no man altogether in his right senses would set himself seriously to oppose.

Meanwhile, Mr. Darwin, as the fruit of a quarter of a century of patient observation and experiment, throws out, in a book whose title at least has by this time become familiar to the reading public, a series of arguments and inferences so revolutionary as, if established, to necessitate a radical reconstruction of the fundamental doctrines of natural history.

Mr. Darwin's mode of accounting for the diversity of the specific forms of organic life, differs alike from the doctrine of LaMarck, (who made transmutation to depend mainly upon the efforts of the animal.) and from the hypothesis of the author of the *Vestiges* (who found the solution of the problem in the idea of consecutive development.)

It differs also, let us say, in having for its basis some undeniable facts.

Not by dubious speculations on the action of "chemico-electric" operation on "germinal vesicles" but by the synthesis of a wide series of appreciable and daylight facts in the structure of animals and plants, does Mr. Darwin sustain his startling theory. Rising from this synthesis, he ascends with the swoop and force of analogy to the august and audacious statement that "all the organic beings which have ever lived on this earth have descended from someone primordial form, into which life was at first breathed!"

It is clear that here is one of the most important contributions ever made to philosophic science and it is at least behooving on scientists, in the light of the accumulation of evidence which the author has summoned in support of his theory, to reconsider the grounds on which their present doctrine of the origin of species is based.

II.

In the construction of his argument, Mr. Darwin takes his *point de depart* from the striking modifications which animals and plants are susceptible of undergoing in domestication. The astonishing results achieved by English breeders in the modification of horses, dogs and cattle, furnish him fruitful exemplifications of the amazing versatility of animal structures. Indeed, breeders habitually speak of an animal's organization as something quite plastic, which they can model almost as they please....

But it is in the wonderful variations which domestic pigeons undergo that he finds his most striking illustrations of variation under domesticity. *In re* pigeons, Mr. Darwin is immense. And really, the diversity of breeds is something astonishing: marked anatomical and physiological characteristics distinguish the various breeds; at least a score of pigeons might be selected which, if shown to an ornithologist, and he were told that they were wild birds, would certainly be ranked by him as well-defined species, and shine even as distinct genera; and yet, great as the differences are between the breeds, it is the universal opinion of naturalists that all have descended from the Rock pigeon!

The key to this production of varieties under domestication is found in man's power of accumulative selection: nature gives successive variations: man adds them up—such is Darwin's striking expression—in certain directions useful to him.

Is it not plain that this capacity for undergoing such modifications exhibits an elasticity of constitution which would equally tend to adapt these animals to varieties of natural conditions, and thus to originate diversified races which would perpetuate themselves without man's interference?

Species, then, are to some extent, and sometimes to a quite remarkable extent, variable—modifiable. There is a certain versatility in animal structures. They are more or less flexible—more or less plastic.

One of the chief powers in this tragedy of transmutation, this new diviner finds in the Struggle for Existence—an expression which he employs in a large and metaphorical sense.

A struggle for existence inevitably follows from the high geometrical ratio of increase which is common to all organic beings. There is no exception to the rule that every organic being naturally increases at so high a rate that, if not destroyed, the earth would soon be covered by the progeny of a single pair. Even slow-breeding man has doubled in twenty-five years, and at this rate, in a few thousand years, there would literally not be standing-room for his progeny. . . .

The law of the Struggle for Existence Mr. Darwin finds the key to many secret facts in the economy of nature. He records many striking instances of how complex and unexpected are the checks and relations between organic beings, which have to struggle together in the same country. By what a web of complex relations are plants and animals most remote in the scale of nature bound together! How, for example, could the domestic cats in any village be imagined to determine the frequency of certain flowers in that district? Mr. Darwin shall make answer in his own words:

"I have reason to believe that humble-bees are indispensable to the fertilization of the heartsease. Hence I have very little doubt that if the whole genus of humble-bees became extinct or very rare in England, the heartsease and red clover would become very rare, or wholly disappear. The number of humble-bees in any district depends in a great degree on the number of field-mice, which destroy their combs and nests; and Mr. H. Newman, who has long attended to the habits of humble-bees, believes that 'more than two thirds of them are thus destroyed all over England.' Now the number of mice is largely dependent, as everyone knows, on the number of cats; and Mr. Newman says, 'Near villages I have found the nests of humble-bees more numerous than elsewhere, which I attribute to the number of cats that destroy the mice!'"

One would think, in reading Mr. Darwin's string of sequences, that the ingenious author of the "House that Jack Built" had intended to imbue the infant mind with a knowledge of high scientific truths: for the whole may be expressed, after the fashion of the immortal legend, in the following concise runic: "This is the old woman, that kept the cat, that ate the mouse, that killed the humble-bees, that fertilized the clover!"

In the midst of this constant and complex struggle for existence, there is a formative principle ever at work modifying the structure of organized beings. This, Mr. Darwin calls the law of Natural Selection. The doctrine of Natural Selection is the radical thought of the book.

We have seen that there is something plastic in all organisms—a certain versatility, susceptible of producing varieties from species, susceptible of modifying the structure to certain extent. We have also seen that man, by selection, can certainly produce great results, and can adapt organic beings to his own uses, through the accumulation of slight but useful variations, given to him by the hand of Nature.

Nature is constantly at work doing the same thing. Whatever variation occurring among the individuals of any species of animals or plants is in any way advantageous in the struggle for existence, will give to those individuals an advantage over their fellows, which will be inherited by their offspring until the modified variety supplants the parent species. By the steady accumulation, during long ages of time, of slight differences, each in some way beneficial to the individual, arise the various modifications of structure by which the countless forms of animal and vegetable life are distinguished from each other. . . .

III.

Shall we frankly declare that, after the most deliberate consideration of Mr. Darwin's arguments, we remain unconvinced?

The book is full of a most interesting and impressive series of minor verifications; but he fails to show the points of junction between these, and no where rises to complete logical statement.

The difficulties, of course, are enormous. This he frankly acknowledges. "Some of them are so grave that to this day I can never reflect on them without being staggered." Such are his own naïve and noble words.

He thinks, however, they are more apparent than real. We fear they are very real. To us insurmountable.

Ten times the space given to this article would not suffice for any adequate treatment of this vast and complicated subject. In a very general way, though, we may touch on a few topics. To this and every hypothesis which assumes the gradual transition from species to species, from genus to genus, geology opposes the irrefragable fact of the utter absence of all transitional links, and the clean and clear identity of specific forms. As on Mr. Darwin's theory an interminable number of intermediate forms must have existed, linking together all the species in each group by gradations as fine as our present varieties, we have a prefect right to ask, why do we not see these linking forms all around us? Why are not all organic beings blended together in an inextricable chaos?

Mr. Darwin answers this difficulty by urging the extreme imperfection of the geological record. That the geological record is imperfect all will admit; but few will be inclined to admit that it is imperfect to the degree Mr. Darwin's doctrine requires.

This draft on the imagination of millionfold transitional organisms to fill up the hiati between fossil species, is succeeded by a hypothesis still more startling. For, unhappily for the theory, when we go back even to the first apparition of life in the palæozoic rocks of the Silurian system, we find ourselves among organic structures specifically just as distinct as those our dredgers now take off our coasts. To obtain a background for this abrupt appearance of distinct types, his theory postulates that "before the lowest Silurian stratum was deposited, long periods elapsed—as long as, or probably far longer than, the whole interval from the Silurian age to the present day; and that during these vast but quite unknown periods of time, the world swarmed with living creatures."

To the question why we do not find records of these vast primordial periods, Mr. Darwin fairly admits that he "can give no satisfactory answer," but tries to satisfy the inquirer by the assurance that his witnesses are all submerged! At a period immeasurably antecedent to the Silurian epoch, he affirms, continents may have existed where oceans are now spread out, and clear and open seas may have existed where our continents now stand. Nor should we be justified in assuming that if, for instance, the bed of the Pacific Ocean were now converted into a continent, we should there find formations older than the Silurian strata, and in those formations, one after another, throughout millions of ages, the successive forms of the primitive fauna and flora were silently entombed.

Amid this pile of unsupported conjecture there is but one consideration that we may profitably keep in mind—that it is an admission in geology that the absence of all organic remains in a particular deposit is no proof whatever that animal life did not abundantly exist during the whole period of that deposit. An azoic *rock* is no necessary proof of an azoic *period*. It is even a doctrine of the most advanced school of English geologists, that the fossils of the Silurian formation are by no means to be regarded as the first apparition of organic life, and the present writer had an opportunity last summer, in company with Sir Wm. Logan, of examining a fossil from the Laurentian rocks of Canada—far older than the Silurian strata—which if it holds its authenticity, must be the intimation of a new and elder world of organic forms, into which research and induction may yet be prolonged.

The fundamental limitation of Mr. Darwin's theory springs from a fact in his own mental structure. He is but a *Naturalist*. Of that lofty series of speculations embracing the doctrine of Homologies, Embryology and Unity of Type, he seems ignorant in any profound sense. It is only, we apprehend, by converging the prophetic omens and intuitions from these grand reaches of science that light can be thrown on the mysterious problem of the Origin of Life. Oken has profoundly remarked that "the animal system cannot be arbitrarily disposed of according to this or that organ, just as it may chance to meet the eye, but only in accordance with the rigid prescripts of the animal body's genesis."

The conception of the Unity of Nature passed like a new spirit into science, at the opening of the present century. It is one of those flaming thoughts the soul projects—splendid prophecies that become the light of all our science and all our day. Applied, within these few years, by St. Hilaire, Oken, Carus, Serres,

Owen, to anatomy, it has revealed a sublime unity of design and composition throughout the whole hierarchy of animate organisms. Geoffroy St. Hilaire formuled it in the grand conception that *there is but one animal*. Embryology has developed the doctrine that all animals resemble each other in their early stages, and Transcendental Anatomy reveals that the higher orders of animals pass through transitory embryonic stages, that are the permanent *states* of the lower classes. Polyp, mollusk, turtle, fish, bird, are successive incarnations of that divine idea that finds its latest vestment in the glorious garb of the human form.

IV.

But it would be to fail to extract the best uses of this book, to expect a finality. Its best suggestion is to show us how far we are from the possibility of any finality. Of how many things are we ignorant. "And we do not know how ignorant we are," adds Mr. Darwin.

No one ought to feel surprise at much remaining as yet unexplained in regard to the origin of life. Think of the baffling problem it presents—the sphinx-riddle of life, the mother-mystery ever environing us! "It is good clearly to understand," said Malebranche, and with a very deep meaning, "that there are some things utterly incomprehensible." Indeed, considering the limitations of science, what is there for the wisest but to work in the spirit of the Shakespearean utterance:

> "In Nature's infinite Book of Secrecy
> A *little* I can read."

But while it seems to our apprehension that, in the very nature of things, the problem of life yet remains to receive epic statement, we at the same time look upon the contribution of Mr. Darwin as a most legitimate and successful attempt to extend the domain of science—as, indeed, the most important of modern contributions to philosophic zoology.

And in respect of the great spinal thought of Mr. Darwin's theory, we are persuaded that the doctrine of progressive modification by Natural Selection, will give a new direction to inquiry into the real genetic relationship of species, existing and extinct, will, in fact, make a revolution in Natural History.

It will give a new and sure basis of classification. Indeed, this grand fact of the grouping of all organic beings seems inexplicable on any other theory. Read the interminable disputes of the naturalists as to what are species and what varieties, and you will see what a scientific chaos classification up to this day is. How the cumbersome catalogues of species increase! Meanwhile the difficulties increase, also, instead of diminishing with the extension of their researches.

That is a great fact, too, which Mr. Darwin so impressively teaches—the imperfection of the geological record—and must go largely to modify existing paleontologic conceptions.

And what a vast background he lights up! What flowing eons mark the ascent from the Silurian mollusks to man—gulfs of time over which the mind grows dizzy in the attempt to gaze, and we feel the shiver of eternity pass over us! It is well to feed the mind with this sense of the amplitude of time, as a counter-agent of our petty and contracted chronologies.

It is the leading idea of modern science that we need not go in search of any other causes than those which are at present in action, for an explanation of the phenomena of Nature. Modern geology has banished the notion of sudden cataclysms, by showing that the same agencies are now at work which have brought about all the wonders of the up-piling of the strata of the globe. We are thus introduced into the grand idea of growth—of the enormous effect of existing agencies, when spread over a great period of time.

Darwin puts himself abreast these same tendencies. And just as Lyell has banished from geology the notion of sudden cataclysms, Darwin threatens to banish from zoology the notion of sudden creations. Together, we feel justified in saying, they have laid the foundation of one of the mightiest changes in philosophical thought. It is certainly more in accordance with our ideas of the philosophy of causation to believe that the entire hierarchy of animate organisms are the result of the continuity of one mode of operation throughout the whole period that has elapsed since life was first introduced into our planet. It harmonizes better with our highest ideas of divine foresight, to believe that the scheme of evolution was originally made so perfect as to require no subsequent interference. We have no sympathy with those who, to use the admirable language of Baden Powell—"behold the Deity more clearly in the dark than in the light—in confusion, interruption and catastrophe, *more* than in order, continuity and progress."

V.

The most important contribution to modern thought is undoubtedly the indirect teachings of physical science. For, magnificent though the direct teachings be, the indirect are perhaps still more wonderful: the former have relation but to the material world, while the latter influence the whole of man's speculative activity. Indeed the only part of science that can ever profoundly touch the great laity are its glorious indirections—its grand out-croppings of truth. In this regard there is much for science yet to attain: science needs literature just as much as literature needs science. He is the master of science who makes his facts but initial, leading to heights where new vistas open in flashes of beauty and repose. . . .

It would seem as if the doctrine of "attractions proportioned to destinies" held equally true in science as in sociology. Harvey's discovery of the circulation of the blood is recorded to have found acceptance from no physician over forty years old. Perhaps Darwin felt that to his own theory some such elective affinities might apply. "I look," says he, towards the close of the volume, "with confidence to the future, to young and rising naturalists, who will be able to view both sides of the question with impartiality."

In that future to which he looks forward, he will not, we apprehend, be regarded as having drawn the cosmic circle of life, but rather as having indicated one of its arcs. At all events, it seems to be a historic law that the greater portion of truths in the theory of nature first appear as purple mirages—ruddy and auroral streaks gilding the matin of man's mind; but the appointed time duly brings up the perfect thought, fraught with the wealth of invisible climes, and flooding the age with the sunlight of science.

—*March 28, 1860*

Is Man Merely an Improved Monkey?

M r. Darwin's recent work on the *Descent of Man* has very naturally attracted considerable attention. It is to be observed that many of Mr. Darwin's disciples think they best serve their master by explaining away, or altogether disowning, the theory upon which reasoning is based. They allege that Mr. Darwin does not say that man is descended from the monkey. In order that there may be no mistake on this point, let us look at Mr. Darwin's own words:

"If the anthropomorphous apes be admitted to form a natural sub-group, then as man agrees with them, not only in all those characters which he possesses in common with the whole Catarhine group, but in other peculiar characters, such as the absence of a tail and of callosities, and in general appearance, *we may infer that some ancient member of the anthropomorphous sub-group gave birth to man.*

"It is therefore probable that Africa was formerly inhabited by extinct apes closely allied to the gorilla and chimpanzee; and as these two species are now man's nearest allies, it is somewhat more probable that our early progenitors lived on the African Continent than elsewhere.

"The early progenitors of man were, no doubt, once, covered with hair, both sexes having beards; their ears were pointed and capable of movement, and their bodies were provided with a tail, having the proper muscles. . . .

"In the class of mammals, the steps are not difficult to conceive which led from the ancient Monotremata to the ancient Marsupials; and from these to the early progenitors of the placental mammals. We may thus ascend to the Lemuridæ; and the interval is wide from these to the Simiadæ. *The Simiadæ then branched off into two great stems, the New World and Old World monkeys; and from the latter, at a remote period, man, the wonder and glory of the universe, proceeded."* . . .

For obvious reasons, we cannot enter here upon any scientific discussion of the subject, but there are some points connected with it which will naturally occur to every reflective reader, and we may briefly touch upon two or three of those points. In the first place, it must strike everybody that in order to maintain his position, Mr. Darwin is obliged to omit from consideration altogether the doctrine that man was created with an immortal soul. This exalted and ennobling belief, which has been the greatest support and encouragement to countless millions of our fellow-creatures, Mr. Darwin only alludes to in an evasive manner in his second volume, while many of his disciples boldly set it down as one of those "old wives' fables" which it is our duty, in this age of progress, to discard. And what do they offer us in place of it? The sublime theory that monkeys were our progenitors, and that we were once gifted with "caudal appendages," which somehow or other have disappeared. Instead of holding fast to the faith that our existence here is but a preparation for a nobler state of being, and that everlasting life is the priceless heritage of man, we are to console ourselves under the innumerable vicissitudes of life with the thought that we were gradually "developed" from monkeys, and that when we die we pass away as the beasts that perish.

We cannot prove existence of the soul—we cannot prove that after we have fulfilled our allotted span here we shall renew our life, although the Divine declaration, "He that believeth in me shall never die," is to myriads of our race proof sufficient. But can Mr. Darwin prove *his* theories? They are as much incapable of proof as any other problem connected with the past or the future. The probabilities are all against them. How is it that, if men sprung from apes, no remains of the intermediate links have ever been found? It is not enough for Mr. Darwin to say, "I cannot find them." We reply, you ought to be able to find them if there is any truth in your "system." . . .

There is nothing in Mr. Darwin's book to shake the faith of any man, unless that faith already rests upon sand. The boundless arrogance of some of Mr. Darwin's followers may lead them to assert that they have proved everything, but those who look closely into their wild speculations will see that their efforts to overthrow the whole structure of the Christian religion have not by any means succeeded.

—March 26, 1871

First-Ever Re-Creation of
New Species' Birth

By CAROL KAESUK YOON

Contemplating the pageant of evolution, biologists have long wondered how much of a role chance played in determining history's course. If researchers could turn back the clock and allow life to evolve again and again, would leaping frogs and blooming orchids always reappear? Or might evolution's path, punctuated by asteroid crashes and random genetic mutations, meander to produce something altogether different?

Now, at last, a team of biologists has found a way to replay one piece of history, re-enacting the birth of a species three separate times and each time tracking the complex genetic changes of speciation. And they have shown that evolution, at least in this case, was much more repeatable and predictable than scientists have imagined.

In their study of the anomalous sunflower, or *Helianthus anomalus*, researchers report in the current issue of the journal *Science* that every time they recreated plants thought to be similar to the original anomalous sunflower, the genes in those plants went through nearly identical changes. Even more surprising, the genetic maps of these laboratory recreations turned out to be very much like the map of the wild sunflower itself.

While researchers have long known that hybrids of any given cross can be quite similar (witness the reliability of producing a mule from a mating between a donkey and a horse), they were greatly surprised that the sunflower hybrids were nearly identical on the fine scale of a genetic map.

"It's a remarkable study," said Dr. Richard Harrison, an evolutionary biologist at Cornell University in Ithaca, N.Y. "What's most remarkable is that you get the same product time after time. It's a very important paper, very elegant and thorough."

Dr. Leslie Gottlieb, a plant evolutionary biologist at the University of California at Davis, said, "This is absolutely unexpected and extremely interesting."

An inhabitant of sand dunes of the Great Basin desert between the Rocky Mountains and the Sierra Nevada, the anomalous sunflower, as its name suggests, is about 100,000 years old as a species. Ethnobotanists say the Hopi

Indians used the powder from dried petals as a facial decoration during traditional ceremonies and as a charm on hunting sticks. Researchers, particularly Dr. Loren Rieseberg, an evolutionary biologist at Indiana University in Bloomington and colleagues, took an interest in the anomalous sunflower for other reasons. It is a natural hybrid. The species, so named because researchers found it so unusual and hard to place, turned out to be the product of the interbreeding of two other distinct sunflower species.

That meant that re-enacting the origin of the species simply required inter-breeding the original parents, the common sunflower, *Helianthus annuus*, and the petioled sunflower, *Helianthus petiolaris*. The resulting offspring would be a modern approximation of hybrid plants that eventually became the anomalous sunflower species.

But what would those first anomalous sunflowers have mated with? The species from which they came, or other hybrids like themselves? To explore the various possibilities, researchers bred the hybrids into three distinct lineages. Each lineage underwent a different regime of mating with other hybrids and parental species to approximate three different situations for the continuing evolution of the species.

After four generations of such breeding, researchers mapped the genomes of the hybrids to see what had happened during the ensuing reorganization and sorting of the original mix of genes.

To do so, Dr. Rieseberg and colleagues borrowed techniques from plant breeders developed for the rapid genetic mapping of crops, using what are known as RAPD markers. The markers, which are used to amplify small portions of DNA for study, allowed researchers to analyze 197 different segments of DNA in each hybrid lineage. Because the researchers already knew what those 197 regions of the genome looked like in the common and in the petioled sunflower species, they were able to identify which regions in the hybrids were descended from which of the original species. Researchers then used the information to con-struct detailed maps showing the patchwork of common sunflower and petioled sunflower DNA in the hybrids.

If chance had played a major role in the evolution of anomalus, then the three different hybrid lineages should each end up with quite different genetic maps, each displaying unique mixtures of common and petioled sunflower genes.

Instead, what researchers found were incredibly similar maps in all three cases. When one lineage carried common sunflower genes on a particular portion of a chromosome, then the other lineages were likely to carry common sunflower genes there as well. Finally researchers compared the maps of experimental hybrids to a genetic map of the wild anomalus. They found, again to their surprise, that the experimental hybrids and the wild species were very similar. Most often, where one had common sunflower genes or petioled sunflower genes, the other did as well.

Researchers say the result indicates that not only were the initial genetic changes in the wild anomalous sunflower the same as those in the experimental lineages, but that even after some 100,000 years of evolution the evidence for those early changes was still clearly detectable.

"I was pretty astonished," Dr. Rieseberg said. "I expected to see some similarities, some concordance, but I didn't expect to see anything like this. I think we'll find that much more about evolution is repeatable and predictable than we think."

The similarities between the experimental hybrids and the wild species are even more striking given that the experimental plants were raised in the wet, rich conditions of the greenhouse quite unlike the extremely hot, dry conditions to which *Helianthus anomalus* became so well adapted.

Dr. Rieseberg said if the researchers had raised the hybrids in conditions more similar to the environment favored by the anomalous sunflower, the closer match of initial environmental pressures for the wild and laboratory sunflowers might have led to an even closer match of the resulting genetic maps.

Dr. Rieseberg and colleagues suggest that the reason the hybrids did end up with the same genetic makeup so consistently is that whether in the greenhouse or the wild, certain combinations of genes from the common sunflower and the petioled sunflower consistently work better together than others. Some of these combinations, it would seem, work so much better that they are essentially always found together in surviving hybrids. If that is the reason, then researchers say the question becomes just what are those particularly potent genes and gene combinations, and what are they doing for the anomalous sunflower?

There may, however, be other explanations for the consistency as well. Researchers suggest the answer could be something much more mechanical. Perhaps, they say, in the inevitable mixing of common sunflower and petioled sunflower chromosomes in the hybrids, the physical interactions among the large gene-carrying structures somehow results each time in certain regions always sorting themselves out together.

Just as the specific cause of the repeatability remains unclear, the bigger question of how widespread over evolution such repeatability is likely to be, remains a matter of debate.

There are those, like Dr. Stephen Jay Gould, an evolutionary biologist at Harvard University, who have previously suggested that any replaying of evolutionary history would probably turn out a quite different result. Dr. Gould said in an interview that the evolution of the anomalous sunflower, while "fascinating," did not suggest that more major events in evolution, for example the evolution of humankind, should be assumed to be deterministic or predictable.

"It's a question of scale," Dr. Gould said. Whenever a person carries three copies of chromosome 21, he said, "you get Down syndrome."

"That's not a general challenge to the notion of unpredictability," he went on. "You put those things together and that's how they're always going to work. But as soon as you get into events whose evolutionary direction depends on interactions with large-scale environments that change unpredictably or the vagaries of interacting with other species through time, that's a different question."

He suggested instead that the genetic machinations of the anomalous sunflower were perhaps more appropriately viewed as a detail in the wider picture. In a wholesale replaying of life, he suggested, perhaps sunflowers themselves would not even reappear.

Dr. Jerry Coyne, an evolutionary biologist at the University of Chicago, cautioned, "You can't extrapolate this to the whole world." Dr. Coyne, who wrote an accompanying article in *Science* suggesting that the new study should perhaps challenge such notions of unpredictability, added, "But in this case, which is the only existing case, you do get these changes that are very repeatable."

One thing scientists do know is that the details of this sunflower's origins are quite unusual.

Dr. Nick Barton, a specialist in the evolution of hybrids at the University of Edinburgh in Scotland, said that in animals new species rarely arose from hybrids. Hybrids and hybrid species are much more common in plants, and some researchers estimate that as many as one in 10 species has hybrid origins. But the most common form of hybrid speciation in plants is very different from the one that gave rise to the anomalous sunflower.

In the more common case, during the crossbreeding of two different species, the number of chromosomes in the hybrid offspring is somehow doubled, making it impossible for the hybrid to breed again with the species from which it came.

In contrast, the incipient anomalous species, when first produced, could still interbreed with its parents. As a result, researchers say, the results of the present study are probably not too relevant to these very different sorts of hybrid origins.

Researchers say at least as important as the finding itself is the successful use of genomic mapping. "What's really important here is the demonstration of these tools," said Dr. Gottlieb of the University of California at Davis. "We are now in a position to ask much more specific questions than were previously possible. My intuition here is that the ability to ask these questions will finally tell us what it means to be a different species and this study is directly on that road."

Dr. Coyne agreed, saying the new study should be emulated, ushering in an era of revitalizing experimentation to evolutionary biology, a field sometimes frustratingly restricted to theorizing and thought experiments.

"You can't bring back the Burgess Shale," he said, referring to an ancient formation of fossils, "but you can do this over and over again and always get the same thing. This is an actual scientific test."

—May 8, 1996

Inherit the Windbags

By MAUREEN DOWD

Do male nipples prove evolution?

Not at all, according to a Web site for a planned Creation Museum devoted to showing that the Bible is literally true.

Nipples may be biologically de trop for men, an "expert" on the site notes, but that doesn't mean they resulted from natural selection. They could just as well be a decorating feature of the Creator's (like a hood ornament). Who are we to question His designs, since we cannot presume to comprehend His mind?

The virtual tour of the museum, to be built in rural Kentucky, says its exhibits will explain many such mysteries, like the claim that *T. rex* lurked around Adam and Eve—"That's the terror that Adam's sin unleashed!"—and how "Noah and his family survive 371 days alone on an animal-filled boat" ("a real *Survivor* story").

The philosophy of the Creation Museum, part of the "Answers in Genesis" ministry, is summed up this way: "The imprint of the Creator is all around us. And the Bible's clear—heaven and earth in six 24-hour days, earth before sun, birds before lizards. Other surprises are just around the corner. Adam and apes share the same birthday. The first man walked with dinosaurs and named them all! God's Word is true, or evolution is true. No millions of years. There's no room for compromise."

Personally, I've decided to stop evolving. No point, really. Evolution is so 20th century.

As with Iraq, President Bush has applied his doctrine of pre-emption on evolution, cutting it off before it can pose a threat to our well-being.

Ever since he observed during his 2000 campaign that "on the issue of evolution, the verdict is still out on how God created the earth," Mr. Bush has been reeling backward as fast as he can toward the Garden of Eden, which, if creationists are to be believed, was really "Jurassic Park."

Seeing the powerful role of evangelicals in getting Mr. Bush re-elected, teachers across the country are quietly ignoring evolution, even when the subject is in their curriculums.

Many teachers take the hint on evolution even without overt pressure, Cornelia Dean wrote this week in Science Times: "Teachers themselves avoid the topic, fearing protests."

On eBay, you can even find replicas of the stickers that a Georgia county put on science textbooks to warn that evolution is "a theory, not a fact." Talk about sticker shock.

So much for the Tree of Knowledge. Mr. Bush gives us the Ficus of Faith.

I knew the president, Dick Cheney and Newt Gingrich wanted to wipe out the psychedelic "if it feels good do it" post-Vietnam '60s and go back to the black-and-white '50s—a meaner *Happy Days*.

They wanted to yank us back in a time machine to a place before Vietnam was lost, free love was found, Roe v. Wade was enacted; they could roll back science to smother stem cells' promise. (Since it was reported last week that all human embryonic lines approved for federally financed research are tainted with a foreign molecule from mice, the administration can't even feign an interest in scientific progress. Who'd a-thunk that science's great hope would turn out to be Arnold Schwarzenegger?)

I misunderestimated this ambitious president. His social engineering schemes in the Middle East and America are breathtakingly brazen.

He doesn't just want to dismantle the '60s. He wants to dismantle the whole century—from the Scopes trial to Social Security. He can shred one of the greatest achievements of the New Deal and then go after other big safety-net Democratic programs, reversing the prevailing philosophy of many decades that our tax and social welfare systems should equalize the distribution of wealth, just a little bit. Barry Goldwater wouldn't have had the brass to take a jackhammer to that edifice.

The White House seems to think Social Security was corrupt from the moment it was enacted in 1935. It wants to replace it with private accounts that will fatten the wallets of stockbrokers and put the savings of Americans who didn't inherit vast fortunes at risk.

Mr. Bush and his crew not only want to scrap the New Deal. By weakening environmental and safety protections and trying to flatten the progressive income tax, they're trying to eradicate not just one Roosevelt but two, going after the progressive legacy of Theodore.

With their brutal assault on history and their sanctimonious manner, they give a whole new meaning to Teddy's philosophy of the presidency. Bully pulpit, indeed.

—*February 3, 2005*

The Dance of Evolution,
or How Art Got Its Start

By NATALIE ANGIER

If you have ever been to a Jewish wedding, you know that sooner or later the ominous notes of "Hava Nagila" will sound, and you will be expected to dance the hora. And if you don't really know how to dance the hora, you will nevertheless be compelled to join hands with others, stumble around in a circle, give little kicks and pretend to enjoy yourself, all the while wondering if there's a word in Yiddish that means "she who stares pathetically at the feet of others because she is still trying to figure out how to dance the hora."

I am pleased and relieved to report that my flailing days are through. This month, in a freewheeling symposium at the University of Michigan on the evolutionary value of art and why we humans spend so much time at it, a number of the presenters supplemented their standard PowerPoint presentations with hands-on activities. Some members of the audience might have liked folding the origami boxes or scrawling messages on the floor, but for me the high point came when a neurobiologist taught us how to dance the hora. As we stepped together in klezmeric, well-schooled synchrony, I felt free and exhilarated. I felt competent and loved. I felt like calling my mother. I felt, it seems, just as a dancing body should.

In the main presentation at the conference, Ellen Dissanayake, an independent scholar affiliated with the University of Washington, Seattle, offered her sweeping thesis of the evolution of art, nimbly blending familiar themes with the radically new. By her reckoning, the artistic impulse is a human birthright, a trait so ancient, universal and persistent that it is almost surely innate. But while some researchers have suggested that our artiness arose accidentally, as a byproduct of large brains that evolved to solve problems and were easily bored, Ms. Dissanayake argues that the creative drive has all the earmarks of being an adaptation on its own. The making of art consumes enormous amounts of time and resources, she observed, an extravagance you wouldn't expect of an evolutionary afterthought. Art also gives us pleasure, she said, and activities that feel good tend to be those that evolution deems too important to leave to chance.

What might that deep-seated purpose of art-making be? Geoffrey Miller and other theorists have proposed that art serves as a sexual display, a means of flaunting one's talented palette of genes. Again, Ms. Dissanayake has other ideas. To contemporary Westerners, she said, art may seem detached from the real world, an elite stage on which proud peacocks and designated visionaries may well compete for high stakes. But among traditional cultures and throughout most of human history, she said, art has also been a profoundly communal affair, of harvest dances, religious pageants, quilting bees, the passionate town rivalries that gave us the spires of Chartres, Reims and Amiens.

Art, she and others have proposed, did not arise to spotlight the few, but rather to summon the many to come join the parade—a proposal not surprisingly shared by our hora teacher, Steven Brown of Simon Fraser University. Through singing, dancing, painting, telling fables of neurotic mobsters who visit psychiatrists, and otherwise engaging in what Ms. Dissanayake calls "artifying," people can be quickly and ebulliently drawn together, and even strangers persuaded to treat one another as kin. Through the harmonic magic of art, the relative weakness of the individual can be traded up for the strength of the hive, cohered into a social unit ready to take on the world.

As David Sloan Wilson, an evolutionary theorist at Binghamton University, said, the only social elixir of comparable strength is religion, another impulse that spans cultures and time.

A slender, soft-spoken woman with a bouncy gray pageboy, a grandchild and an eclectic background, Ms. Dissanayake was trained as a classical pianist but became immersed in biology and anthropology when she and her husband moved to Sri Lanka to study elephants. She does not have a doctorate, but she has published widely, and her books—the most recent one being *Art and Intimacy: How the Arts Began*—are considered classics among Darwinian theorists and art historians alike.

Perhaps the most radical element of Ms. Dissanayake's evolutionary framework is her idea about how art got its start. She suggests that many of the basic phonemes of art, the stylistic conventions and tonal patterns, the mental clay, staples and pauses with which even the loftiest creative works are constructed, can be traced back to the most primal of collusions—the intimate interplay between mother and child.

After studying hundreds of hours of interactions between infants and mothers from many different cultures, Ms. Dissanayake and her collaborators have identified universal operations that characterize the mother-infant bond. They are visual, gestural and vocal cues that arise spontaneously and unconsciously between mothers and infants, but that nevertheless abide by a formalized code: the calls and responses, the swooping bell tones of motherese, the widening of the eyes, the exaggerated smile, the repetitions and variations, the laughter of the baby met by the mother's emphatic refrain. The rules of engagement have a pace and a set of expected responses, and should the rules be violated, the pitch prove too jarring, the delays between coos and head waggles too long or too short, mother or baby may grow fretful or bored.

To Ms. Dissanayake, the tightly choreographed rituals that bond mother and child look a lot like the techniques and constructs at the heart of much of our art. "These operations of ritualization, these affiliative signals between mother and infant, are aesthetic operations, too," she said in an interview. "And aesthetic operations are what artists do. Knowingly or not, when you are choreographing a dance or composing a piece of music, you are formalizing, exaggerating, repeating, manipulating expectation and dynamically varying your theme." You are using the tools that mothers everywhere have used for hundreds of thousands of generations.

In art, as in love, as in dancing the hora, if you don't know the moves, you really can't fake them.

—*November 27, 2007*

A Leading Mystery of Life's Origins Is Seemingly Solved

By NICHOLAS WADE

An English chemist may have found the hidden gateway to the RNA world, the chemical milieu from which the first forms of life are thought to have emerged on earth some 3.8 billion years ago.

The discovery seemingly solves a problem that for 20 years has thwarted researchers trying to understand the origin of life: how the building blocks of RNA, called nucleotides, could have spontaneously assembled themselves in the primitive earth's conditions.

The finding, if correct, should set researchers on the right track to solving many other mysteries about the origin of life. It will also mean that for the first time, a plausible explanation exists for how an information-carrying biological molecule could have emerged through natural processes from chemicals on the primitive earth.

In an interview, the author, John D. Sutherland of the University of Manchester, likened his work to addressing a crossword puzzle in which solving the first clues makes the others easier. "Whether we've done one across is an open question," Dr. Sutherland said. "Our worry is that it may not be right."

But other researchers think he has made a major advance in prebiotic chemistry, the study of the natural chemical reactions that preceded the first living cells.

"It is precisely because this work opens up so many new directions for research that it will stand for years" as a great advance, Jack W. Szostak of Massachusetts General Hospital writes in a commentary in *Nature*.

Scientists had long suspected that the first forms of life carried their biological information not in DNA but in RNA, its close chemical cousin. But despite 20 years' work, they had found no plausible way in which nucleotides could have been assembled.

A nucleotide consists of a chemical base, a sugar molecule called ribose and a phosphate group. Chemists quickly found plausible natural ways for each of these constituents to form from natural chemicals. But there was no natural way for them all to join together.

The spontaneous appearance of nucleotides on the primitive earth "would have been a near miracle," two leading researchers, Gerald F. Joyce and Leslie E.

Orgel, wrote in 1999. Others were so despairing that, believing some other molecule must have preceded RNA, they started looking for a pre-RNA world.

But the miracle seems now to have been explained. In an article being published Thursday in *Nature*, Dr. Sutherland and his colleagues Matthew W. Powner and Béatrice Gerland report that they have taken the same starting chemicals used by others but have made them react in an order and in combinations different from those of previous experiments. They discovered their recipe, which is far from intuitive, after 10 years of working through every possible combination of plausible chemicals.

The starting chemicals, they found, will naturally form a compound that is half-sugar and half-base. When another half-sugar and half-base are added, the RNA nucleotide called ribocytidine phosphate emerges. A second nucleotide is created if ultraviolet light is shone on the mixture.

Dr. Sutherland said he had not yet found natural ways to generate the two other types of nucleotide found in RNA molecules, but synthesis of the first two was thought to be harder to achieve.

If all four nucleotides formed naturally, then they zipped together easily to form an RNA molecule with a backbone of alternating sugar and phosphate groups. The bases attached to the sugar constitute a four-letter alphabet in which biological information can be represented.

"My assumption is that we are here on this planet as a fundamental consequence of organic chemistry," Dr. Sutherland said. "So it must be chemistry that wants to work."

The reactions he has described look convincing to most other chemists. "The chemistry is very robust: all the yields are good, and the chemistry is simple," said Dr. Joyce, an expert on the chemical origin of life at the Scripps Research Institute in San Diego.

In Dr. Sutherland's reconstruction, phosphate plays a critical role not only as an ingredient but also as a catalyst and in regulating acidity. Dr. Joyce said he was so impressed by the role of phosphate that "this makes me think of myself not as a carbon-based life form but as a phosphate-based life form."

Dr. Sutherland's proposal has not convinced everyone. Robert Shapiro, a chemist at New York University, said the recipe "definitely does not meet my criteria for a plausible pathway to the RNA world."

If the proposal is correct, it will set conditions that should help solve the many other problems in reconstructing the origin of life. Darwin, in a famous 1871 letter to the botanist Joseph Hooker, surmised that life began "in some warm little pond, with all sorts of ammonia and phosphoric salts." But the warm little pond has given way in recent years to the belief that life began in some exotic environment like the fissures of a volcano or the deep sea vents that line the ocean floor.

Dr. Sutherland's report supports Darwin. His proposed chemical reactions take place at moderate temperatures, though one does best at 60 degrees Celsius. "It's consistent with a warm pond evaporating as the sun comes out," he said.

And because his proposal requires ultraviolet light, it would rule out deep sea vents as the place where life originated.

—*May 14, 2009*

THE PREHISTORIC RECORD

Finder Says Fossil Links Ape and Man

Dr. Louis S. B. Leakey, a leading British anthropologist, reported today that a recently discovered skull was the link between man and the South African ape man.

He said he and his wife made the discovery July 17 in the Olduvai Gorge, Tanganyika, after an exploration that began there 27 years ago. He believes the skull is between 600,000 and 1,000,000 years old.

Dr. Leakey said the skull, which was almost intact, was the oldest yet discovered of tool-making man and a convincing piece of evidence of truth of Darwin's view that man and apes evolved from common stock.

Features of the skull, he said, were its enormous teeth, small brain cavity and very large face. It was that of a youth of about 18 who had lived mainly on nuts and had just begun to eat mice, snakes and lizards, the anthropologist said.

The youth died of causes other than violence, possibly pneumonia, and had probably been covered with thorn bushes by comrades to prevent his being eaten by hyenas.

Soon afterward, a lake rose in the gorge, covering and preserving him.

Dr. Leakey said at a news conference that he announced his discovery last week at the fourth Pan-African Congress of Prehistory at Leopoldville, Belgian Congo, but that the world outside scientific circles was still unaware of its importance. He said leading authorities at the congress had agreed that the skull was authentic and were "jubilant" at the discovery.

"For years," he said, "many scientists have been trying to find the connecting link between the South African near-men or ape men—*Australopithecus* and *Paranthropus*—and true man as we know him, from the primitive *Pithecanthropus* in Java and China on the one hand, to the more advanced humans—*Atlanthropus* of North Africa and the skulls from Steinheim and Swanscombe in Europe—on the other. Now at last we have got this link.

"We discovered the Olduvai skull on his living floor, with examples of the very primitive stone culture called Oldowan. On the same living floor were the bones of the animals, birds and reptiles that formed part of his diet."

Dr. Leakey said the characters of the skull separated it clearly from *Australopithecus* and *Paranthropus*. In some of these characters, the creature seemed even more primitive, while in others he was much closer to *Pithecanthropus* and *Homo sapiens*. The teeth and palate were far bigger than the largest previously known on a man or man-like fossil.

—September 4, 1959

Two New Theories Offered on Mass Extinctions in Earth's Past

By WALTER SULLIVAN

Two proposals involving catastrophes that enveloped the earth in sun-shading debris have been advanced to account for past episodes of widespread extinction and climatic change.

One was a direct hit on the earth by an asteroid, causing an explosion whose debris cut off all or most sunlight for several years. This could explain the extinction of the dinosaurs and numerous other creatures some 65 million years ago.

The other was a volcanic eruption on the moon, from which the debris remained in orbit around the earth, forming an equatorial ring like that of Saturn. Such a ring would have partially shaded the winter hemisphere of the earth, chilling the climate radically. Part of the debris, in the form of solidified droplets of molten material, was spread around at least half the earth, according to recently found evidence.

Many explanations have been proposed for the mass extinction that occurred at the end of the Cretaceous period, some 65 million years ago, including radiation from a nearby supernova, or stellar explosion.

Among those who favored the supernova hypothesis were Drs. Luis W. Alvarez and Walter Alvarez at the University of California in Berkeley. They have now proposed that the extinctions were caused by the impact of an asteroid some six miles in diameter. The resulting explosion would have filled the stratosphere with sufficient dust to cut off most sunlight for a few years.

The scientists, with Frank Asaro and Helen V. Michel of the Lawrence Berkeley Laboratory, reported their hypothesis in the latest issue of the journal *Science*.

Land plants and small animals would not have become extinct, as indicated by the fossil record, since seeds and roots could have survived a few years without sunlight. Sea plants would have perished, as well as large animals dependent on plant food. Smaller animals could have lived on insects and decaying vegetation.

The evidence cited for an asteroid impact is an abnormal abundance of iridium in sedimentary rocks laid down at that time. Iridium is far more abundant in meteorites (and presumably asteroids) than in earth rocks.

The more recent encirclement of the earth by volcanic debris has been proposed by Dr. John A. O'Keefe of the National Aeronautics and Space

Administration's flight center in Greenbelt, Md. Dr. O'Keefe has specialized in the study of the glassy fragments known as tektites that have fallen on the earth at various times in the past.

In a recent issue of the British journal *Nature* he pointed out that three catastrophes affected the earth some 34 million years ago: a large part of its surface was bombarded by at least a billion tons of tektites, winter temperatures in the northern latitudes plunged and five abundant ocean-dwelling species became extinct.

The explanation, Dr. O'Keefe believes, was a lunar eruption that placed about 25 billion tons of volcanic debris in orbit around the earth. Roughly a billion tons of it fell through the atmosphere. The rest, within about a year, formed a ring extending 4,000 to 12,000 miles above the equator. In the northern winter, when the sun is over the Southern Hemisphere, such a ring would shadow high latitudes in the Northern Hemisphere.

On several occasions falling tektites have formed large "strewn fields" across the earth. Some scientists, such as Dr. Billy P. Glass of the University of Delaware in Newport, believe they were thrown up by explosions produced by giant meteorite impacts.

As noted by Dr. O'Keefe, Dr. Glass and his colleagues have recently delineated a strewn field that extends halfway around the world from the United States and the Caribbean across the Pacific to the Indian Ocean.

The tektites occur in a deep layer of sea-floor sediment, laid down 34 million years ago, that reveals the coincident extinction of five abundant species of warm-water radiolaria. An implication is that the sea suddenly became too cold for them. Radiolaria are single-celled organisms whose skeletons are spiked or highly perforated spheres.

Evidence that high northern latitudes suddenly became cold has been compiled by Dr. Jack A. Wolfe of the United States Geological Survey offices in Menlo Park, Calif. He has found, from the fossil record, that prior to some 34 million years ago vegetation in the Puget Sound area resembled a "marginally tropical rain forest."

Then, according to Dr. Wolfe, in middle to high latitudes of the Northern Hemisphere the vegetation "changed drastically." The broad-leaved evergreens gave way to cool-climate deciduous trees. Species diversity was also reduced.

Dr. Wolfe estimates that the drop in mean annual temperature in Alaska was about 25 degrees Fahrenheit. Furthermore, the difference between average summer and winter temperatures increased radically—from about 8 degrees to as high as 50 degrees.

It has long been known that some great change occurred at this time, marking the end of Eocene epoch. It had been assumed that the process was gradual, but Dr. Wolfe believes it must have been swift, geologically speaking, caused by an increase in the tilt of the earth's spin axis.

Dr. O'Keefe rules out this possibility on the grounds that the earth's equatorial bulge is highly stabilizing, preventing any rapid changes in spin axis.

While it has been proposed that climate cooling and ice ages occur when the solar system passes through a cloud of interstellar dust, this cannot explain the tektites, Dr. Glass said last week. They do not carry the scars of exposure to space radiation (cosmic rays) and could not have been there long.

The recent extension of the "North American" tektite field to the Indian Ocean is based on the discovery of tektites of the same age and composition in sediment extracted at four sites across the equatorial Pacific and one in the Indian Ocean. Such clues tell little about the full extent of the field, but Dr. Glass believes it may represent a fall of at least 10 billion tons.

It overlaps another field, laid down some 700,000 years ago, that covers Southeast Asia, Australia and the Indian Ocean. This fall, which Dr. Glass and his colleagues estimate at 100 million tons, coincided with a reversal of the earth's magnetic field.

A smaller fall, on and near the Ivory Coast of Africa a million years ago, seems to have been associated with another such reversal. Dr. Glass believes that some—but not all—reversals occur when the earth is struck by a large object. In the past 35 million years there have been about 105 reversals in polarity of the earth's magnetism, whereas the remains of only six craters large enough to suggest catastrophic impacts (six miles wide or greater) are known from that period.

Others may lie hidden beneath the sea but not enough to account for all the reversals. Some theorists believe that the earth's magnetic field flips over spontaneously.

The tektites of each fall are distinctive in composition and age. The tektites that fell 34 million years ago occur in three forms: transparent; opaque-white with high calcium content and little iron, and opaque-black, rich in iron and magnesium.

Dr. O'Keefe believes a ring of this material extending from 4,000 to 12,000 miles above the Equator would have cut off 75 percent of sunlight.

—June 10, 1980

New Theory on Dinosaurs: Multiple Meteorites Did Them In

By WILLIAM J. BROAD

For more than a decade, most scientists have believed that the extinction of the dinosaurs was caused by a single event: the crash of an immense body from outer space, its explosive force like a hundred million hydrogen bombs, igniting firestorms and shrouding the earth in a dense cloud of dust that blocked sunlight and sent worldwide temperatures plummeting.

The theory gained wide acceptance in 1991, after the discovery of a crater buried under the tip of the Yucatán Peninsula. The giant gash stretched 110 miles from rim to rim, and its age was found to be 65 million years, the same time as the death of the dinosaurs.

Now, however, scientists working in Ukraine have discovered that a well-known but smaller crater, some 15 miles wide, had been inaccurately dated and is actually 65 million years old, making the blast that created it a likely contributor to the end of the dinosaurs.

So too, a British team has recently found a crater at the bottom of the North Sea dating to the same era and stretching over 12 miles in a series of concentric rings.

The discoveries are giving new support to the idea that killer objects from outer space may have sometimes arrived in pairs or even swarms, perhaps explaining why the extinctions seen in the fossil record can be messy affairs, with species reeling before a final punch finishes them off.

"It's so clear," said Dr. Gerta Keller, a geologist and paleontologist at Princeton, who studies the links between cosmic bombardments and life upheavals. "A tremendous amount of new data has been accumulated over the past few years that points in the direction of multiple impacts."

But Dr. Keller added that many scholars had staked their reputations on the idea of a single dinosaur-ending disaster and were reluctant to consider the new evidence. "Old ideas," she said, "die hard."

Her own research, Dr. Keller added, suggests the reality of multiple strikes and raises doubts that the Yucatán rock, whose crater is known as Chicxulub, was the event that sealed the dinosaurs' fate. Instead, she said, the main killer "has yet to be found."

The ferment is prompting scientists around the globe to look for new craters and to reassess the ages of old ones in search of clues to the wave of global extinction that did in thousands of species—not only the dinosaurs but many plants, fish and plankton—at the end of the Cretaceous period.

"There are over 170 confirmed craters on earth and we know the precise impact age of only around half," said Dr. Simon P. Kelley of the Open University in Britain, who found the dating error on the Ukraine crater, along with Dr. Eugene P. Gurov of the Institute of Geological Sciences in Ukraine. Even in the United States, he added, several craters are poorly dated.

"In the U.K., we have a phrase, 'You wait an hour for a bus, then three come along all at once,'" he remarked in an interview. "Maybe impacts are like that."

The idea that a giant intruder from outer space killed off the dinosaurs was proposed in 1980 by Dr. Luis W. Alvarez; his son, Dr. Walter Alvarez; and their colleagues at the University of California at Berkeley. It was met with great skepticism at first, but in time became the standard belief.

In his 1997 book, *T. Rex and the Crater of Doom*, Dr. Walter Alvarez, a geologist, said he had considered the possibility of multiple impacts until 1991 and the discovery of the huge Yucatán crater, which seemed big enough to solve the mystery on its own.

Dr. Kelley and Dr. Gurov presented their findings from Ukraine in the August issue of the journal *Meteoritics & Planetary Science*. In geologic time, the twin birth throes of the Ukraine and Yucatán craters, they note, suggest rather than prove "that they combined to lead to the mass extinction" at the end of the Cretaceous period and raise questions of other possible cosmic killers.

Known as Boltysh, the newly dated crater lies in eastern Ukraine in the basin of the Tyasmin River, a tributary of the Dnieper. Though just 15 miles wide, the buried crater, whose presence is revealed by deep jumbled masses of melted and broken rocks, is surrounded by a ring of rocky debris that extends over many hundreds of square miles, conjuring up a fiery cataclysm. The two scientists say in their report that this kind of crash today would have devastated a densely populated nation.

Over the years, scientists had analyzed rocky samples from the Boltysh crater and found ages ranging from 88 million to 105 million years.

The new dating of the crater by Dr. Kelley and Dr. Gurov used a highly accurate method that carefully measures the ratio of two isotopes of the element argon, a colorless, odorless gas that makes up about 1 percent of the earth's atmosphere. Argon-argon dating works because the isotopes decay at different rates. By measuring the ratio, it is possible to estimate how long ago the sample melted to trap atmospheric argon.

Dr. Kelley and Dr. Gurov report that seven samples of melted rock from the depths of the Boltysh crater yielded an average age of 65.2 million years, with an accuracy of plus or minus 600,000 years.

By contrast, Chicxulub (pronounced CHEEK-soo-loob) has been dated to 65.5 million years, plus or minus 600,000. Given the range of dating uncertainty, the two impacts that made the craters may have occurred simultaneously or been separated by thousands of years.

Scientists have recently looked more favorably at the idea that comets can travel in packs. In the 1980s, a few speculated that comet showers might produce strikes on the earth over a period of a million years or so to bring on extinctions. The idea gained support in 1994 when the comet Shoemaker-Levy 9 was fractured by the gravitational pull of Jupiter into 21 discernible pieces that then, one by one, bombarded the planet.

Dr. Kelley and colleagues at the University of Chicago and the University of New Brunswick, writing in the journal *Nature* in 1998, gave precise dating evidence to argue that a similar kind of celestial barrage hit the earth 214 million years ago. Spread over Europe and North America, the chain of five craters, they wrote, indicated that a large comet or asteroid had broken up and struck the earth in a synchronized assault.

Today, Dr. Kelley said, the odds of the Boltysh and Chicxulub craters' having formed simultaneously, like the chain, are not great. Still, even if their times of impact prove to have been only close, experts say, the one-two punch could still have added to the global turmoil that did in the dinosaurs and other creatures.

Beneath the North Sea, two British oil geologists have found another crater, buried under hundreds of feet of ooze, that may have contributed to the chaos. Writing in the Aug. 1 issue of *Nature*, Simon A. Stewart and Philip J. Allen said they were able to date the 12-mile structure to a period 60 million to 65 million years ago. They named it Silverpit, after a nearby sea-floor channel.

Experts say the new finds may answer an old criticism of the single-impact theory. Critics, especially the paleontologists who specialize in dinosaur extinction rates, had long noted that the fossil record of the late Cretaceous shows a slow decline of many life forms rather than a single vast die-off. That seemed inconsistent with a cosmic catastrophe.

But now, the emerging family ties among the Boltysh, Silverpit and Chicxulub craters suggest that a series of impacts may have driven or contributed to this slow decline.

Dr. Keller of Princeton and her colleagues have found signs of other intruders from outer space that hit at slightly different times about 65 million years ago, strengthening the gradualist idea.

Working in northeastern Mexico, they discovered that glass spheres of melted rock once thought to have been thrown out by the Chicxulub impactor were more likely the result of at least two separate disasters, about 300,000 years apart. They recently presented their findings in a paper for the Geological Society of America.

Moreover, Dr. Keller said, the evidence suggests that the earlier of the two cataclysms formed the Chicxulub crater, making its arrival too early to account for the killer punch of the dinosaur extinction.

Geologic clues that she and her colleagues are collecting from Mexico, Guatemala, Haiti and Belize, Dr. Keller said, suggest that a barrage of cosmic bodies hit the earth over the course of 400,000 years. The first was the Chicxulub event, the second the unlocated impactor at the end of the Cretaceous period and then a straggler some 100,000 years later.

Strong evidence exists for three impacts at the end of the Cretaceous era, Dr. Keller said, followed by wide climate shifts that lasted through the turbulent period.

While geologists hunt for other craters and impact events, they say the most compelling evidence of all may have vanished. Since the earth's surface is more than 70 percent water, it is likely that most signs of speeding rocks from space disappeared long ago in the churning geological processes that constantly renew the seabed. The North Sea, being relatively shallow, is an exception.

Despite the inherent difficulties of the research, Dr. Kelley of the Open University said he planned to redouble his hunt to "try to solve this problem."

—November 5, 2002

A Prehistoric Giant Is Resurrected, if Only in Name

By JAMES GORMAN

For anyone who has ever been told by a smarty-pants seven-year-old, "There's no such thing as *Brontosaurus*; it's called *Apatosaurus*"—it is payback time.

"We have good evidence now for the resurrection of *Brontosaurus*," Emanuel Tschopp, a paleontologist with the New University of Lisbon, said on Tuesday.

He was referring to the name, not to the creature itself, of course, and to an exhaustive study of 80 or so fossils in a group of long-necked giants called the Diplodocidae, familiar from natural history museums the world over. These plant-eaters grew to lengths of more than 100 feet and weighed thousands of pounds, and it is thought they could crack their long tails like bullwhips, creating sonic booms to scare away predators.

The name "Brontosaurus" was first used in the late 1800s to describe fossils of a dinosaur now on display at the Yale Peabody Museum, but by 1905 it had been reclassified as *Apatosaurus*, because it was so similar to another sauropod dinosaur of that name.

Dr. Tschopp and his colleagues Octávio Mateus at the New University and Roger B. J. Benson at the University of Oxford in England decided that although *Brontosaurus* and *Apatosaurus* are similar, there are actually two different genera and the Yale specimen is really a *Brontosaurus* after all. So are several other museum specimens, they said, including one at the University of Wyoming, and a baby *Apatosaurus* at the Carnegie Museum of Natural History in Pittsburgh.

Their paper, released online on Tuesday in the journal *Peer J*, with all of its nearly 300 pages freely available to anyone, will not be the last word on whether the *Brontosaurus* name should come back into scientific use. Names of species and genera are matters of expert opinion. There is no national or international board of official dinosaur names that decides who is right.

"What's interesting to me is that *Brontosaurus* and *Apatosaurus* are still extremely close," said Matthew T. Carrano, a curator at the Smithsonian Institution's National Museum of Natural History. "This provides a lot of new information for the argument, but the argument will continue."

Dr. Tschopp and his colleagues looked at 477 distinct traits that could be identified on the fossil bones and then analyzed them in several ways to look for differences.

Dr. Benson said they set a standard based on the differences between two well-known species of similar long-necked dinosaurs, *Diplodocus* and *Barosaurus*. "*Brontosaurus* is at least as different from *Apatosaurus* as *Diplodocus* is from *Barosaurus*," he said.

Dr. Benson pointed out that "the names are just handles" that help scientists study how life evolves into different forms.

Most of the differences are highly technical and noticeable only to anatomists, said Dr. Tschopp. To pick one understandable example, he said, "*Apatosaurus* has a relatively wider neck than *Brontosaurus*." But it is the number of differences that is important, he said.

Jeffrey A. Wilson, a paleontologist at the University of Michigan, said that he did not have a position on whether the name should be resurrected but that he found the criteria for distinguishing between *Apatosaurus* and *Brontosaurus* arbitrary.

It was a matter of judgment that involved deciding how different was different enough to justify two distinct genera. "It's as if they had a pizza," he said, and "cut it in six pieces." Why not cut it into four pieces? he asked.

The name game is played not only by scientists. There are pitfalls for parents and children doing their obligatory museum visits. The *Apatosaurus* on display at the American Museum of Natural History is probably still an *Apatosaurus*, said Dr. Tschopp, as is the adult skeleton at the Carnegie Museum.

In the end, it may be too complicated for parents to remember which is which. So the seven-year-olds may continue to rule.

—*April 8, 2015*

THE WORLD OF ANIMALS

Computer Helps Chimpanzees Learn to Read, Write and "Talk" to Humans

By BOYCE RENSBERGER

Timothy Gill peered into a Plexiglas room where a chimpanzee named Lana lives with a computer console. Lana pushed a series of symbol-coded buttons on the console and, outside her room, the computer typed out a translation of the symbols. "Please Tim move into room."

Mr. Gill, who read the message in symbols on a display panel above Lana's console, reached to his own console and pushed a button marked "yes." The symbol for "yes" flashed onto Lana's display panel and she excitedly rushed over to the door.

Mr. Gill, a graduate student at the Yerkes Primate Research Center in Atlanta, who is Lana's best friend, opened the door and went in. The chimpanzee took Mr. Gill's hand and they walked to the computer console, which acts as their medium of communication.

Mr. Gill pushed some buttons and Lana watched the display panels to see what he said. An automated typewriter monitored the conversation.

"Please, Lana, groom Tim."

"Yes," the chimpanzee answered, and immediately Lana began picking through Mr. Gill's hair, carrying out a friendly social behavior common among chimpanzees.

Read, Write and Talk

Such exchanges are typical of half a dozen chimpanzees in research centers around the country that are demonstrating that chimps can learn languages approximating English well enough to read and write and even to converse with human beings.

Although efforts to teach apes to use human language were largely given up as impossible some two decades ago, renewed efforts using new methods over the last five years have shown that the animals are capable of learning hundreds of words and of chaining them into rudimentary but meaningful sentences.

In just the last year, some of the chimpanzees have achieved still more remarkable language skills, such as the mastery of a rigorous grammar and an apparent understanding of conceptual and abstract terms.

These and other recent developments suggest that behind the sometimes comical face of the chimpanzees there lies an intellectual capacity vastly more sophisticated than even the most ardent anthropomorphists had dared to suppose.

The accomplishments of the chimps are growing so rapidly that it may be seriously questioned whether man can long continue to claim to be the only animal that uses language.

Defining Language

The issue is not likely to be resolved conclusively very soon, for there is no generally accepted definition of language against which to measure the chimps' achievements. Some skeptics contend that the chimpanzees are exhibiting nothing more than rote learning or stimulus-response conditioning. While the chimps' teachers cannot prove this to be untrue, neither can the critics demonstrate that human language itself is not based simply on memorization or conditioning.

One possible application of the new technique could be in efforts to communicate with retarded or autistic children who fail to learn language under conventional teaching methods. If the researchers can devise ways to unlock the chimpanzee's limited intellect, such methods might also work with human beings of limited or isolated mental abilities.

The breakthrough came in the late 1960s when Dr. Allan Gardner and his wife, Beatrice, psychologists at the University of Nevada, hit upon a way to circumvent the chimpanzee's lack of a pharynx, the space just above the voice box that changes shape to help produce the varying sounds needed in speech. The Gardners tried the sign language of the deaf and found that their chimp, named Washoe (for the country in which the Gardners lived), picked it up readily.

Washoe, who lived in a trailer behind the Gardner home near Reno, learned sign language well enough that visiting deaf people understood her and she them.

Earliest Words

Washoe's earliest words in the order she learned them, were: come-gimme, more, up, sweet, go, hear-listen, tickle, toothbrush, hurry, out, funny, drink, sorry, please, food-eat, flower, cover-blanket, you, in.

Early on, Washoe began to combine her words into sentence-like strings such as "come-gimme sweet" and "out please."

She also engaged in little conversations such as:

Washoe: Gimme, gimme.

Human being: What do you want?

Washoe: Sweet.

After Washoe learned the word "open" in connection with opening a house door, she quickly generalized it to ask for the opening of refrigerator doors, car doors, cupboards and jars.

Whenever she sustained a cut or bruise, she learned to sign the gesture for "hurt" or "pain." Later, when she saw people with red stains on their bodies, she volunteered, "hurt."

In most cases she easily generalized a sign for the name of a concrete object to a picture of the object and even to the thought of the object.

Shows Fear of Dogs

Washoe, for example, does not like dogs. Once the Gardners tried to fool her, in sign language, by saying there was a big dog outside. The chimpanzee immediately became agitated and nervous. Her hair bristled up and she behaved as if in danger of attack.

The rate at which Washoe learned new signs steadily increased during her training. In her first seven months, beginning when she was a year old and the equivalent of a human child of about the same age, Washoe learned only four words. During the second seven months she learned nine and during the third such period she picked up 21 new signs. At the end of three years she could understand and "speak" 85 words. After only one more year, her vocabulary nearly doubled to 160 words.

At about this point the Gardners changed their plans and wanted to start anew with younger chimps. Washoe was becoming too large and unruly. They turned her over to a student, Roger Fouts, who, upon earning his doctorate in psychology, took Washoe to the Institute for Primate Studies in Norman, Okla.

The institute is a privately owned facility run by Dr. William B. Lemmon of the University of Oklahoma.

"Dirty Monkey"

Dr. Fouts taught Washoe the sign for monkey and she readily used it for squirrel monkeys and gibbons, but when it came to a particularly nasty rhesus monkey, she invented the term "dirty monkey." Previously she had used the word dirty only in reference to feces or soiled objects. Now she regularly uses the term for the offending rhesus and has even made "dirty" into a kind of swear word, repeating "dirty, dirty" whenever some human being annoys her.

Another instance of her inventing names for new things has impressed some experts as a genuine indication of a human-like language ability—the combining of previously unrelated words (or signs of symbols) to signify a new meaning.

On seeing her first swans, Washoe inverted the term "water bird," combining two words she knew independently.

These days Washoe, now eight years old, is in semi-retirement, living in an indoor cage with several other chimps with which she no longer tries to strike up conversations in sign language. (They never responded.) Attention has turned to Lucy, another eight-year-old female trained by Dr. Fouts, who lives with a human family near the institute.

Lucy knows 93 words and is learning new ones at the rate of one to three a week, sometimes after only five minutes of demonstration.

Holds Interview

Recently Lucy sat for her first interview with this reporter, who knows sign language. Actually, she didn't sit very much. She jumped, rolled, climbed, walked and ran. But she did pay attention enough for brief sign language conversations.

Reporter (holding up a key): What this?

Lucy: Key.

R: (holding a comb) What this?

L: Comb (takes comb and combs reporter's hair, then hands comb to reporter). Comb me.

R: O.K. (combs Lucy).

My longest exchange with Lucy was this:

R: Lucy, you want go outside?

L: Outside, no. Want food, apple.

R: I have no food. Sorry.

The conversation may not have been especially deep, but it certainly was communication.

When she started to walk away, I signed, "Lucy, sit," and she sat. When I signed, "Where Roger?" she pointed to Dr. Fouts.

After each exchange, Lucy and I would stare into each other's eyes for a few seconds. I don't know how she felt, but I was nervous. I was participating in something extraordinary. I was conversing in my own language with a member of another species of intelligent beings. What was she thinking about me? What should I say to her?

What Does She Think

Lucy is only eight years old, and because chimpanzees have a life span of 50 to 60 years, she is really still a child. What will she know and say 10 years from now? Will she be able to tell us what life is like for a chimpanzee? What does she think?

Dr. Fouts, who talks not only to Washoe and Lucy but also to three other similarly trained chimps at the institute, has given such questions much thought.

"First of all," he said, "chimps are going to express themselves as chimpanzees. They aren't humans and the differences are likely to show up in language as in other things."

Dr. Lemmon puts it differently: "I suspect we'll come to recognize that the chimp, like the human, has a nature."

Quite possibly Lucy thought it not at all unusual to converse with people. It is about all she has known. But Jane Temerlin, in whose home Lucy lives, tells of one incident that gives cause to wonder.

Lucy had a pet cat. One day when Lucy was alone with her pet, she was seen to sit down on the floor, place the cat between her legs, facing her, and hold up a book. Lucy pointed several times to the book and, signing so the cat could see her, made the sign for "book."

In Other Ways

When news of the Gardners' success spread, a number of other researchers thought up alternate ways of communicating without speech. One was Dr. David Premack of the Center for Advanced Study in the Behavioral Sciences in Stanford, Calif. He constructed a number of distinctively shaped and colored pieces of plastic, each signifying an English word, and taught chimps both to arrange them in sensible sequences and to read the meaning of sequences he assembled.

His prize pupil was Sarah, a young chimp who knows over 130 words and can construct such simple sentences as "Ann give apple Sarah" and can read and obey such sentences as "Sarah insert apricot red dish," selecting the correct fruit and dish from several possibilities.

Sarah, now considered too big to handle safety, is in retirement in a cage, and Dr. Premack has taken on two younger chimpanzees that are learning quickly.

Still another alternative to signing is the computer-controlled language being learned by Lana, which is under study by Dr. Duane M. Rumbaugh of Georgia State University. Dr. Rumbaugh hopes that through use of the artificial language programed into the computer, Lana will learn to adhere to a rigorous syntax— something that some observers feel is missing from the signing chimps' language.

"Too Easy to Accept"

"Sign language is fine," Dr. Rumbaugh said, "but it's just too easy to accept an ungrammatical sentence from a signing chimpanzee and take it to mean what you want it to mean."

To eliminate any ambiguity in what is being said, Dr. Rumbaugh collaborated with Ernst von Glasersfeld and Pier P. Pisani of the University of Georgia to develop Lana's computerized language, called Yerkish in honor of the primate center's founder, Dr. Robert M. Yerkes.

Rules of Yerkish grammar are programed into the computer and, if Lana is trying to command the operation of any of the automated food dispensing devices in her room, the computer will accept and relay only messages that are in correct Yerkish.

Thus, for example, if Lana pushes the word buttons in the following sequence, "Please, machine, give milk," an automated dispenser with a straw will fill with milk. If Lana says, "Please, machine, make milk," the computer will reject the sentence.

Nonsensical Sequence

While it may be tempting to accept the sentence as a good try, it is, in strict Yerkish, nonsensical. The only things the machine can "make" are "window open," "music" or "movie."

After a year and a half in training, Lana is now 3½ years old. She has learned 71 words, but Dr. Rumbaugh and Tim Gill both say they have deliberately concentrated not on expanding her vocabulary but on teaching more sophisticated concepts and uses of words.

For example, Lana knows the names of colors. If a picture of a blue ball is projected on her room wall and Lana is asked, "What color of this," she answers, "Blue color of this." If, on the next exchange, she is asked, "What name of this," she readily composes a new sentence, pressing the button for ball instead of the one for blue.

Asks for Music

All of Lana's words and sentences are recorded by the computer-controlled typewriter. Because the computer is left running every night, Dr. Rumbaugh has an opportunity to see what Lana tries with language when she is alone. She often asks the machine to play her a movie or recorded music.

"We feel," Dr. Rumbaugh said, "she is about to convince us that she has language. What we're looking for is to have continuing and meaningful conversation with her."

Dr. Rumbaugh looks forward to the day when Lana can become a partner in the behavioral study of other chimps, reporting in Yerkish the meaning of various things chimps do in their own societies. Toward that end he expects that Lana's training will continue for many years if adequate research funds can be found.

For the moment, Lana's interests are simpler. At night, when Tim Gill has gone home and Lana is alone in her room, she has typed out the sentences, "Please, machine, move into room" and "Please, machine, tickle Lana."

—*May 29, 1974*

Mating for Life?
It's Not for the Birds or the Bees

By NATALIE ANGIER

A h, romance. Can any sight be as sweet as a pair of mallard ducks gliding gracefully across a pond, male by female, seemingly inseparable? Or better yet, two cygnet swans, which, as biologists have always told us, remain coupled for life, their necks and fates lovingly intertwined.

Coupled for life, with just a bit of adultery, cuckoldry and gang rape on the side.

Alas for sentiment and the greeting card industry, biologists lately have discovered that, in the animal kingdom, there is almost no such thing as monogamy. In a burst of new studies that are destroying many of the most deeply cherished notions about animal mating habits, researchers report that even among species assumed to have faithful tendencies and to need a strong pair bond to rear their young, infidelity is rampant.

Biologists long believed, for example, that up to 94 percent of bird species were monogamous, with one mother and one father sharing the burden of raising their chicks. Now, using advanced techniques to determine the paternity of offspring, biologists are finding that, on average, 30 percent or more of the baby birds in any nest were sired by someone other than the resident male. Indeed, researchers are having trouble finding bird species that are not prone to such evident philandering.

Faithless Females

"This is an extremely hot topic," said Dr. Paul W. Sherman, a biologist at Cornell University in Ithaca, N.Y. "You can hardly pick up a current issue of an ornithology journal without seeing a report of another supposedly monogamous species that isn't. It's causing a revolution in bird biology."

In related studies of creatures already known to be polygamous, researchers are finding their subjects to be even more craftily faithless than previously believed. And to the astonishment, perhaps disgruntlement, of many traditional animal behaviorists, much of that debauchery is committed by females.

Tracking rabbits, elk and ground squirrels through the fields, researchers have learned that the females of these species will copulate with numerous males in a single day, each time expelling the bulk of any partner's semen to make room for the next mating. Experts theorize that the female is storing up a variety of semen, perhaps so that different sperm will fertilize different eggs and thus assure genetic diversity in her offspring.

Males Retaliate

Most efficiently energetic of all may be the queen bee, who on her sole outing from her hive mates with as many as 25 accommodating, but doomed drones.

Scientists also have gathered evidence of many remarkable instances of attempts by males to counteract philandering by females. Among Idaho ground squirrels, a male will stick unerringly by a female's side whenever she is fertile, sometimes chasing her down a hole and sitting on top of it to prevent her from cavorting with his competitors. Other squirrels simply use a rodent's version of a chastity belt, topping an ejaculation with a rubber-like emission that acts as a plug.

The new research, say scientists, gives the lie to the old stereotype that only males are promiscuous. "It's all baloney," said Dr. Sherman. "Both males and females seek extra-pair copulations. And what we've found lately is probably just the tip of the iceberg." Even mammals, which have never been paragons of virtue, are proving to be worse than expected, and experts are revising downward the already pathetic figure of 2 percent to 4 percent that represented, they thought, the number of faithful mammal species.

"It was believed that field mice, certain wolf-like animals and a few South American primates, like marmosets and tamarins, were monogamous," said Dr. David J. Gubernick, a psychologist at the University of Wisconsin in Madison who studies monogamy in mammals. "But new data indicate that they, too, engage in extra-pair copulations."

Scientists say their new insights into mating and the near-universality of infidelity are reshaping their ideas about animal behavior and the dynamics of different animal social systems.

"It's been a bandwagon," said Dr. Susan M. Smith, a biologist at Mt. Holyoke College in South Hadley, Mass. "Nobody can take monogamy for granted anymore, in any species they look at, so we're all trying to rewrite the rules we once thought applied."

Old Assumptions, Darwin's Misconceptions

Biologists say their new research suggests that many animal social systems might have developed as much to allow animals to selectively cheat as they did out of a need for animals to divide into happy couples. They propose that pair bonds among animals might be mere marriages of convenience, allowing both partners enough stability to raise their young while leaving a bit of slack for the occasional dalliance.

More than anything else, say biologists, they are increasingly impressed by the complexity of animal sexuality. "It seems that all our old assumptions are incorrect, and that there's a big difference between who's hanging out with whom and who's actually mating with whom," said Dr. Patricia Adair Gowaty, a biologist at Clemson University in South Carolina and one of the first to question the existence of fidelity among animals. "For those of us in the field, this is a tremendously exciting time."

Researchers say that many of the misconceptions about monogamy and infidelity began in Darwin's day, when he and other naturalists made presumptions, perhaps understandable, about mating based on field observations of coupled animals. Nearly all birds form pairs during the breeding season, and biologists assumed that the pair bond was necessary for the survival of the young. Without the contributions of both males and females to feed and protect the young, experts thought, few offspring would make it to the fledgling stage. And that demand for stability, biologists assumed, likely included monogamy as well.

But as field researchers became more sophisticated in their observation techniques, they began spotting instances in which one member of a supposedly monogamous avian couple would flit off for a tête-à-tête with a paramour.

"Extra-pair copulations are called sneakers, and they really are," said Dr. Robert Montgomerie, a biologist at Queens University in Kingston, Ontario. "They're not easy to observe because the birds are very surreptitious about such behavior."

Such sightings inspired biologists to apply DNA fingerprinting and other techniques used in paternity suits to help determine the parentage of chicks. They discovered that between 10 percent and 70 percent of the offspring in a nest did not belong to the male caring for them.

Explanations, Females Look Up

Redoubling their efforts in the field, biologists began to seek explanations for the infidelities. In some cases, the female clearly was the one seeking outside liaisons.

Dr. Smith has studied the familiar black-capped chickadee of North America. She had found that, during winter, a flock of chickadees forms a dominance hierarchy in which every bird knows its position relative to its fellows, as well as the ranking of the other birds.

In the spring breeding season, says Dr. Smith, the flock breaks up into pairs, with each pair defending a territorial niche and breeding in it. Though she has determined that infidelity is rare among the chickadees, it does occur "and in a very interesting way," she said. On occasion, a female mated to a low-ranking male will leave the nest and sneak into the territory of a higher-ranking male nearby.

"In every single case of extra-pair copulations, the female wasn't moving randomly, but very selectively," said Dr. Smith. "She was mating with a bird ranked above her own mate."

Dr. Smith suggests that the cheating chickadee may have the best of both worlds: a stable mate at home to help rear the young, along with the chance to bestow on at least one or two of her offspring the superior genes of a dominant male. "This fits into the idea that the female is actively attempting to seek the best-quality genes," she said.

Selectivity of Barn Swallows

In similar studies, Dr. Anders Moller, a biologist at the University of Uppsala in Sweden, has found that female barn swallows likewise are very finicky about their adulterous encounters. When cheating, he said, the females invariably copulate with males endowed with slightly longer tails than those of their mates. Dr. Moller has learned that, among barn swallows, a lengthy tail appears to be evidence that the birds are resistant to parasites; this trait would be beneficial to a female's young. "Females mated to very short-tailed males engage in these extra-marital affairs the most," he said. "Short-tailed males attempt to have affairs themselves, but they're rarely successful."

Some females that mate promiscuously may be gaining not so much the best genes as enough genetic diversity to assure that at least some of their offspring thrive. Biologists studying honeybees have found that the queen bee will leave her hive only once, to mate with as many as 25 drones patrolling nearby. Tabulating her wantonness is easy: to complete intercourse, the poor drone must explode his genitals onto the queen's body, dying but leaving behind irrefutable evidence of an encounter.

And while the queen bee does have considerable reproductive demands, needing enough sperm to fertilize about four million eggs, researchers have determined that any one of the drones could provide enough sperm to accommodate her. They, therefore, suspect that her profligate behavior is intended to insure genetic diversity in her brood.

The Devious Males, Strategies for Success

But biologists say there are evolutionary counterbalances that can keep cheating in check. Females that actively seek outside affairs might risk losing the devotion of their own mates. Researchers have found that among barn swallows, a male that observes his mate copulating with other males responds by reducing his attention to her babies. Of course, males themselves are always attempting to philander, say biologists, whether or not they are paired to a steady mate at home. In an effort to spread their seed as widely as possible, some males go to exquisitely complicated lengths.

A Kind of Betrayal

Studying the purple martin, the world's largest species of swallow, Dr. Gene S. Morton, a research zoologist at the National Zoo in Washington, has found that older males will happily betray their younger counterparts. An older martin will first establish his nest, attract a mate and then quickly reproduce, both parents again being needed for the survival of the young.

His straightforward business tended to, the older bird will start singing songs designed to lure a younger male to his neighborhood. That inexperienced yearling moves in and croons a song to attract his own mate, who is promptly ravished by the elder martin. A result is that a yearling male manages to fertilize less than 30 percent of his mate's eggs, although he is the one who ends up caring for the brood.

"The only way for the older males to get the younger females is to attract the young males first," said Dr. Morton. "The yearlings end up being cuckolded."

Older males often try to appropriate a younger male's partner. Studying mallards and related ducks, Dr. Frank McKinney, curator of ethology at the Bell Museum of Natural History at the University of Minnesota in Minneapolis, has found that males often try to force sex on females paired to other males. The females struggle mightily to avoid these copulations, he said, by flying away, diving underwater or fighting back.

"Our finding is that it's usually the older, experienced males that are successful in engaging in forced copulation," he said. They have more skills, and capturing females is a skillful business."

Guarding the Females

Driven by evolutionary pressures, males have developed an impressive array of strategies to fend off competitors and keep their females in line as well. "In almost any animal you look at, males do things in order to be certain of paternity," said Dr. David F. Westneat, a biologist at the University of Kentucky in Lexington. Mate-guarding is one widespread strategy, he said, with males staying beside females during her fertile times. But other strategies result in what biologists have called "sperm wars," a battle by males to give their sperm the best chance of success.

Among many species of rodents, the last male's sperm is the sperm likeliest to inseminate the female, for reasons that remain mysterious.

Hence, several males may engage in an exhausting round robin, as each tries, repeatedly, to be the last one to copulate with the female.

In studies of the damselfly, Dr. Jonathan Waage, a biologist at Brown University in Providence, R.I., has learned that the male has a scoop at the end of his genitals that can be used before copulating to deftly remove the semen of a previous mate.

In other species, natural selection seems to have favored males with the most generous ejaculation. Over evolutionary time, researchers say, this has resulted in the development of some formidable testicles.

The more likely a female is to mate with more than one male, they say, the bigger the sperm-producing organs will be.

Comparing the dimensions of testes relative to body size among several species of primates, biologists have found that gorillas have the smallest. Among the great apes, a dominant silverback male manages to control a harem of females with little interference from other males, biologists say.

Chimpanzees have the largest testes of the primates relative to body size. They are the ones that live in troupes with multiple males, multiple females, and considerable mating by all.

Human beings have mid-sized testicles, further evidence, biologists say, that our species is basically monogamous, but that there are no guarantees.

One Exception, A Paragon of Fidelity

But lest everybody cynically conclude that nothing and nobody can be trusted, a study has unearthed at least one example of an irrefutably monogamous animal: *Peromyscus californicus*, or the California mouse, found in the foothills of the Sierra Nevada.

Dr. David Ribble, of the University of California at Berkeley, and Dr. Gubernick of the University of Wisconsin have performed extensive tests to prove the rodent's fidelity. DNA analysis has shown that, in 100 percent of the time, the pups are fathered by a female's lifelong mate.

The scientists also have coated the female in fluorescent pigment powders to see with whom the female has contact. "The powder only shows up on her mate and offspring," said Dr. Gubernick. Mother and father split child-rearing duties 50-50 he says.

"This is an extremely unusual animal," said Dr. Gubernick. "It may be one of the only truly monogamous species in the world."

The Human Urge to Cheat

The most intrepid biologists are trying to apply the new insights about infidelity among animals to the study of humans. Some say that we are basically a monogamous species, but that the urge to cheat might have an evolutionary basis.

Babies need long-term care, which probably led to pair-bonding among humans early in our evolution, biologists say. But they suggest that a man might be driven to stray from his partner to slip a few more of his genes into the pool. For her part, a woman might philander to mate with a man who has hardier genes than those of her husband.

Dr. Robert L. Smith of the University of Arizona in Tucson, believes that lapses in monogamy helped spawn male sexual jealousy.

"There are nasty cultural manifestations of male jealousy," he said. "Female genital disfiguration, foot-binding in China—these are mechanisms by which males have controlled female opportunities to run off and mate with other males."

But women are not entirely helpless, Dr. Smith says. He suggests that evolution has provided them with ways of avoiding being too closely monitored by men, for example, by giving no clue of when they are fertile.

"If males don't know when their mates are ovulating, they can't be so diligent about guarding their partners during that time," said Dr. Smith. "That allows women to exercise their reproductive options."

Another way that women may exercise such options, Dr. Smith suggests, is by having breasts. "In great apes, conical breasts are a signal that a female is lactating and thus has low reproductive value," he said. "By having perennially enlarged breasts, women make it ambiguous to males when they're fertile and when they're lactating," again confusing men about when to guard their partners.

—August 21, 1990

Slow Is Beautiful

By NATALIE ANGIER

This was no euphemistic brushoff, no reptilian version of "Sorry, I'll be busy that night washing my hair." Paddling around in a tropically appointed pool at the National Aquarium in Baltimore, the husky female Gibba turtle from South America made all too palpable her disdain for the petite male Gibba that pursued her. He crawled onto the parqueted hump of her bark-brown shell. She shrugged and wriggled until he slipped off. He looped around to show her his best courtship maneuvers, bobbing his head, quivering his neck. She kicked him aside like a clot of algae and kept swimming.

"I feel sorry for the little guy," said Jack Cover, a turtle specialist and the general curator of the aquarium. "He's making no progress, she's got zero interest in him, yet he just keeps coming back for more."

And why not? The male Gibba may be clueless, he may at the moment have the sex appeal of a floating toupee, but he is a turtle, and, as a major new book and a wealth of recent discoveries make abundantly clear, turtles are built for hard times. Through famine, flood, heat wave, ice age, a predator's inspections, a paramour's rejections, turtles take adversity in stride, usually by striding as little as possible. "The tale of the tortoise and the hare is the turtle's life story," said Mr. Cover, who calls himself a card-carrying member of the "turtle nerds" club. "Slow and steady wins the race."

With its miserly metabolism and tranquil temperament, its capacity to forgo food and drink for months at a time, its redwood burl of a body shield, so well engineered it can withstand the impact of a stampeding wildebeest, the turtle is one of the longest-lived creatures Earth has known. Individual turtles can survive for centuries, bearing silent witness to epic swaths of human swagger. Last March, a giant tortoise named Adwaita said to be as old as 250 years died in a Calcutta zoo, having been taken to India by British sailors, records suggest, during the reign of King George II. In June, newspapers around the world noted the passing of Harriet, a Galapagos tortoise that died in the Australia Zoo at age 176—171 years after Charles Darwin is said, perhaps apocryphally, to have plucked her from her equatorial home.

Behind such biblical longevity is the turtle's stubborn refusal to senesce—to grow old. Don't be fooled by the wrinkles, the halting gait and the rheumy gaze.

328

Researchers lately have been astonished to discover that in contrast to nearly every other animal studied, a turtle's organs do not gradually break down or become less efficient over time.

Dr. Christopher J. Raxworthy, the associate curator of herpetology at the American Museum of Natural History, says the liver, lungs and kidneys of a centenarian turtle are virtually indistinguishable from those of its teenage counterpart, a Ponce de Leonic quality that has inspired investigators to begin examining the turtle genome for novel longevity genes.

"Turtles don't really die of old age," Dr. Raxworthy said. In fact, if turtles didn't get eaten, crushed by an automobile or fall prey to a disease, he said, they might just live indefinitely.

Turtles have the power to almost stop the ticking of their personal clock. "Their heart isn't necessarily stimulated by nerves, and it doesn't need to beat constantly," said Dr. George Zug, curator of herpetology at the Smithsonian Institution. "They can turn it on and off essentially at will."

Turtles resist growing old, and they resist growing up. Dr. Zug and his co-workers recently determined that among some populations of sea turtles, females do not reach sexual maturity until they are in their forties or fifties, which Dr. Zug proposes could be "a record in the animal kingdom."

Turtles are also ancient as a family. The noble chelonian lineage that includes all living turtles and tortoises extends back 230 million years or more, possibly predating other reptiles like snakes and crocodiles, as well as birds, mammals, even the dinosaurs.

The turtle's core morphology has changed little over time, and today's 250 or so living species all display an unmistakable resemblance to the earliest turtle fossils. Yet the clan has evolved a dazzling array of variations on its blockbuster theme, allowing it to colonize every continent save Antarctica and nearly every type of biome nested therein: deserts; rainforests; oceans; rivers; bogs; mountains; New Brunswick, Canada; New Brunswick, N.J.

"Turtles can persist in habitats where little else can survive," said Dr. J. Whitfield Gibbons, a professor of ecology at the University of Georgia in Athens.

Troubles Foreseen

The iconic turtle likewise has colonized the human heart. People may despise cats or fear dogs, but practically everybody has a soft spot for turtles. "Turtles are by far the most popular reptile," said Peter C. H. Pritchard, director of the Chelonian Research Institute in Oviedo, Fla. "Unlike snakes, which may threaten you and which move like a flash, turtles are benign and slow, and you can't dislike or distrust the clumsy."

Yet such warm and fuzzy feelings have proved cold comfort for turtles, and herpetologists fear that in humans the stalwart survivors from the Mesozoic era may at last have met their mortician. Turtle habitats are fast disappearing, or are being fragmented and transected by roads on which millions of turtles are crushed each year. "There's no defense against that predator known as the automobile," Dr. Gibbons said.

Researchers estimate that at least half of all turtle species are in serious trouble, and that some of them, like the Galapagos tortoise, the North American bog turtle, the Pacific leatherback sea turtle and more than a dozen species in China and Southeast Asia, may effectively go extinct in the next decade if extreme measures are not taken. "People love turtles, people find them endearing, but people take turtles for granted," Mr. Cover said. "They have no idea how important turtles are to the ecosystems in which they, and we, live."

Researchers are also impressed by the turtle's many sensory talents. Box turtles and other forest-dwelling species can spot a lake or pond a mile in the distance, possibly by detecting polarized light glinting off the surface of the water. Female sea turtles migrate across entire oceans every breeding season, unerringly making their way from far-flung feeding grounds right back to the beach where they were born, and where they are instinctively driven to lay their own eggs.

Instinctive does not mean inflexible, however. Should a weary wayfarer arrive at her natal beach in the dead of night and find it has eroded away, Dr. Pritchard said, she can adapt, swimming down the coast until she locates a suitably sandy nesting site.

Turtles, it seems, are all ears, all the time. Dr. Ray Ashton, who runs the Finca de la Tortuga biological preserve in Archer, Fla., has highly preliminary evidence that some turtle species may communicate subsonically, just as elephants do, transmitting and detecting ultralow frequency sound waves as vibrations in the ground.

In their new book, *Turtles of the World* (Johns Hopkins Press), Franck Bonin, Bernard Devaux and Alain Dupre seek to loft turtles into the limelight by showcasing the group's diversity—its beauties, its goofies, its gargoyles.

There is the Indian star tortoise, its shell a vivid basket weave of dark and light veins that dance like spattered sunlight as the tortoise crosses the forest floor; and the Matamata turtle of the Amazon basin, with a flattened, ragged head and neck that look like dead leaves and a bumpy shell that mimics an old log—just try to spot that Matamata at the bottom of a stream, awaiting passing prey; and the massive alligator snapping turtle of the south-central United States, which lures fish right into its open jaw with a red bleb of flesh on the floor of its mouth that jiggles like a chubby worm.

Some turtles have serpentine necks twice the length of their shells; others sport sweet little snorkeling snouts that look like double-barreled cocktail straws; still others have beaks so fiercely hooked their bearers could easily serve, in the authors' words, as "adornment of the upper reaches of Notre Dame."

Among the most common questions leveled at turtle researchers is, What is the difference between a turtle and a tortoise? It depends on where you live, researchers reply. In the United States, any reptile with a shell is referred to as a turtle, and the term tortoise is reserved for those turtle species that have elephantine feet and live entirely on land, like the desert tortoise of the American Southwest. In Australia, by contrast, the word tortoise often applies to aquatic side-necked species—bizarre beasts with necks that cannot be drawn into the shell for protection but instead must be tucked on the side, under the shell's eavelike overhang.

Whatever their group identity badge, turtles vary considerably in size, from the tiny speckled padloper tortoise of South Africa, which in adulthood is no bigger than a computer mouse, to the great leatherback sea turtle, which can measure seven feet long and weigh 2,000 pounds.

Menu plans vary as well. Many turtles are omnivores, happily consuming fruits, leaves, insects, mollusks, fish, frogs, ice cream. Dr. Gibbons told of a friend whose his pet box turtle would respond to the sound of a spoon being tapped on a glass ice cream bowl by emerging from behind the couch, walking over to its owner, rearing up on its hind legs and waiting to be spoon-fed its just dessert. "Had I not seen this a few times myself," he said, "I would not have believed it."

A few turtles have highly specialized palates. Green sea turtles prize the tender tips of sea grass, and will clip away and discard tough, older grass to stimulate the sprouting of fresh buds beneath. Leatherback sea turtles dine only on jellyfish, or what they think are jellyfish. "Plastic bags look like jellyfish," said

Dr. Joseph Mitchell, an ecologist and turtle specialist in Richmond, Va., "and quite a few leatherbacks have stomachs impacted with plastic bags."

Some turtles, conversely, seek out the world's detritus. Scavenger turtles that live in the Ganges River devour human corpses, making it possible for devout Hindus to deposit their loved ones' remains in the waters they deem sacred.

An Iconic Feature

Whether they wrest it from sea grass, shellfish or Haagen-Dazs, all turtles need a substantial amount of calcium in their diet, to sustain the structure that marks them as turtles and that remains among the most extraordinary architectural achievements in vertebrate evolution: the shell. A number of invertebrates have shells, of course, and so, too, do a few vertebrates, most notably the armadillo. But whereas the armadillo's shell is built of bony segments slapped down over its muscle tissue and is distinct from the mammal's underlying skeletal frame, in the turtle the skeleton has become the shell.

During embryonic development, the bones of the turtle's rib cage grow straight out, rather than curving toward one another as they do in other vertebrates. Those ribs, spinal vertebrae and other skeletal bones are then fused to form the upper shell, called the carapace, the lower shell, or plastron, and the bony bridges that join upstairs with down. In many turtle species, the bony shell is in turn plated over with tough fingernail-like structures called scutes.

As a result of the osteotic overhaul, not only can a turtle not crawl out of its shell, it has trouble crawling, period. "Its legs stick out at bizarre angles, and the only reason it can walk at all is through sheer strength," Dr. Pritchard said. "The turtle has enormously strong muscles and extremely thick leg bones." A clumsy gait proved a small price to pay, however, for the acquisition of body armor that protects adult turtles against a panoply of jaws and claws.

Geneticists have proposed that the turtle shell may have appeared quite suddenly in the distant past, rather than emerging slowly through modest, mincing modifications of pre-existing structures. They suggest that the dramatic innovation could have arisen from just a few key mutations in master genes like the so-called homeobox genes, which help specify an animal's basic body plan. If the shell did burst on the reptilian stage more or less fully formed, they said, that would explain the lack of "intermediary" fossils or prototurtles in the paleontological record.

The shell very likely helps explain the turtle's elongated storyline. It takes time to consolidate a large, thick shell, but upon reaching adult stature, the turtle is close to invulnerable. At that point, it can compensate for its Darwinically unproductive youth with a very prolonged and zealously fecund adulthood. A female turtle will continue laying eggs until she dies, and a male turtle will just as mulishly pursue her.

—December 12, 2006

CHAPTER 8

Mathematics: Reality to Infinity

Chaos Is Defined by New Calculus

A new mathematical definition of chaos, which brings "utter confusion" for the first time under the control of man, was reported to the Fourth International Congress for Applied Mechanics here [in Cambridge, Mass.] today.

The definition is a new form of calculus. It enables scientists to predict what will happen in states of complete confusion. Practical uses are many. Examples are the solving of air turbulence which hampers airline flights and the flow of liquids in pipes.

This calculus was reported by Dr. Norbert Wiener of Massachusetts Institute of Technology. The mathematics first card indexes sample kinds of chaos. An engineer can select the sample which most nearly resembles his problem.

The steps of a drunken person illustrate the samples. The wabbly walk was one of the problems in chaos which first interested Dr. Wiener. Each step has no relation to previous steps. But calculus can show far the "drunk" is likely to go in a given time.

Problem in Planets' Motions

"Perturbations" of the planets, their slight, irregular wanderings from their orbits, are a practical example of chaotic movement. Astronomers have been figuring them for centuries, and an explanation by Dr. Albert Einstein of one unexpected motion helped build his fame.

The new definition, Dr. Wiener explained, is entirely mathematical. It defines "pure" chaos. Practical examples of the "pure" state are difficult, he said, but closest to 100 percent chaos are the noises you here in a vacuum tube, like the radio. Electrons make them, by striking a target. But no single electron makes a sound. The racket comes from "mass," that is, when a lot of electrons hit simultaneously.

There are three steps in the new calculus, said Dr. Wiener. The first was the "ergotic" (from the Greek word *ergon*, meaning "work") theorem of Willard Gibbs, American scientist of the last century. It declared that a chaotic system would run through all possible phases of confusion.

Scientists, said Dr. Wiener, found that the systems did not run the entire gamut.

The next step was by Dr. G. D. Birkhoff of Harvard, with new mathematics by which the time when chaotic events would happen could be averaged.

Adds Factor of Space

Dr. Wiener's word adds "space" to chaos, so as to tell not only the time but the place where a change is likely.

In airplane flights one of the "chaos" troubles is to figure out when and where the smooth flow of air over the top of a streamlined wing will bring up into eddies; and what kind of eddies. The eddies interfere with the plane's lifting power.

Engineers by experiment measure some of these eddies. But it is impossible to measure them all. Designers have to guess on the basis of their sample eddies. The new mathematics is designed to take the guesswork out of these experiments by showing whether the samples are well chosen.

—September 14, 1938

Puzzle on Map Conjecture:
Color It Solved

By TOM FERRELL

One of the best-known problems in mathematics, the so-called four-color conjecture, has at last been solved. The problem has engaged both amateur and professional attention for many years because it is so easily stated: Prove that four colors are enough to color any conceivable map, such that no two adjacent areas will be colored the same.

The four-color conjecture has long been considered probably true, because no one has ever been able to devise an imaginary map that violated it. However, such failure is not considered proof. For proof to exist, mathematicians must show that the conjecture is true for all possible maps that might ever be imagined.

The new proof has been carried out by Kenneth Appel and Wolfgang Haken of the University of Illinois, using about 1,200 hours of computer time. A brief and incomplete statement of their method, in nontechnical terms, might be this:

They took advantage of the fact that maps (divisions of a plane into regions) can be, by standard mathematical procedures, converted into graphs (points connected by lines). They found that 1,936 graphs could be used to represent all the possible configurations of maps. The computer analysis began by supposing the four-color conjecture is false; if so, a "bad map," one that would require five colors, could exist. The computer showed for each of the 1,936 configurations that no "bad map" exists. Therefore, since the possibilities of falsehood are exhausted, the conjecture must be true.

Mathematicians are still in search of a simpler proof of the conjecture, one that will not lean so heavily on the computer but will be accessible to the techniques of more or less ordinary geometry.

There is some anxiety in mathematical circles that because the four-color conjecture has been broken at last, amateur mathematicians will now be encouraged to proceed with such projects as squaring the circle and trisecting the angle. Both, according to professionals, have been proven to be simply impossible.

—September 26, 1976

But Aren't Truth and Beauty
Supposed to Be Enough?

By JAMES GLEICK

Could the mathematicians, winners of the most prestigious awards of their discipline, please tell the audience what their work is good for?

Flush with pleasure, these four young men, carrying home three Fields Medals and a Nevanlinna Prize, were telling a lay audience what their work was about. Two had discovered astonishing facts about shapes in four-dimensional space. One had developed important insights into what makes hard problems hard. One had proved Mordell's conjecture, the idea that a large class of equations can have only a finite number of rational solutions.

To the nearly 4,000 mathematicians who gathered here [in Berkeley, Calif.] for the International Congress of Mathematicians, which ended Monday, these were an astonishing set of breakthroughs demonstrating new vitality in the purest of sciences. But a reporter-cameraman for a local television station wanted at least one of the prizewinners to address a basic question: How would their achievements improve life for the viewers at home?

Embarrassed silence. The mathematicians suddenly seemed to have remembered pressing engagements elsewhere. They looked at one another. Gerd Faltings, a boyish, blond West German who became one of mathematics' great men when he proved the Mordell conjecture, gave an awkward smile and flatly refused to speak.

It was Michael Freedman, a topologist in California, who rose in the end to say what all the mathematicians felt: That theirs is a way of thinking that thrives by disdaining the need for practical applications. Let the applications come later by accident—they always do. A weird, curved parody of Euclidean geometry turns out to be just the framework a physicist needs to invent the General Theory of Relativity. Notoriously unpractical techniques of number theory turn out to be just what the National Security Agency needs to make efficient, secure codes.

Usually unspoken, but always present, is the faith that doing mathematics purely, following an internal compass, seeking elegance and beauty in a strange abstract world, is the best way in the long run to serve practical science. As physics or biology progress, they will inevitably find that the way ahead has been cleared by some odd piece of pure mathematics that was thought dead and buried for many decades.

"We're a part of a gigantic enterprise that has gone on for hundreds of years, and interacts in interesting ways with science, and operates on a very low budget," Dr. Freedman said, "and we've learned that it's hard to prophesy what piece of mathematics will have what particular applications. Mathematics has to advance as an organic whole, in ways that seem right to the people inside it."

An older mathematician, Sir Michael Atiyah of Oxford University, who won a Fields Medal himself in 1966, offered one correction. "It's been thousands of years," he said. "So we're in business on a long-term basis."

Yet the meaning of mathematical purity is changing—has changed, many mathematicians said, even since the last Congress in Warsaw in 1983. Questions about the nature of that change, and what it might mean for the future, hovered in the air through 16 "plenary addresses," scores of 45-minute lectures and hundreds of 10-minute "short communications" in the nine-day conference.

Mostly unrepresented was the somewhat less exalted discipline known as applied mathematics, the traditional route for mathematical ideas to filter down to engineering and other sciences. A few mathematicians could not help noticing, though, that recently physical scientists have been plucking ideas directly from the heart of pure mathematics, bypassing applied mathematics altogether. Many unexpected connections have arisen—between knot theory and genetic processes in DNA strands, for example—but the most important has been the use of geometric ideas in the theory of cosmic strings, the hottest new game in the physics of fundamental forces and particles.

The suspicion of a few mathematicians here was that biologists and chemists can no longer be relied on to be naive about the arcana of number theory or topology. That will take some getting used to.

And in the case of strings, the physics has begun feeding back into mathematics, meaning that the up-to-date pure mathematician may now have to learn some unpure science. This state of affairs was highlighted by two unusual talks by physicists, Edward Witten of Princeton University and Aleksandr Polyakov of the Soviet Union's Landau Institute for Theoretical Physics.

Dr. Polyakov, an intense man with long sandy hair, paced nervously before his lecture, a red knapsack on his back. He was worried that his mathematician audience would be put off by having to hear a foreign language—physics, not Russian.

"I apologize if you are irritated by the reckless manner of a physicist," he told his audience. Reckless, because the two disciplines have different standards of proof: A physicist is content to say that the earth orbits the sun; a mathematician will say only that there is convincing evidence.

Dr. Freedman, in the work that won his Fields Medal, proved that certain exotic four-dimensional spaces exist. Another medalist, Simon Donaldson of Oxford, meanwhile, used tools from physics to prove that these same spaces could not exist.

"So the conclusion a mathematician would draw," Dr. Freedman said, "is that physics doesn't exist."

To some mathematicians, purity has always meant a certain degree of inscrutability. That, at least, has not changed.

Sometimes inscrutability comes with the territory—for example, when the territory has four dimensions or more. A mathematician needs to be comfortable with shapes in many dimensions, but not everyone can actually visualize more than the usual three. That is one reason geometry relies, for the sake of purity, on rigorous proofs using numbers and symbols. Visual imagination cannot be trusted.

One dimension is a line. The second dimension comes when you add a second line at right angles to the first, so that now you have east-west and north-south. The third dimension requires a new line at right angles to the others, so you must leave the flat plain and draw one up-down. To imagine a fourth dimension, it is necessary to imagine a fourth line at right angles to all the others, and this most mortals cannot do.

Yet some kind of inner vision led John Milnor of the Institute for Advanced Study in Princeton, in describing the four-dimensional discoveries of Dr. Freedman to an audience of several thousand, to start gesturing with his hands.

"The problem is," he was saying, "when you try to embed a two-dimensional disk inside a four-dimensional manifold, it will usually intersect itself." His hands formed loops and handles in the air, as though he were describing some new kind of suitcase.

Sometimes inscrutability is just a matter of style.

It has been said that the ideal mathematics talk has three parts. The first part should be understood by most of your audience. The second part should be understood by four or five specialists in your field. The third part should be understood by no one—because how else will people know you are serious?

Some speakers seemed to follow these guidelines, mathematicians felt. Others, perhaps to save time, skipped directly to part three.

There was one question that Fields medalists could not wait to answer, and that was whether they used computers, the unloved child of mathematics and an object whose influence was more in evidence at the congress than ever before.

"No," Dr. Freedman said. Dr. Donaldson: "No." Dr. Faltings said, "Perhaps it could reduce some sorts of tedious work for us, but it doesn't do the thinking." Personally he doesn't use one.

Eyes turned to Leslie G. Valiant of Harvard University, winner of the recently established Nevanlinna Prize for Information Science, whose work centered on computer algorithms. "Maybe I should clarify my own position," Dr. Valiant said. "I don't use computers either."

If the mathematicians were inclined toward parable and metaphor—and they most definitely are not—they might describe a vast wilderness, and in it a small society of men and women whose business it is to lay railroad track. This has become an art, and they have become artists—artists of track, lovers of track, connoisseurs of track.

Almost perversely, they ignore the landscape around them. A network of track may head to the northeast for many years and then be abandoned. An old, nearly forgotten line to the south may sprout new branches, heading toward a horizon that the tracklayers seem unable or unwilling to see.

As long as each new piece of track is carefully joined to the old, so that the progression is never broken, an odd thing happens. People come along hoping to explore this forest or that desert, and they find that a certain stretch of track takes them exactly where they need to go. The tracklayers, for their part, may have long since abandoned that place. But the track remains, and track, of course, is the stuff on which the engines of knowledge roll forward.

—August 12, 1986

At Last, Shout of "Eureka!" in Age-Old Math Mystery

By GINA KOLATA

More than 350 years ago, a French mathematician wrote a deceptively simple theorem in the margins of a book, adding that he had discovered a marvelous proof of it but lacked space to include it in the margin. He died without ever offering his proof, and mathematicians have been trying ever since to supply it.

Now, after thousands of claims of success that proved untrue, mathematicians say the daunting challenge, perhaps the most famous of unsolved mathematical problems, has at last been surmounted.

The problem is Fermat's last theorem, and its apparent conqueror is Dr. Andrew Wiles, a 40-year-old English mathematician who works at Princeton University. Dr. Wiles announced the result yesterday at the last of three lectures given over three days at Cambridge University in England.

Within a few minutes of the conclusion of his final lecture, computer mail messages were winging around the world as mathematicians alerted each other to the startling and almost wholly unexpected result.

Dr. Leonard Adelman of the University of Southern California said he received a message about an hour after Dr. Wiles's announcement. The frenzy is justified, he said. "It's the most exciting thing that's happened in—geez—maybe ever, in mathematics."

Impossible Is Possible

Mathematicians present at the lecture said they felt "an elation," said Dr. Kenneth Ribet of the University of California at Berkeley, in a telephone interview from Cambridge.

The theorem, an overarching statement about what solutions are possible for certain simple equations, was stated in 1637 by Pierre de Fermat, a 17th-century French mathematician and physicist. Many of the brightest minds in mathematics have struggled to find the proof ever since, and many have concluded that Fermat, contrary to his tantalizing claim, had probably failed to develop one despite his considerable mathematical ability.

343

With Dr. Wiles's result, Dr. Ribet said, "the mathematical landscape has changed." He explained: "You discover that things that seemed completely impossible are more of a reality. This changes the way you approach problems, what you think is possible."

Dr. Barry Mazur, a Harvard University mathematician, also reached by telephone in Cambridge, said: "A lot more is proved than Fermat's last theorem. One could envision a proof of a problem, no matter how celebrated, that had no implications. But this is just the reverse. This is the emergence of a technique that is visibly powerful. It's going to prove a lot more." Remember Pythagoras?

Fermat's last theorem has to do with equations of the form $xn + yn = zn$. The case where n is 2 is familiar as the Pythagorean theorem, which says that the squares of the lengths of two sides of a right-angled triangle equal the square of the length of the hypotenuse. One such equation is $32 + 42 = 52$, since $9 + 16 = 25$.

Fermat's last theorem states that there are no solutions to such equations when n is a whole number greater than 2. This means, for instance, that it would be impossible to find any whole numbers x, y and z such that $x3 + y3 = z3$. Thus $33 + 43 (27 + 64) = 91$, which is not the cube of any whole number.

Mathematicians in the United States said that the stature of Dr. Wiles and the imprimatur of the experts who heard his lectures, especially Dr. Ribet and Dr. Mazur, convinced them that the new proof was very likely to be right. In addition, they said, the logic of the proof is persuasive because it is built on a carefully developed edifice of mathematics that goes back more than 30 years and is widely accepted.

Experts cautioned that Dr. Wiles could, of course, have made some subtle misstep. Dr. Harold M. Edwards, a mathematician at the Courant Institute of Mathematical Sciences in New York, said that until the proof was published in a mathematical journal, which could take a year, and until it is checked many times, there is always a chance it is wrong. The author of a book on Fermat's last theorem, Dr. Edwards noted that "even good mathematicians have had false proofs."

Luring the World's "Cranks"

But even he said that Dr. Wiles's proof sounded like the real thing and "has to be taken very seriously."

Despite the apparent simplicity of the theorem, proving it was so hard that in 1815 and in again 1860, the French Academy of Sciences offered a gold medal and 300 francs to anyone who could solve it. In 1908, the German Academy of Sciences offered a prize of 100,000 marks for a proof that the theorem was correct. The prize, which still stands but has been reduced to 7,500 marks, about $4,400, has attracted the world's "cranks," Dr. Edwards said. When the Germans said the proof had to be published, "the cranks began publishing their solutions in the vanity press," he said, yielding thousands of booklets. The Germans told him they would even award the prize for a proof that the theorem was not true, Dr. Edwards added, saying that they "would be so overjoyed that they wouldn't have to read through these submissions."

But it was not just amateurs whose imagination was captured by the enigma. Famous mathematicians, too, spent years on it. Others avoided it for fear of being sucked into a quagmire. One mathematical genius, David Hilbert, said in 1920 that he would not work on it because "before beginning I should put in three years of intensive study, and I haven't that much time to spend on a probable failure."

Mathematicians armed with computers have shown that Fermat's theorem holds true up to very high numbers. But that falls well short of a general proof.

Tortuous Path to Proof

Dr. Ribet said that 20th-century work on the problem had begun to grow ever more divorced from Fermat's equations. "Over the last 60 years, people in number theory have forged an incredible number of tools to deal with simple problems like this," he said. Eventually, "people lost day-to-day contact with the old problems and were preoccupied with the objects they created," he said.

Dr. Wiles's proof draws on many of these mathematical tools but also "completes a chain of ideas," said Dr. Nicholas Katz of Princeton University. The work leading to the proof began in 1954, when the late Japanese mathematician Yutaka Taniyama made a conjecture about mathematical objects called elliptic curves. That conjecture was refined by Dr. Goro Shimura of Princeton University a few years later. But for decades, Dr. Katz said, mathematicians had no idea that this had any relationship to Fermat's last theorem. "They seemed to be on different planets," he said.

In the mid-'80s, Dr. Gerhard Frey of the University of the Saarland in Germany "came up with a very strange, very simple connection between the Taniyama conjecture and Fermat's last theorem," Dr. Katz said. "It gave a sort of rough idea that if you knew Taniyama's conjecture you would in fact know Fermat's last theorem," he explained. In 1987, Dr. Ribet proved the connection. Now, Dr. Wiles has shown that a form of the Taniyama conjecture is true and that this implies that Fermat's last theorem must be true.

"One of the things that's most remarkable about the fact that Fermat's last theorem is proven is the incredibly roundabout path that led to it," Dr. Katz said.

Another remarkable aspect is that such a seemingly simple problem would require such sophisticated and highly specialized mathematics for its proof. Dr. Ribet estimated that a tenth of one percent of mathematicians could understand Dr. Wiles's work because the mathematics is so technical. "You have to know a lot about modular forms and algebraic geometry," he said. "You have to have followed the subject very closely."

The general idea behind Dr. Wiles's proof was to associate an elliptic curve, which is a mathematical object that looks something like the surface of a doughnut, with an equation of Fermat's theorem. If the theorem were false and there were indeed solutions to the Fermat equations, a peculiar curve would result. The proof hinged on showing that such a curve could not exist.

Dr. Wiles, who has told colleagues that he is reluctant to speak to the press, could not be reached yesterday. Dr. Ribet, who described Dr. Wiles as shy, said he had been asked to speak for him.

Dr. Ribet said it took Dr. Wiles seven years to solve the problem. He had a solution for a special case of the conjecture two years ago, Dr. Ribet said, but told no one. "It didn't give him enough and he felt very discouraged by it," he said.

Dr. Wiles presented his results this week at a small conference in Cambridge, England, his birthplace, on "P-adic Galois Representations, Iwasawa Theory and the Tamagawa Numbers of Motives." He gave a lecture a day on Monday, Tuesday and Wednesday with the title "Modular Forms, Elliptic Curves and Galois Representations." There was no hint in the title that Fermat's last theorem would be discussed, Dr. Ribet said.

"As Wiles began his lectures, there was more and more speculation about what it was going to be," Dr. Ribet said. The audience of specialists in these arcane fields swelled from about 40 on the first day to 60 yesterday. Finally, at the end of his third lecture, Dr. Wiles concluded that he had proved a general case of the Taniyama conjecture. Then, seemingly as an afterthought, he noted that that meant that Fermat's last theorem was true. Q.E.D.

People raised their cameras and snapped pictures of this historic moment, Dr. Ribet said. Then "there was a warm round of applause, followed by a couple of questions and another warm round of applause," he added.

"I had to give the next lecture," Dr. Ribet said. "It was something incredibly mundane." Since mathematicians are "a pretty well behaved bunch," they listened politely. But, he said, it was hard to concentrate. "Most people in the room, including me, were incredibly shell-shocked," he said.

—June 24, 1993

The Spies' Code and How It Broke

By GEORGE JOHNSON

To the human brain, with its insatiable hunger for order, nothing is more disorienting than randomness. Soviet cryptographers knew this well when they set out to devise a code for communicating with the spy ring that included Julius and Ethel Rosenberg. With a system as simple as it was ingenious, they tried to insure that any message intercepted by United States intelligence agents would seem as meaningless as the snow on a television set tuned to an empty station.

But pure, unadulterated randomness can be extremely difficult to manufacture. As was revealed last week after decades of secrecy, in a ceremony at the Central Intelligence Agency headquarters, the Russians suffered from a lapse in quality control. They inadvertently let some pattern find its way into their scrambled codes, a loose thread that allowed American code breakers to unravel the scheme slowly.

The fine details of the Soviet encryption remain among the secrets of the National Security Agency, America's premier decoding service. But the principle behind the system, called a "one-time pad," has been known to cryptologists for years.

One begins with an alphabetical list of words or phrases likely to be needed in messages. These are numbered sequentially. Suppose that "Antenna," an early code name for Julius Rosenberg (later changed to "Liberal"), was assigned the number 2222. If the next item on the list was "anti-tank," it would be 2223, and "Anton," the code name for the Rosenbergs' KGB handler, Leonid Kvasnikov, would be 2224. In such a system, a message can be converted into a sequence of numbers and decoded by anyone with a copy of the translation table.

Names not on the list could be spelled out. A code number—9953, say— would tell the recipient that the following numbers stood for individual letters encrypted according to some agreed-upon scheme. Then 9954 could be a signal to stop spelling and pop back up to the dominant system in which whole words or phrases are assigned numbers.

If that is all there were to the code, even someone without the table might be able to crack it. One might profitably assume that the most frequent pattern of numbers probably represented the period or full stop. Articles like "a" and "the" would be among the next most common patterns. In many languages, the next pattern after a period would probably be a noun. With some lucky guesses, sophisticated statistical analysis and a lot of trial and error, meaning could be squeezed from the noise.

Code upon Code

Hoping to guard against this possibility, the Russian cryptographers added another layer of obfuscation. After using the table to translate the message into a string of digits, they disguised it further by adding to it a long random number. The result would also be a random number, patternless and theoretically indecipherable.

The message would be decodable by the Russians because sender and receiver each knew the random number used in the encoding scheme. If the sender wanted to encrypt "Antenna," he would translate it into 2222, then take out a pad imprinted with the random number key and copy down the first four digits, perhaps 3913. Adding the two numbers would produce 6135. Then he would move onto the next part of the message, adding it to the next digits on the pad. Once they had been used, the random numbers would be discarded—hence the name "one-time pad."

To decode the message, the recipient would take out his random number pad, copy the appropriate digits and subtract them from the message to recover the original number string. Proceeding like this, always carefully keeping their place on the pad, sender and receiver would be able to read dispatches that to anyone intercepting them would look like pure noise. While "Antenna" might be 3913 in one sentence, in the next it might be 4710. Since there is no structure to the key, there are none of the patterns cryptanalysts need to get a statistical foothold.

"Given a pure, perfect one-time system, you're not going to break it," said David Kahn, visiting historian at the NSA's Center for Cryptologic History and author of *The Codebreakers* (Macmillan, 1967). Even if the message were short enough for an intelligence agent to systematically subtract from it every possible number string, the result would be meaningless. "You would simply find that you had generated every possible message in every possible language with no way of telling which one was correct," Mr. Kahn said.

But no system is foolproof. First, generating a truly random number is harder than it sounds. Flipping a coin produces a random pattern containing an equal number of heads and tails—but only if the coin is perfectly balanced. More likely, differences in the engravings could make one side heavier than the other. The result would be deviations from randomness that might allow an observer to tell from the record of a coin toss whether it was more likely generated by a nickel or a dime. Similarly, if the code sender's random number generator is flawed, there might be enough order in the message for cryptanalysts to reconstruct the key.

Asking for Infinity

As revealed by American cryptologists last week, the Russians' crucial flaw was much more trivial. The problem with the one-time pad is that it depends on generating a number that is, as Mr. Kahn put it, not only absolutely random but endless. In practice, the number cannot be infinite, of course, but it must be long enough to encode every possible message that will conceivably be sent over a channel.

As traffic between Moscow and the KGB office in New York increased in volume, the Russians apparently ran out of numbers and committed the cryptographer's cardinal sin. They repeated themselves, betraying details of Soviet espionage efforts on American soil. The spell of randomness was broken, and meaning began seeping in.

—July 16, 1995

The Mighty Mathematician
You've Never Heard Of

By NATALIE ANGIER

Scientists are a famously anonymous lot, but few can match in the depths of her perverse and unmerited obscurity the 20th-century mathematical genius Amalie Noether.

Albert Einstein called her the most "significant" and "creative" female mathematician of all time, and others of her contemporaries were inclined to drop the modification by sex. She invented a theorem that united with magisterial concision two conceptual pillars of physics: symmetry in nature and the universal laws of conservation. Some consider Noether's theorem, as it is now called, as important as Einstein's theory of relativity; it undergirds much of today's vanguard research in physics, including the hunt for the almighty Higgs boson.

Yet Noether herself remains utterly unknown, not only to the general public, but to many members of the scientific community as well.

When Dave Goldberg, a physicist at Drexel University who has written about her work, recently took a little "Noetherpoll" of several dozen colleagues, students and online followers, he was taken aback by the results. "Surprisingly few could say exactly who she was or why she was important," he said. "A few others knew her name but couldn't recall what she'd done, and the majority had never heard of her."

Noether (pronounced NER-ter) was born in Erlangen, Germany, 130 years ago this month. So it's a fine time to counter the chronic neglect and celebrate the life and work of a brilliant theorist whose unshakable number love and irrationally robust sense of humor helped her overcome severe handicaps—first, being female in Germany at a time when most German universities didn't accept female students or hire female professors, and then being a Jewish pacifist in the midst of the Nazis' rise to power.

Through it all, Noether was a highly prolific mathematician, publishing groundbreaking papers, sometimes under a man's name, in rarefied fields of abstract algebra and ring theory. And when she applied her equations to the universe around her, she discovered some of its basic rules, like how time and energy are related, and why it is, as the physicist Lee Smolin of the Perimeter Institute put it, "that riding a bicycle is safe."

Ransom Stephens, a physicist and novelist who has lectured widely on Noether, said, "You can make a strong case that her theorem is the backbone on which all of modern physics is built."

Noether came from a mathematical family. Her father was a distinguished math professor at the universities of Heidelberg and Erlangen, and her brother Fritz won some renown as an applied mathematician. Emmy, as she was known throughout her life, started out studying English, French and piano—subjects more socially acceptable for a girl—but her interests soon turned to math. Barred from matriculating formally at the University of Erlangen, Emmy simply audited all the courses, and she ended up doing so well on her final exams that she was granted the equivalent of a bachelor's degree.

She went on to graduate school at the University of Göttingen, where she earned her doctorate summa cum laude and met many of the leading mathematicians of the day, including David Hilbert and Felix Klein, who did for the bottle what August Ferdinand Möbius had done for the strip. Noether's brilliance was obvious to all who worked with her, and her male mentors repeatedly took up her cause, seeking to find her a teaching position—better still, one that paid.

"I do not see that the sex of the candidate is an argument against her," Hilbert said indignantly to the administration at Göttingen, where he sought to have Noether appointed as the equivalent of an associate professor. "After all, we are a university, not a bathhouse." Hilbert failed to make his case, so instead brought her on staff as a more or less permanent "guest lecturer"; and Noether, fittingly enough, later took up swimming at a men-only pool.

At Göttingen, she pursued her passion for mathematical invariance, the study of numbers that can be manipulated in various ways and still remain constant. In the relationship between a star and its planet, for example, the shape and radius of the planetary orbit may change, but the gravitational attraction conjoining one to the other remains the same—and there's your invariance.

In 1915 Einstein published his general theory of relativity. The Göttingen math department fell "head over ear" with it, in the words of one observer, and Noether began applying her invariance work to some of the complexities of the theory. That exercise eventually inspired her to formulate what is now called Noether's theorem, an expression of the deep tie between the underlying geometry of the universe and the behavior of the mass and energy that call the universe home.

What the revolutionary theorem says, in cartoon essence, is the following: Wherever you find some sort of symmetry in nature, some predictability or homogeneity of parts, you'll find lurking in the background a corresponding conservation—of momentum, electric charge, energy or the like. If a bicycle wheel is radially symmetric, if you can spin it on its axis and it still looks the same in all directions, well, then, that symmetric translation must yield a corresponding conservation. By applying the principles and calculations embodied in Noether's theorem, you'll see it is angular momentum, the Newtonian impulse that keeps bicyclists upright and on the move.

Some of the relationships to pop out of the theorem are startling, the most profound one linking time and energy. Noether's theorem shows that a symmetry of time—like the fact that whether you throw a ball in the air tomorrow or make the same toss next week will have no effect on the ball's trajectory—is directly related to the conservation of energy, our old homily that energy can be neither created nor destroyed but merely changes form.

The connections that Noether forged are "critical" to modern physics, said Lisa Randall, a professor of theoretical particle physics and cosmology at Harvard. "Energy, momentum and other quantities we take for granted gain meaning and even greater value when we understand how these quantities follow from symmetry in time and space."

Dr. Randall, the author of the newly published *Knocking on Heaven's Door,* recalled the moment in college when she happened to learn that the author of Noether's theorem was a she. "It was striking and even exciting and inspirational," Dr. Randall said, admitting, "I was surprised by my reaction."

For her part, Noether left little record of how she felt about the difficulties she faced as a woman, or of her personal and emotional life generally. She never married, and if she had love affairs she didn't trumpet them. After meeting the young Czech math star Olga Taussky in 1930, Noether told friends how happy she was that women were finally gaining acceptance in the field, but she herself had so few female students that her many devoted pupils were known around town as Noether's boys.

Noether lived for math and cared nothing for housework or possessions, and if her long, unruly hair began falling from its pins as she talked excitedly about math, she let it fall. She laughed often and in photos is always smiling.

When a couple of students started showing up to class wearing Hitler's brown shirts, she laughed at that, too. But not for long. Noether was one of the first Jewish scientists to be fired from her post and forced to flee Germany. In 1933, with the help of Einstein, she was given a job at Bryn Mawr College, where she said she felt deeply appreciated as she never had been in Germany.

That didn't last long, either. Only 18 months after her arrival in the United States, at the age of 53, Noether was operated on for an ovarian cyst, and died within days.

—March 27, 2012

The Life of Pi, and Other Infinities

On this day that fetishizes finitude, that reminds us how rapidly our own earthly time share is shrinking, allow me to offer the modest comfort of infinities.

Yes, infinities, plural. The popular notion of infinity may be of a monolithic totality, the ultimate, unbounded big tent that goes on forever and subsumes everything in its path—time, the cosmos, your complete collection of old Playbills. Yet in the ever-evolving view of scientists, philosophers and other scholars, there really is no single, implacable entity called infinity.

Instead, there are infinities, multiplicities of the limit-free that come in a vast variety of shapes, sizes, purposes and charms. Some are tailored for mathematics, some for cosmology, others for theology; some are of such recent vintage their fontanels still feel soft. There are flat infinities, hunchback infinities, bubbling infinities, hyperboloid infinities. There are infinitely large sets of one kind of number, and even bigger, infinitely large sets of another kind of number.

There are the infinities of the everyday, as exemplified by the figure of pi, with its endless post-decimal tail of nonrepeating digits, and how about if we just round it off to 3.14159 and then serve pie on March 14 at 1:59 p.m.? Another stalwart of infinity shows up in the mathematics that gave us modernity: calculus.

"All the key concepts of calculus build on infinite processes of one form or another that take limits out to infinity," said Steven Strogatz, author of the recent book *The Joy of x: A Guided Tour of Math, From One to Infinity* and a professor of applied mathematics at Cornell. In calculus, he added, "infinity is your friend."

Yet worthy friends can come in prickly packages, and mathematicians have learned to handle infinity with care.

"Mathematicians find the concept of infinity so useful, but it can be quite subtle and quite dangerous," said Ian Stewart, a mathematics researcher at the University of Warwick in England and the author of *Visions of Infinity*, the latest of many books. "If you treat infinity like a normal number, you can come up with all sorts of nonsense, like saying, infinity plus one is equal to infinity, and now we subtract infinity from each side and suddenly naught equals one. You can't be freewheeling in your use of infinity."

355

Then again, a very different sort of infinity may well be freewheeling you. Based on recent studies of the cosmic microwave afterglow of the Big Bang, with which our known universe began 13.7 billion years ago, many cosmologists now believe that this observable universe is just a tiny, if relentlessly expanding, patch of space-time embedded in a greater universal fabric that is, in a profound sense, infinite. It may be an infinitely large monoverse, or it may be an infinite bubble bath of infinitely budding and inflating multiverses, but infinite it is, and the implications of that infinity are appropriately huge.

"If you take a finite physical system and a finite set of states, and you have an infinite universe in which to sample them, to randomly explore all the possibilities, you will get duplicates," said Anthony Aguirre, an associate professor of physics who studies theoretical cosmology at the University of California, Santa Cruz.

Not just rough copies, either. "If the universe is big enough, you can go all the way," Dr. Aguirre said. "If I ask, will there be a planet like Earth with a person in Santa Cruz sitting at this colored desk, with every atom, every wave function exactly the same, if the universe is infinite the answer has to be yes."

In short, your doppelgangers may be out there and many variants, too, some with much better hair who can play Bach like Glenn Gould. A far less savory thought: There could be a configuration, Dr. Aguirre said, "where the Nazis won the war."

Given infinity's potential for troublemaking, it's small wonder the ancient Greeks abhorred the very notion of it.

"They viewed it with suspicion and hostility," said A. W. Moore, professor of philosophy at Oxford University and the author of *The Infinite* (1990). The Greeks wildly favored tidy rational numbers that, by definition, can be defined as a ratio, or fraction—the way 0.75 equals ¾ and you're done with it—over patternless infinitums like the square root of 2.

On Pythagoras' Table of Opposites, "the finite" was listed along with masculinity and other good things in life, while "the infinite" topped the column of bad traits like femininity. "They saw it as a cosmic fight," Dr. Moore said, "with the finite constantly having to subjugate the infinite."

Aristotle helped put an end to the rampant infiniphobia by drawing a distinction between what he called "actual" infinity, something that would exist all at once, at a given moment—which he declared an impossibility—and "potential" infinity, which would unfold over time and which he deemed perfectly intelligible.

As a result, Dr. Moore said, "Aristotle believed in finite space and infinite time," and his ideas held sway for the next 2,000 years.

Newton and Leibniz began monkeying with notions of infinity when they invented calculus, which solves tricky problems of planetary motions and accelerating bodies by essentially breaking down curved orbits and changing velocities into infinite series of tiny straight lines and tiny uniform motions. "It turns out to be an incredibly powerful tool if you think of the world as being infinitely divisible," Dr. Strogatz said.

In the late 19th century, the great German mathematician Georg Cantor took on infinity not as a means to an end, but as a subject worthy of rigorous study in itself. He demonstrated that there are many kinds of infinite sets, and some infinities are bigger than others. Hard as it may be to swallow, the set of all the possible decimal numbers between 1 and 2, being unlistable, turns out to be a bigger infinity than the set of all whole numbers from 1 to forever, which in principle can be listed.

In fact, many of Cantor's contemporaries didn't swallow, dismissing him as "a scientific charlatan," "laughable" and "wrong." Cantor died depressed and impoverished, but today his set theory is a flourishing branch of mathematics relevant to the study of large, chaotic systems like the weather, the economy and human stupidity.

With his majestic theory of relativity, Einstein knitted together time and space, quashing old Aristotelian distinctions between actual and potential infinity and ushering in the contemporary era of infinity seeking. Another advance came in the 1980s, when Alan Guth introduced the idea of cosmic inflation, a kind of vacuum energy that vastly expanded the size of the universe soon after its fiery birth.

New theories suggest that such inflation may not have been a one-shot event, but rather part of a runaway process called eternal inflation, an infinite ballooning and bubbling outward of this and possibly other universes.

Relativity and inflation theory, said Dr. Aguirre, "allow us to conceptualize things that would have seemed impossible before." Time can be twisted, he said, "so from one point of view the universe is a finite thing that is growing into something infinite if you wait forever, but from another point of view it's always infinite."

Or maybe the universe is like Jorge Luis Borges's fastidiously imagined Library of Babel, composed of interminable numbers of hexagonal galleries with polished surfaces that "feign and promise infinity."

Or like the multiverse as envisioned in Tibetan Buddhism, "a vast system of 10^{59} universes, that together are called a Buddha Field," said Jonathan C. Gold, who studies Buddhist philosophy at Princeton.

The finite is nested within the infinite, and somewhere across the glittering, howling universal sample space of Buddha Field or Babel, your doppelganger is hard at the keyboard, playing a Bach toccata.

—January 1, 2013

Don't Expect Math to Make Sense

By MANIL SURI

Each year, March 14 is Pi Day, in honor of the mathematical constant. Saturday is the once-in-a-century event when the year, '15, brings the full date in line with the first five digits of pi's decimal expansion—3.1415. Typical celebrations revolve around eating pies and composing "pi-kus" (haikus with three syllables in the first line, one in the second and four in the third). But perhaps a better way to commemorate the day is by trying to grasp what pi truly is, and why it remains so significant.

Pi is irrational, meaning it cannot be expressed as the ratio of two whole numbers. There is no way to write it down exactly: Its decimals continue endlessly without ever settling into a repeating pattern. No less an authority than Pythagoras repudiated the existence of such numbers, declaring them incompatible with an intelligently designed universe.

And yet pi, being the ratio of a circle's circumference to its diameter, is manifested all around us. For instance, the meandering length of a gently sloping river between source and mouth approaches, on average, pi times its straight-line distance. Pi reminds us that the universe is what it is, that it doesn't subscribe to our ideas of mathematical convenience.

Early mathematicians realized pi's usefulness in calculating areas, which is why they spent so much effort trying to dig its digits out. Archimedes used 96-sided polygons to painstakingly approximate the circle and showed that pi lay between 223/71 and 22/7. By the time Madhava (in India, around 1400) calculated pi to over 10 decimal places using his groundbreaking infinite series (which regrettably bears Leibniz's name), it was already more than accurate enough to address all practical applications. Pursuing pi further had essentially become a mathematical challenge.

With the advent of computers, pi offered a proving ground for successively faster models. But eventually, breathless headlines about newly cracked digits became less compelling, and the big players moved on. Recent records (currently in the trillions of digits) have mostly been set on custom-built personal computers. The history of pi illustrates how far computing has progressed, and how much we now take it for granted.

So what use have all those digits been put to? Statistical tests have suggested that not only are they random, but that any string of them occurs just as often as any other of the same length. This implies that, if you coded this article, or any other, as a numerical string, you could find it somewhere in the decimal expansion of pi. Of course, that's relatively meaningless, since you don't know where to find the material you want. An apt metaphor for an age when we are being asphyxiated by mushrooming clouds of information.

But pi's infinite randomness can also be seen more as richness. What amazes, then, is the possibility that such profusion can come from a rule so simple: circumference divided by diameter. This is characteristic of mathematics, whereby elementary formulas can give rise to surprisingly varied phenomena. For instance, the humble quadratic can be used to model everything from the growth of bacterial populations to the manifestation of chaos. Pi makes us wonder if our universe's complexity emerges from similarly simple mathematical building blocks.

Pi also opens a window into a more uncharted universe, the one consisting of transcendental numbers, which exclude such common irrationals as square and cube roots. Pi is one of the few transcendentals we ever encounter. One may suspect that such numbers would be quite rare, but actually, the opposite is true. Out of the totality of numbers, almost all are transcendental. Pi reveals how limited human knowledge is, how there exist teeming realms we might never explore.

The combination of utility and mystery makes pi a perfect symbol for all of mathematics. Surely the ancients, had they understood pi better, would have deified it, just as they did the moon and the sun. They would have praised pi's immutability: Pi = 3.14159 . . . is one of the few absolutes that remain.

Or is it? The ratio of circumference to diameter might not be as fixed as we think. To understand why, imagine a circle drawn on the surface of a sphere. Its diameter, as measured along the bulging surface, will be greater than if the same circle is traced out on a flat sheet of paper. This observation might have been of only academic interest except for our inability, so far, to definitively determine whether the geometry of our universe is flat. If there is even a little curvature, then the value of pi, as defined by this ratio, is not what we think.

But pi, on cue, reminds us that it is an abstraction, like all else in mathematics. The perfect flat circle is impossible to realize in practice. An area calculated using pi will never exactly match the same area measured physically.

This is to be expected whenever we approximate reality using the idealizations of math.

Many decades ago, a teacher had me memorize another approximation, pi = 22/7. Except she told my class this was pi's exact value. Perhaps she feared we'd be as traumatized as Pythagoras by the idea of a non-fractional universe; more likely, she didn't want to confuse us. I wish we'd had Pi Day then, because pi = 22/7 was a misconception I carried all the way to college. Since then I've learned that it's only when we try to stretch our minds around mathematics' enigmas that true understanding can set in.

—*March 14, 2015*

CHAPTER 9

Medicine: Outbreaks and Breakthroughs

Progress of the Germ Theory

The cable announced recently that Robert Koch, who went from Germany to study the Egyptian cholera epidemic, has found what many medical men of several nations have long looked for in vain—the cause of the disease. A predecessor of the same nationality during the last epidemic injected choleraic blood into his veins for the purposes of study. Within six hours he died while examining his own blood with the microscope. Only a few weeks ago the telegraph reported a similar fate regarding one of the commission dispatched under the management of Pasteur, the famous Frenchman, but science was none the wiser. Now, however, a man's eye has again seen the parasite, a living thread-like organism, which has slain greater numbers of men in brief periods of time than, perhaps, any other known cause.

It is surprising to consider the number of similar discoveries recently. It is believed that the parasite of yellow fever has been seen, and regarding this it is ominously asserted that it flourishes even in the earth, as is the case with the disease germ most fatal to animals. An offset against this persistent vitality is the discovery of alternative methods of inoculation. Pasteur has familiarized the medical world with the preventive virtues of an attenuated virus. The Brazilian commission has successfully applied this method to the *Micrococcus xanthogenicus*, and it has further discovered a distinct germ fatal to the yellow fever germ. This killing of one parasite with another is a new departure in the germ theory, and opens to the study of microscopists a fertile field. During the last epizooty in Paris, Pasteur found in the horses' nostrils a living speck which produced pure typhoid in rabbits. The same brilliant scientist has demonstrated that hydrophobia is due to a parasite peculiar to the saliva. All these are too recent to be found in the books. Back of them are a number of equally striking and perfectly well-established discoveries. Abscesses in general, and malignant pustules in particular, are due to the deadly microbes. Malarial and puerperal fevers, leprosy, diphtheria, and erysipelas are suspected to belong to this class of diseases. And Koch, who has found the cholera germ, previously found the cause of the death of one-seventh of humanity, i.e., the microscopic cause of tubercular diseases. Pasteur's discoveries regarding anthrax, the silk-worm disease, and chicken cholera are worth mentioning, not especially because of their economic worth, but because they strengthen the analogy of such diseases in men. Briefly and generally speaking, all epidemic and contagious diseases are under suspicion of being caused by organisms visible only under the microscope, and only then to a trained observer.

To unprofessional people it will seem that the doctors would do better to discuss such a theory as this than quarrel about questions of medical ethics. If the theory be established treatment will be revolutionized. For instance, Koch's bacillus of consumption can live only between the temperatures of 86° and 104°, and like limitations may readily be supposed regarding other bacteria. Now, within bounds the temperature of the human body can be raised or lowered. For the imagination—which usually precedes calm discovery—it is, then, an easy step to believe that here is a hint as to how disease germs may be enfeebled and crippled, if not killed. It is not necessary to suggest to the sick, or to those who profess to heal them, what an immense field is opened here—if the theory is established, and there's the rub.

Dr. Rollin R. Gregg, of Buffalo, after laboratory experiments extending over many months, has recently boldly challenged the theory in toto. And not one, but several, have asserted that Koch's tubercular germ is characteristic of all saliva, even the saliva of perfect health. Dr. Gregg declares that the so-called bacteria are merely fragments of fibrine. He boiled pure fibrine. He boiled, baked, and pulverized healthy blood. He rotted other blood under warmth. All the processes resulted in minute forms which moved violently and were identical. Now, boiling heat, it is agreed, is positively fatal to bacteria. Dr. Gregg's conclusion is that the bacteria of disease are delusions of the observer, and he asks them to repeat their experiments. They will scarcely ask him to repeat his, for all he says may not be inconsistent with their position. But they will probably ask him to supplement his studies by others similar to theirs. They will request him to produce a disease with them. And they will request him to inquire regarding the unfailing existence of that species of bacteria in spontaneous cases of the disease. All this they have done. If he fails to do it, their response will be that he never saw the bacteria of disease, but mistook for them forms which he declared "identical."

—October 28, 1883

Ehrlich's Remedy a Medical Wonder

The first authentic information given to an American newspaper by those associated with Dr. Paul Ehrlich, Director of the Royal Institute for Experimental Therapeutics at Frankfort-on-the-Main, concerning Prof. Ehrlich's remarkable discovery of a specific for a widespread blood disease of great antiquity, deemed for centuries well-nigh unconquerable, has been furnished to *The Times* by Dr. Lewis Hart Marks. Dr. Marks, who is an American, is first assistant to Prof. Ehrlich.

The chemical name of this curative agent is dioxydiamidoarsenobenzol. It is popularly called "606" by Prof. Ehrlich, for the reason that it was the 606th preparation compounded by him in an effort to find a cure. The first 605 preparations proved to be failures.

A small supply of the drug has been placed in the hands of Dr. Simon Flexner, director of the Rockefeller Institute for Medical Research, in this city, and in the hands of Prof. L. P. Barker of Johns Hopkins University, Baltimore. Within a few days, however, limited supplies of the curative agent, which is not a serum but a definite chemical compound, will be furnished to Prof. Anders of the University of Pennsylvania, Philadelphia; Prof. Simon, Baltimore, and Dr. William Kohlmann, New Orleans. Prof. Ehrlich and his associates request prospective patients in this country to address themselves by letter to the nearest of these doctors, as they will receive more prompt and satisfactory replies than by addressing the physicians at Frankfort.

For the reason that Prof. Ehrlich's relations with America have been of the most friendly nature, and also because he owes a debt of gratitude to John D. Rockefeller, who made him a grant of $10,000 last year, and to the Rockefeller Institute he intends, as soon as possible, to send a large quantity of the drug to the Rockefeller Institute to be distributed among the foremost physicians in the United States.

How soon Prof. Ehrlich will be able to carry out this plan is not known, as the demand for "606" exceeds the supply by something like 1,000 percent, and the clinics in Europe which are trying out the drug scientifically must be supplied first.

The drug will be put on the market in this country about Nov. 1 by the American representatives of the German firm which manufactures "606." After that date it can be obtained by any physician. At present, cases are being treated in all of the large European clinics, but Dr. Marks advises prospective patients in this country to await the liberal supply of the drug in America.

Dr. Marks desires to take this opportunity to correct certain misrepresentations due to the sudden publicity given to Prof. Ehrlich's discovery of the new drug, and so save many persons from grave disappointment. Prof. Ehrlich and his assistants also hope to gain much valuable time for themselves by this announcement as it is becoming daily more impossible for them to reply to all the letters of inquiry they receive.

In the first place Dr. Marks wishes to impress upon the public the fact that "606" is purely and only a specific for those diseases caused by the presence in the blood of the parasites known as spirilla [syphilis, caused by spirochete bacteria]. Of these, the most important and widely spread is the blood disease, unconquerable for centuries. . . . The drug is not a cure for various blood diseases, but is a specific for the one mentioned. It will not cure paralytics of long standing, or cancer, or cases of pernicious anemia, or anything except the disease described and such closely related diseases as, for example, recurrent fever, which is common in Russia and Africa.

The effect on the disease for which it is a specific is described by Dr. Marks as wonderful. The treatment consists simply of one hypodermic injection, but this must be made by a physician. It is impossible for patients to treat themselves.

No one can say yet if the cure resulting from the one injection is permanent. It will be years before that can be definitely determined. It is not yet known exactly how large a dose a man can stand. The doses given so far have been relatively small.

It can be said, Dr. Marks asserts, that "606" is the most remarkable drug known to the medical profession, because it acts with unprecedented and amazing rapidity and because it immediately cures certain lesions which all previous modes of treatment were unable to influence at all.

Prof. Ehrlich has records, up to the present, of nearly 6,000 cases which have been treated. In 95 percent of the cases the results are all that could be desired. Some of the few failures were due to the use of too small doses at the beginning. There are no ill effects at all to be feared from the use of the drug if it is properly used by an experienced physician. Prof. Ehrlich has given away 17,000 doses already.

Dr. Marks wishes the public to understand that it is impossible for Prof. Ehrlich and his assistants to reply to all the letters received from physicians and prospective patients in America and elsewhere.

—*September 11, 1910*

The Mold That Fights for the Life of Man

By DANIEL SCHWARZ

U sually the first thing doctors say when the subject of penicillin is brought up is, "Please don't call it a 'wonder drug.'" But then they proceed to grow enthusiastic in spite of themselves: "A remarkably effective antibacterial agent" . . . "better than the sulfa drugs" . . . "a few months ago in a case like this we'd have been helpless."

Penicillin inspires such enthusiasm partly because it has come on the medical scene at exactly the right time. Army doctors have always searched for a drug that would cure infections in open wounds. The sulfa drugs help, but they aren't always completely effective against pus-forming bacteria. Penicillin has cleared infected wounds that had defied all the usual treatments.

In addition, penicillin has proved extremely useful in treating types of pneumonia that have resisted all other treatment: boils and abscesses, infected burns, the bone disease called osteomyelitis and dozens of other less familiar illnesses. Given in large doses it even conquers the dread sub-acute bacterial endocarditis, formerly "100 percent fatal." Gonorrhea can't stand up against it—all traces of the germ vanish within a few days—and it shows promise of being effective in treating syphilis, too, though experiments in that field haven't gone far enough to justify positive statements. Moreover, unlike the sulfa drugs, which are sometimes hard on the kidneys, penicillin has the great advantage of being practically nontoxic.

What is penicillin? A rare drug secreted by a greenish-blue mold similar to the familiar mold that forms on cheese, oranges that have spoiled, bread, etc. Molds are fungous growths. Fungi are nonflowering plants, of which mushrooms are the most familiar example. The particular kind of mold from which penicillin is secreted is called technically *Penicillium Chrysogenum notatum*. Its spores (reproductive bodies) are microscopic; they are carried by the wind.

This particular kind of *Penicillium* differs from other fungi in one important respect, as Dr. Alexander Fleming, an English bacteriologist, discovered by accident in 1928. While busy with research on influenza he left a culture plate full of *Staphylococcus* colonies out in the open. The next time he looked at the plate he noticed a spot of mold on it, not an uncommon experience, but instead of discarding the contaminated specimen, he studied it more closely. All around the mold, like a moat in a field of bacteria, was a clear ring. He looked at the ring under the microscope. It was completely free of bacteria. For some reason the bacteria died when they entered it. Evidently the *Penicillium* was giving off some substance that was death to bacteria.

What were Dr. Fleming's first thoughts? He himself reported them just the other day. "Nothing is more certain," he said, "than that when I saw the bacteria fading away, I had no suspicion that I had got a clue to the most powerful therapeutic substance yet used to defeat bacterial infections in the human body. But the appearance of that culture plate was such that I thought it should not be neglected."

Nor did Fleming neglect it. He made pure cultures of the mold, and most of the penicillin produced up to now has been obtained from the descendants of his original colony. He made test-tube experiments on many other types of bacteria, using the strongest solutions he could get of the strange factor *Penicillium* secreted, and got mixed results. But the thing that surprised him most of all was that the secretion did not damage blood corpuscles. He had been testing all kinds of antiseptics in human blood for years and had never before come across one that did not do more damage to blood corpuscles than it did to bacteria. "Here," as he said with great satisfaction, "was something to bite on."

Fleming gave the name "penicillin" to the mysterious X that *Penicillium* produced, published his results and hoped that someone would purify the active substance. But his work attracted little attention because it came at the wrong time—just at the period when the potentialities of the sulfa drugs had first begun to be understood and attention had turned back to Ehrlich's enthralling idea of fighting bacteria with man-made chemicals.

It was not until 1939, eleven years after Fleming's original discovery, that a group of British scientists at Oxford, led by Dr. H. W. Florey, professor of pathology, began work on penicillin in earnest. They knew that if they grew *Penicillium* in a liquid under favorable conditions something that killed bacteria was somehow added. So they grew *Penicillium* by the square yard and after a week or two poured off the liquid in which it grew and tried to extract from it the essential compound.

Extraction of one compound from a mixture of several can be very simple: For example, to extract the salt from salt and water all that has to be done is to boil off the water. Or to extract the fat from a piece of meat (a silly idea in times like these) you simply pour ether over the meat, let the ether dissolve the fat and then evaporate the ether from the fat-and-ether mixture. The problem of extracting penicillin was, and is, far more complex. Its properties were unknown, it proved to be exceedingly unstable, and it was mixed with a number of other organic materials any of which might have been a part of it or a necessary ally in its work.

So the job was difficult—probably it seemed hopeless at times—but, with the aid of a grant from the Rockefeller Foundation, the Oxford group of bacteriologists pushed it along and began to get results. In time they were able to make a concentrate over 1,000 times the strength of the original fluid. They tried it on mice and it had no poisonous effects. They shot deadly doses of bacteria into other mice, produced infections that were "incurable"—and then cured those infections with their concentrated penicillin. By 1941 they had a pure enough form of penicillin and enough evidence that it was nontoxic to justify experiments with it on human cases, people desperately, hopelessly ill, and it produced amazing results. They came to America and demonstrated their technique here.

From all this work some of the advantages of penicillin became clear. Other drugs that killed bacteria worked much faster than penicillin, but they had some nasty toxic effects. Even the sulfa drugs, despite their amazing results, had the disadvantage that a number of people were apparently allergic to the forms of sulfonamides known at the time and some persons even suffered permanent harm. But penicillin singled out the bacteria and left the patient alone. Furthermore, penicillin worked against some bacteria that the sulfa drugs could not cope with, in particular the pus-forming bacteria that gather in open cuts and wounds.

Then came Pearl Harbor. Medicine, like everything else, was mobilized and penicillin research and production, up to that time conducted only on a laboratory scale, was given an A-1 priority. Manufacture was strictly controlled; a number of chemical houses were encouraged to set up plants for the mass production of penicillin (there are now about twenty such plants in operation); twenty-two groups of investigators were promised supplies of penicillin to carry out experiments in hospitals; and the Army and Navy created penicillin units to do similar work in the armed forces.

Complete authority over every drop of penicillin produced for civilian use was given to a sort of penicillin director, Dr. Chester S. Keefer of Boston. Now requests come to him from doctors all over the country and he doles out the precious stuff on the basis of a set of principles set up by a group of doctors. If previous experience has shown that penicillin won't fight a particular kind of sickness, he is in duty bound to refuse a request for the slightest bit of it. But if penicillin seems likely to help, he rushes the necessary supply and keeps a check on the result.

While penicillin has probably been the most carefully rationed war essential of all, the output is being stepped up so rapidly that a surplus for civilian doctors may be available by next spring or summer. Details of how penicillin is being mass-produced are a military secret, but it isn't revealing anything to Hitler's technicians to say that three methods of growing the temperamental mold are in use: surface culture, bran culture (in which the mold is grown on bran moistened by a liquid nutrient) and submerged culture.

In submerged culture the mold is grown in huge, covered vats holding hundreds and perhaps thousands of gallons of nutrient fluid (the preferred culture medium is corn steep liquor). The necessary air is pumped through the fluid. After several days, when the mold has produced a sufficient supply of penicillin, the fluid is led through pipes to other containers, where it is concentrated at low temperatures and under high vacuums by a technique somewhat similar to that of drying blood plasma. It has been estimated that the cost of producing penicillin is $18,000 a pound, but the doses needed are so small that each costs only $2.

Penicillin must be concentrated some 20,000 times, or until it is completely dry, before it can be stored without deteriorating. It arrives at the hospital as a yellow or yellow-brown powder and is stored in refrigerators until it is needed. Then it is diluted and jabbed with a hypodermic needle into the muscles (not necessarily near the point of infection) or allowed to drip into a vein in a glucose mixture.

According to an estimate made early this year five gallons of culture fluid were needed to yield a single gram of penicillin. Current production is a secret, but it probably is still a relatively low percentage of the fluid treated. Fortunately the potency of penicillin is extremely high: it requires only one part in 25,000,000 to stop the growth of the pus-forming germ, *Staphylococcus aureus*, which causes boils.

Since penicillin is so powerful, why not simply lay some of the mold on a wound and let it secrete its healing drug? For several reasons. One is that the mold is dirty and would carry more infection than the penicillin could destroy. Another is that the mold is temperamental; it will grow almost anywhere, but it will not secrete penicillin unless conditions are exactly right. Even under what scientists consider perfect conditions, it sometimes balks. It grows, but it refuses to produce the drug.

One object of current penicillin research, and a most important one, is to synthesize the drug with known chemicals, as some of the vitamins have been synthesized. The first step, obviously, is to decide what penicillin is made of, itself an exceedingly difficult thing, and progress has been made on that line. Then, since many houses can be made with the same set of bricks, the particular molecular structure of penicillin will have to be worked out. Biological chemists are convinced that both these things will be done before very long.

Another thing that nobody knows yet is how penicillin does its job. In the case of the sulfa drugs one theory is that the drug is absorbed by the bacteria and that it then (somehow) prevents them from digesting an essential element in their diet (para-aminobenzoic acid). So they starve to death. Other antibacterial agents seem to work by upsetting the osmotic balance of the bacteria with this dire result: they absorb so much liquid that they get too big for their skins, like the bullfrog that tried to blow himself up to the size of a bull, and explode.

As for penicillin, it is known to work relatively slowly—in a matter of hours rather than minutes, as some other bactericides do. Therefore one hypothesis is that it doesn't kill the bacteria but simply upsets their process of reproduction. That is enough to do the job. If the total number of bacteria can be kept constant, the white blood cells can be relied on to clean them out. The trouble is that bacteria multiply exceedingly fast (it has been estimated that under proper conditions a single bacterium produces 1,000,000,000 descendants in fifteen hours) and if they aren't stopped they soon become so numerous that the body cannot cope with them.

The potentialities of penicillin seem even greater than its achievements. It hasn't yet been possible to explore all of penicillin's possibilities because the supply has been so limited and the need so great that even qualified researchers have had difficulty in getting it. Moreover, after penicillin is synthesized, it seems likely that variations of it will be worked out, just as they were in the case of the sulfa drugs, and these derivatives may do miracles as yet unthought of. Finally, the success of penicillin seems certain to lead to extensive research on similar properties of other fungi, a field whose possibilities have hardly been scratched.

But the doctors' warning—"Please don't call it a 'wonder drug'"—is worth remembering. The reason for the warning was explained the other day by a doctor in the "penicillin ward" at Halloran Hospital, the Army's impressive base hospital on Staten Island. "Penicillin is an amazing drug," he said, "but the danger is that people may think of it as a panacea. It isn't. It's only another medical tool.

"Look at the boys in this ward," he went on. "They've all had gunshot or shrapnel wounds of one sort or another. They were all brought here because their wounds became infected and the usual treatments, including the sulfa drugs, didn't work. As soon as they were given penicillin their wounds cleared up, their temperatures went down to normal and they began to eat again. Remarkable? Certainly. But that was only the beginning of the job. A four-inch piece of bone in this fellow's right tibia was shot away; we had to graft a new piece of bone there. Penicillin alone couldn't have made it possible for that boy to walk again.

"Don't misunderstand me. Without penicillin we might have had to wait six months before that wound cleared up. Meanwhile that boy's leg would have got weaker and weaker from disuse. With penicillin we could operate quickly so the results were far better. But you see now what I mean when I say it's no panacea."

—*January 2, 1944*

Pills Are Tested in Birth Control

By ROBERT K. PLUMB

Two tests of oral contraceptive tablets over the last three years by 1,000 women were reported yesterday to have given a higher degree of conception control than contraceptives now used. However, a third test, using different pills, did not produce the same unequivocal results.

Although the two tests were said to indicate that the tablets were effective, they are not regarded as the ideal method of birth control.

The contraceptive is a combination of two hormones. In the tests, women took the tablets for twenty days of their monthly cycle. Menstruation was normal. Pregnancies occurred when the medication was discontinued.

The compound inhibits the anterior pituitary gland from production of the gonadotrophic hormone. This hormone stimulates the ova to ripen.

Two compounds, Enovid of G. D. Searle & Co., and Norlutin of Parke Davis & Co., were used in the study.

830 Women Tested

In the journal *Science* for July 10, Dr. Gregory Pincus and his associates at the Worcester Foundation for Experimental Biology in Shrewsbury, Mass., reported on the use of tablets by 830 women in Puerto Rico and Haiti over a time interval covering 8,133 menstrual cycles.

When the tablets were taken as directed, from the fifth to the twenty-fourth day of the cycle, the pregnancy rate for 100 woman-years was found to be 0.2 percent.

Dr. Alan F. Guttmacher, chairman of the medical committee of the Planned Parenthood Federation of America, said in comment on the results:

"At this point, we regard the new tablets as the first important step in the development of a physiological method of conception control. We do not feel that the present tablets are ideal. In ten years a better conception control method will be available, however."

The reported pregnancy rate when women did not miss taking a tablet is lower than the rate in conventional contraception, Dr. Guttmacher noted.

Dr. Pincus summarized his study by saying:

"For the period studied, the data appear to us to answer the following questions in the manner indicated:

"Is the method contraceptively effective? Yes.

"Does it cause any significant abnormalities of the menstrual cycle? No.

"Does it adversely affect the reproductive tract and adnexae [adjuncts]? No.

"Does it have physiologically adverse effects generally? No.

"Does it affect the sex life of the subjects adversely? No.

"Does it impair fertility upon cessation? No."

Better Agent Sought

Dr. Guttmacher noted, however, that a tablet that had to be taken daily could not be considered an ideal contraceptive. The present cost of the tablets is about 50 cents each, he said. What is sought is a pill that can be taken once or twice a year, and a compound that will not alter the body hormone chemistry in a major way.

Meanwhile, Dr. Joseph Goldzieher of the Southwest Foundation, San Antonio, Tex., said that a third combination tablet, manufactured by the Syntex Corporation, had produced "remarkable" results in a recent test with 188 women over a total of 600 months. None became pregnant and there were no side effects, Dr. Goldzieher said yesterday by telephone.

A third and different study of five tablets in contraceptive use was reported in the April 18 issue of the *Journal of the American Medical Association*. Patients experienced good and bad side effects in a study directed by Dr. Edward T. Tyler of the University of California at Los Angeles School of Medicine. Dr. Tyler found a pregnancy rate of 8.6 percent compared with about 4 percent for other contraceptive methods, he said.

—*July 10, 1959*

Brady's Recovery: Doctors Describe Dramatic Sequence of Lucky Moments

By LAWRENCE K. ALTMAN, M.D.

Six weeks ago today, White House Press Secretary James S. Brady was critically wounded by a shot in the head during an attempt on the life of President Reagan. The bullet destroyed a large segment of the right side of Mr. Brady's brain, and even his neurosurgeon, Dr. Arthur I. Kobrine, was initially pessimistic about Mr. Brady's chances for survival. Indeed, for a time television networks erroneously had declared Mr. Brady dead—as indeed he might be, were it not for a web of fortunate circumstances that Dr. Kobrine and others close to the case detailed as they reconstructed in recent days the crucial moments that contributed to what has so far been a remarkable recovery.

"Overwhelming luck" ranks highest in the list of those circumstances, according to Dr. Kobrine. First, critical time was saved because Mr. Brady did not have to wait for an ambulance. Secret Service agents sped him to a trauma center within minutes of the injury, applying a lesson that doctors learned from wars— get the injured patient to the doctor as quickly as possible. Mr. Brady would have died had he reached the hospital a few minutes later, Dr. Kobrine said.

Second, even more valuable time was saved because a brain surgeon happened to be in the emergency room and began appropriate therapy immediately upon Mr. Brady's arrival.

Third, the injury occurred in the early afternoon when all doctors and technicians needed for Mr. Brady's care were present in the hospital and not involved in the care of other patients. As a result, they were free to do the necessary tests instantly.

Fourth, under the circumstances, Mr. Brady was fortunate in being right-handed. The bullet damaged that section of the brain governing the motor functions of left-handed persons. The nerve impulses that control motor function normally cross over, in the pathways in the spinal cord, to activate muscles on the opposite side of the body. The bullet caused very little damage to the left side of Mr. Brady's brain, which also contain the centers governing his speech and thinking processes.

Accordingly, Dr. Kobrine said that if the bullet had traveled in the opposite direction—entering through his right eyebrow and going into the left brain, instead of entering through the left eyebrow as it did to pierce his right brain and

376

to stop near the ear—Mr. Brady might still have survived. But in that event, he probably "wouldn't be a functional human being," Dr. Kobrine said.

Recent medical advances were additional factors. Because Mr. Brady had just eaten lunch, he might have developed a fatal pneumonia if he had vomited and inhaled stomach contents during surgery. But Mr. Brady's doctors used the techniques of so-called "crash" anesthesia, designed to put the patient under in a hurry without vomiting.

Death also can occur suddenly when swelling from an injury forces the brain stem to bulge through the small opening at the base of the skull into the area where the spinal cord begins. The centers controlling heartbeat and breathing are in the brain stem.

Now doctors can reduce the amount of such swelling by immediately injecting mannitol, a drug that reduces brain pressure, steroids and other drugs, and by making such a patient hyperventilate. To do so, a tube is inserted into the windpipe and connected to a mechanical respirator. Those steps bought time while Dr. Kobrine's team operated to relieve the pressure.

Dr. Kobrine said that throughout the operation all four members of his neurosurgical team wore magnifying glasses, called loupes, so that they could use microsurgical techniques to help stop the bleeding from two arteries in Mr. Brady's brain and to preserve vital brain tissue. The bullet destroyed about 20 percent of his right cerebral hemisphere.

Technology played another role. CAT scan X-rays—computerized axial tomography machines, which offer a cross-sectional view of body parts and were unknown a decade ago, allowed Dr. Kobrine's team to assess the extent of Mr. Brady's head injuries with greater precision than possible with standard X-rays before the first operation and during the complications in the post-operative period.

Today, after Mr. Brady had read newspapers, watched television and sat up for the first time in two weeks in his room at George Washington University Hospital, Dr. Kobrine described his patient's recovery as "extraordinary"—although that recovery has been hampered by several life-threatening complications. The most serious indication of the favorable chances for Mr. Brady's recovery is that Mr. Reagan is holding his job open for his return. The offer is understood to be based partly on the medical opinion that Dr. Kobrine said he gave to White House officials. "Nothing now precludes Mr. Brady from eventually resuming his duties" as press secretary, Dr. Kobrine said in an interview, although he emphasized that he was not offering a prediction.

Mr. Brady "is not out of the woods yet," Dr. Kobrine said. After his most recent examination, he said that the best prospects are that the 40-year-old press secretary will "walk with a limp and a cane, his arm in a sling, and have essentially normal mental function."

Mr. Brady's memory for events in the distant past "is excellent—absolute and quick," Dr. Kobrine said. The neurosurgeon can test Mr. Brady's distant recall more accurately than in other patients because by chance both men grew up and went to colleges in Illinois at about the same time.

However, Mr. Brady's recall of more recent events was "fuzzy" at times, usually when he was fatigued, Dr. Kobrine said. Although Mr. Brady's brain was severely damaged by the bullet, it seems to have retained everything stored in the past. "His computer didn't crash," Dr. Kobrine said, adding that "for a while he may have some trouble storing new information. That doesn't bother me because that situation is almost always reversible. Part of the problem is that any patient who stays in the same room for so long loses track of time. The brain needs references that his hasn't had a chance to get since the shooting."

The bullet caused greatest damage to the brain cells that control the movements of Mr. Brady's left hand, and Dr. Kobrine said it was still too early to judge precisely how much function will return. One encouraging note is that during the past week Mr. Brady regained additional movement in his left leg. He has some voluntary motion in his left shoulder.

Mr. Brady's speech, vision, hearing, and thought processes were reported to be normal. However, his senses of smell and taste might be impaired.

No one knows just how much of the brain someone can spare without functional impairment. Clearly, however, the degree of recognizable impairment depends on the area of the brain destroyed—some regions being more crucial than others—and many people have survived bullet injuries to the brain.

Despite the severity of Mr. Brady's brain injury, he now usually begins his day by keeping abreast of the news. "He understands what he reads," Dr. Kobrine said, basing his opinion on his own frequent tests of Mr. Brady's memory of complex reports of budgetary and other issues.

Mr. Brady jokes with his doctors and nurses with the same sense of humor that Mrs. Brady says her husband had before the shooting, according to Dr. Kobrine. As the day passes, Mr. Brady's strength is sapped by the recuperative processes, and he naps frequently.

At other times, he talks with visitors. Any patient who stays in a hospital for long periods gets bored, and to relieve his boredom Mr. Brady plays with small hand-held electronic football games. His spirits are buoyed by looking at the 50-odd stuffed bears sent by strangers after learning of his nickname, "Bear."

He ends the day by watching late evening news programs, even those that report on his progress and complications. Among the problems: leaks of air and spinal fluid from a bullet hole that that Dr. Kobrine said he could not repair in the first operation. His goals then were to stop the bleeding, to remove dead brain tissue, the bullet and bone fragments, and to prevent infections. He was cautious to touch the left side of the brain as little as possible to avoid further loss of function.

Mr. Brady's mood fluctuates according to his medical course. When he expressed concern about new complications such as the one that threatened his life on April 22, Dr. Kobrine said he offered reassurance. That morning, Mr. Brady had been less alert than usual and "he just seemed a little different to me," Dr. Kobrine said. A CAT scan X-ray revealed an emergency condition called tension pneumocephalus, showing a large air space in the ventricles, the chambers normally filled with spinal fluid.

Dr. Kobrine released the air trapped in the ventricles by inserting long needles through holes that he had drilled in Mr. Brady's skull during the first operation just in case such a problem developed later. "He improved dramatically within minutes and for the first time since the original operation I knew he had a hole between his brain and the outside world," Dr. Kobrine said.

That night, Mr. Brady underwent a six-hour operation. It was three weeks since the shooting, enough time for the swelling of the left side of Mr. Brady's brain to go down, and it allowed Dr. Kobrine to examine an area he did not dare touch during the first operation. The leak occurred through a large hole in the dura, the lining of the brain, and a skull fracture, which Dr. Kobrine covered with tissue taken from elsewhere in Mr. Brady's body.

However, on April 27, Mr. Brady suffered another complication—several drops of spinal fluid leaked from his brain through his nose. In the hope that the leak would heal without another operation, from then until today Mr. Brady constantly rested in bed with his head elevated at a 20-degree angle. In that position, the pressures outside and inside Mr. Brady's skull were balanced and if no more fluid escaped, the body's normal healing mechanisms should have sealed the leak.

Yet another complication occurred a week ago; Mr. Brady complained of a brief episode of sweating and discomfort in his chest. Tests showed that emboli, or blood clots, had lodged in his lungs after traveling through the veins from his legs or pelvis.

From the time of Mr. Brady's first operation, his doctors had been concerned about such emboli because he was overweight and immobile. Doctors often prescribe blood-thinning drugs, such as heparin, to reduce the chances of dangerous blood clots. But the use of such drugs increases the patient's risk of bleeding. Dr. Kobrine said he did not want to take that step in Mr. Brady's case, because such an adverse reaction could destroy more brain tissue and threaten the patient's life.

To prevent further potentially fatal emboli from reaching Mr. Brady's lungs, Dr. Hugh H. Trout did an operation on May 4 to insert a device that resembles a bundle of bobby-pins in the shape of a missile nose cone. The umbrella-like device, which permits the normal flow of blood while straining out clots, was inserted into the inferior vena cava, the large vein that carries blood back to the heart from the lower part of the body. The operation was similar to the one that President Nixon had in 1974.

After examining new X-ray pictures today, Dr. Kobrine said that he expects that Mr. Brady will sit up for increasingly longer periods. However, if more spinal fluid leaks, Mr. Brady may face yet more surgery.

"We have had many forks in the road with Mr. Brady, and, knock on wood, so far we have gone down the right paths," Dr. Kobrine said. Other complications could develop any time in the next year. "Statistically that is unlikely, but it's not zero," Dr. Kobrine concluded.

—May 12, 1981

Rare Cancer Seen in 41 Homosexuals

By LAWRENCE K. ALTMAN

Doctors in New York and California have diagnosed among homosexual men 41 cases of a rare and often rapidly fatal form of cancer. Eight of the victims died less than 24 months after the diagnosis was made.

The cause of the outbreak is unknown, and there is as yet no evidence of contagion. But the doctors who have made the diagnoses, mostly in New York City and the San Francisco Bay area, are alerting other physicians who treat large numbers of homosexual men to the problem in an effort to help identify more cases and to reduce the delay in offering chemotherapy treatment.

The sudden appearance of the cancer, called Kaposi's sarcoma, has prompted a medical investigation that experts say could have as much scientific as public health importance because of what it may teach about determining the causes of more common types of cancer.

First Appears in Spots

Doctors have been taught in the past that the cancer usually appeared first in spots on the legs and that the disease took a slow course of up to 10 years. But these recent cases have shown that it appears in one or more violet-colored spots anywhere on the body. The spots generally do not itch or cause other symptoms, often can be mistaken for bruises, sometimes appear as lumps and can turn brown after a period of time. The cancer often causes swollen lymph glands, and then kills by spreading throughout the body.

Doctors investigating the outbreak believe that many cases have gone undetected because of the rarity of the condition and the difficulty even dermatologists may have in diagnosing it.

In a letter alerting other physicians to the problem, Dr. Alvin E. Friedman-Kien of New York University Medical Center, one of the investigators, described the appearance of the outbreak as "rather devastating."

Dr. Friedman-Kien said in an interview yesterday that he knew of 41 cases collated in the last five weeks, with the cases themselves dating to the past 30 months. The Federal Centers for Disease Control in Atlanta is expected to publish the first description of the outbreak in its weekly report today, according to a spokesman, Dr. James Curran. The report notes 26 of the cases—20 in New York and six in California.

There is no national registry of cancer victims, but the nationwide incidence of Kaposi's sarcoma in the past had been estimated by the Centers for Disease Control to be less than six one-hundredths of a case per 100,000 people annually, or about two cases in every three million people. However, the disease accounts for up to 9 percent of all cancers in a belt across equatorial Africa, where it commonly affects children and young adults.

In the United States, it has primarily affected men older than 50 years. But in the recent cases, doctors at nine medical centers in New York and seven hospitals in California have been diagnosing the condition among younger men, all of whom said in the course of standard diagnostic interviews that they were homosexual. Although the ages of the patients have ranged from 26 to 51 years, many have been under 40, with the mean at 39.

Nine of the 41 cases known to Dr. Friedman-Kien were diagnosed in California, and several of those victims reported that they had been in New York in the period preceding the diagnosis. Dr. Friedman-Kien said that his colleagues were checking on reports of two victims diagnosed in Copenhagen, one of whom had visited New York.

Viral Infections Indicated

No one medical investigator has yet interviewed all the victims, Dr. Curran said. According to Dr. Friedman-Kien, the reporting doctors said that most cases had involved homosexual men who have had multiple and frequent sexual encounters with different partners, as many as 10 sexual encounters each night up to four times a week.

Many of the patients have also been treated for viral infections such as herpes, cytomegalovirus and hepatitis B as well as parasitic infections such as amebiasis and giardiasis. Many patients also reported that they had used drugs such as amyl nitrite and LSD to heighten sexual pleasure.

Cancer is not believed to be contagious, but conditions that might precipitate it, such as particular viruses or environmental factors, might account for an outbreak among a single group.

The medical investigators say some indirect evidence actually points away from contagion as a cause. None of the patients knew each other, although the theoretical possibility that some may have had sexual contact with a person with Kaposi's sarcoma at some point in the past could not be excluded, Dr. Friedman-Kien said.

Dr. Curran said there was no apparent danger to nonhomosexuals from contagion. "The best evidence against contagion," he said, "is that no cases have been reported to date outside the homosexual community or in women."

Dr. Friedman-Kien said he had tested nine of the victims and found severe defects in their immunological systems. The patients had serious malfunctions of two types of cells called T- and B-cell lymphocytes, which have important roles in fighting infections and cancer.

But Dr. Friedman-Kien emphasized that the researchers did not know whether the immunological defects were the underlying problem or had developed secondarily to the infections or drug use.

The research team is testing various hypotheses, one of which is a possible link between past infection with cytomegalovirus and development of Kaposi's sarcoma.

—July 3, 1981

New Type of Drug for Cholesterol Approved and Hailed as Effective

By JANE E. BRODY

The Food and Drug Administration yesterday approved for marketing a new type of drug that experts have hailed as the most effective remedy yet devised for lowering cholesterol in the blood.

The drug, lovastatin, is expected to revolutionize treatment of high cholesterol levels, which heart specialists regard as the most important underlying cause of atherosclerosis and coronary heart disease.

About 20 million American adults have cholesterol levels that put them at very high risk of developing coronary heart disease, and experts predict that lovastatin will be prescribed for many of these people, as well as millions of others whose risk is not quite as high.

Treatment to Be Costly

But the treatment, which should be available in two to three weeks, will be costly. At a news conference yesterday, representatives of the developer, Merck & Company, said that the price of lovastatin to the pharmacist would be $1.25 for a 20-milligram dose, and that some patients would need four 20-milligram tablets a day. Thus, with the retail markup, the annual cost for patients at highest risk could exceed $3,000.

Lovastatin, which will be sold by prescription under the trade name Mevacor, is expected to become a pharmaceutical gold mine for Merck, perhaps reaching sales of more than $1 billion a year. In heavy trading yesterday, Merck stock jumped as much as four points, then settled down to close at 205, a gain of just one point above Monday's closing.

Adjunct to Dietary Changes

Experts cautioned yesterday that lovastatin should not be considered an alternative to a cholesterol-lowering diet but rather an adjunct to dietary changes. In many people, they said, diet alone is effective in bringing cholesterol down to desirable levels. The drug agency approved the drug for use only when diet and exercise alone have not achieved the desired reduction in cholesterol.

Furthermore, "the drug works best when taken in conjunction with a low-fat, low-cholesterol diet," said Dr. Michael S. Brown, whose Nobel Prize-winning studies with Dr. Joseph L. Goldstein provided the scientific basis for the development of lovastatin. Dr. Brown and Dr. Goldstein, both at the University of Texas Health Science Center in Dallas, deciphered the body's natural mechanism for controlling cholesterol levels in the blood.

"Unlike most drugs, which are discovered by screening large numbers of compounds in animal tests, lovastatin was purposely designed to take advantage of the mechanism of cholesterol metabolism elucidated through basic research," Dr. Goldstein said.

New Class of Medications

Lovastatin is the first to be approved of a new class of medications, called HMG-CoA reductase inhibitors, that work by blocking the liver enzyme needed to manufacture cholesterol in the body.

Seventy percent of the body's cholesterol is manufactured in the liver. When liver production is blocked, the liver is forced to remove cholesterol from the blood to meet its metabolic needs. This in turn lowers the level of cholesterol circulating in the blood. Several other inhibitors of the liver's cholesterol-producing enzyme are still being tested, including another Merck product and one by Squibb.

"This class of drugs is very exciting," said Dr. Basil Rifkind, chief of the lipid metabolism and atherogenesis branch of the National Heart, Lung and Blood Institute in Bethesda, Md. "Lovastatin is powerful in its ability to lower cholesterol and it is easy to take in contrast to other cholesterol-lowering drugs. It should make a big difference in getting people to stay on treatment for elevated cholesterol."

"More Than Just a New Drug"

Dr. Rifkind said he believed that lovastatin "will be more than just a new drug," adding, "It will serve as an educational instrument in its own right because it comes along at a time when a lot of emphasis is being placed on the importance of lowering cholesterol, but until now the tools for doing this weren't too good."

The average cholesterol level of middle-aged adults in this country is currently about 215 milligrams for each 100 milliliters of blood serum. Goals recently announced by federal health officials call for keeping cholesterol levels under 200 in people 30 and older and under 180 for those 20 to 29. For those in their 30s, a level of 220 to 239 is regarded as moderately risky and a level of 240 or more is considered high risk. For those 40 and older, a cholesterol level of 240 to 259 is considered a moderate risk and a level of 260 or more is high risk.

In four years of clinical testing sponsored by Merck Sharp & Dohme Research Laboratories in West Point, Pa., lovastatin was shown to be simpler to take, easier to tolerate and less likely to cause unpleasant or serious side effects than other available cholesterol-lowering drugs. It was also found to be far more potent, capable of reducing harmful cholesterol levels by 30 to 40 percent when used alone and by as much as 50 percent when combined with other remedies.

How the New Drug Works

The F.D.A. said yesterday that in clinical trials lovastatin lowered total cholesterol levels by 18 to 34 percent, depending on the dosage used. More importantly, it reduced levels of damaging low-density-lipoprotein (LDL) cholesterol by 19 to 39 percent without lowering the protective high-density-lipoprotein (HDL) cholesterol that is thought to cleanse the body of artery-clogging cholesterol. In fact, in most patients taking lovastatin, HDL-cholesterol rises by about 10 percent, researchers have reported.

Nonetheless, the drug agency did not give lovastatin a blanket approval. Rather, the agency advised that patients for whom lovastatin is prescribed should have blood tests every six weeks to check on liver function, as well as annual eye examinations. An increase in liver enzymes has been noted in about 1 percent of patients taking lovastatin, which could mean their livers are being overworked. Other patients have experienced changes in the lens of the eye that could suggest an increased risk of cataracts.

"The real test will be the next year or so, when a lot of people are taking the drug," said Dr. Brown. "Will there be serious side effects, and will we begin to see a drop in heart attacks?" Dr. Goldstein predicted it would take 5 to 10 years to see an effect on the nation's coronary death rate, adding that lovastatin is the first drug to make it possible to test nationally whether lowering cholesterol can significantly forestall the nation's leading cause of death.

"With this drug, we now have the ability to control the three major risk factors for heart disease: cigarette smoking, high blood pressure and elevated cholesterol," Dr. Brown noted.

—September 2, 1987

Poor Black and Hispanic Men
Are the Face of HIV

By DONALD G. McNEIL JR.

The AIDS epidemic in America is rapidly becoming concentrated among poor, young black and Hispanic men who have sex with men.

Despite years of progress in preventing and treating HIV in the middle class, the number of new infections nationwide remains stubbornly stuck at 50,000 a year—more and more of them in these men, who make up less than 1 percent of the population.

Giselle, a homeless 23-year-old transgender woman with cafe-au-lait skin, could be called the new face of the epidemic.

"I tested positive about a year ago," said Giselle, who was born male but now has a girlish hair spout, wears a T-shirt tight across a feminine chest and identifies herself as a woman. "I don't know how, exactly. I was homeless. I was escorting. I've been raped."

"Yes, I use condoms," she added. "But I'm not going to lie. I slip some-times. Trust me—everyone here who says, 'I always use condoms'? They don't always."

Besides transgender people like Giselle, the affected group includes men who are openly gay, secretly gay or bisexual, and those who consider themselves heterosexual but have had sex with men, willingly or unwillingly, in shelters or prison or for money. (Most of those interviewed for this article spoke on the condition that only their first names be used.)

Nationally, 25 percent of new infections are in black and Hispanic men, and in New York City it is 45 percent, according to the Centers for Disease Control and Prevention and the city's health department.

Nationally, when only men under 25 infected through gay sex are counted, 80 percent are black or Hispanic—even though they engage in less high-risk behavior than their white peers.

The prospects for change look grim. Critics say little is being done to save this group, and none of it with any great urgency.

"There wasn't even an ad campaign aimed at young black men until last year—what's that about?" said Krishna Stone, a spokeswoman for GMHC, which was founded in the 1980s as the Gay Men's Health Crisis.

Phill Wilson, president of the Black AIDS Institute in Los Angeles, said there were "no models out there right now for reaching these men."

Federal and state health officials agreed that it had taken years to shift prevention messages away from targets chosen 30 years ago: men who frequent gay bars, many of whom are white and middle-class, and heterosexual teenagers, who are at relatively low risk. Funding for health agencies has been flat, and there has been little political pressure to focus on young gay blacks and Hispanics.

Reaching those men "is the Holy Grail, and we're working on it," said Dr. Jonathan Mermin, director of HIV prevention at the CDC. His agency created its Testing Makes Us Stronger campaign—the one Ms. Stone referred to—and has granted millions of dollars to local health departments and community groups to pay for testing.

But he could not name a city or state with proven success in lowering infection rates in young gay minority men.

"With more resources, we could make bigger strides," he said.

Reaching Out

Gay black youths are hard to reach, experts say. Few are out to their families. Many live in places where gays are stigmatized and cannot afford to move. Few attend schools with gay pride clubs or gay guidance counselors.

"When we talked about HIV in sex ed, the class started freaking out," said Alex, 20, who was born in St. Croix but raised in New York. "One guy said, 'We ain't no faggots; why do we have to learn this stuff?' So the teacher stopped and moved on to another topic."

When those who are poor and homeless go to traditional gay hangouts, they become prey.

Kwame, a 20-year-old from Philadelphia, said that on his first day wandering around New York last year, he was propositioned by an older homeless man and by an older transgender person. The homeless man later admitted that he was infected, and added: "If you sit here long enough, you're going to get some propositions—and that's where you're going to sleep tonight. It happened to me, and it's going to happen to you."

Kwame said he had sex that night—with a man he met at a gay services center, where he had gone in search of emergency housing. "I wore a condom," he said. "I did it sort of out of guilt, or pity. It's how I was raised. I didn't want him to think I thought less of him. Also, I needed someplace to stay."

According to a major CDC-led study, a male-male sex act for a young black American is eight times as likely to end in HIV infection as it is for his white peers.

That is true even though, on average, black youths in the study took fewer risks than their white peers: they had fewer partners, engaged in fewer acts of sex while drunk or high, and used condoms more often.

They had other risk factors. Lacking health insurance, they were less likely to have seen doctors regularly and more likely to have syphilis, which creates a path for HIV.

But the crucial factor was that more of their partners were older black men, who are much more likely to have untreated HIV than older white men.

Among the poor, untreated or inadequately treated HIV is the norm, not the exception, said Perry N. Halkitis, a professor of psychology and public health at New York University. According to the CDC, 79 percent of HIV-infected black men who have sex with men and 74 percent of Hispanics are not "virally suppressed," meaning they can transmit the infection, either because they are not yet on antiretroviral drugs or are not taking them daily.

Giselle admitted to sometimes skipping days. "The medicine gets you sick," she said. "It messes up your mental state. Or it can be freezing and I'm sweating."

Missed doses let the virus rebound, sometimes in drug-resistant strains, experts said.

Other risk factors include depression and fatalism. In a 2012 project by the National Youth Pride Services, an advocacy organization for gay black youths, more than half of the young gay black people questioned said they feared their friends or families would disown them if they came out as gay, and about 4 in 10 said they had contemplated suicide over being gay.

"The image of a black gay man almost doesn't exist," said Shariff Gibbons, 25, who works with other young men at GMHC. "In the black community, the image that 'gay men are sissies' is amped up a billion times. And we all have an aunt who goes to church and says, 'Being gay is wrong.' That makes young men hide."

Fighting Isolation

Roderick, 22, said his aunt, who took him in after his parents were arrested on drug charges, became furious after he told her at age 15 that he was gay. Later he attended a small New Jersey university and studied to be a veterinarian. But when his aunt learned that he was dating a white man, she demanded that he return home and go to a local community college.

She and his cousins called him an "Oreo" and even viler names, he said. "It got to where I felt I was going to snap, and kill myself or kill them. I didn't want to do either, so one night I took my cousin's bike and I left, and took a train to New York. I'm just basically dead to my family now."

In New York, he found housing through the Ali Forney Center, which is named after a young gay rights advocate murdered in 1997 and which shelters gay minority youths, who are often abused in regular shelters. Roderick briefly supported himself by having sex for money at parties organized through Craigslist. But he gave that up, he said, has one partner and is applying for veterinary scholarships.

Several young men described having felt isolated and scared as teenagers, and so depressed that they hardly cared if they lived or died, which left them indifferent about using condoms, especially when they were offered money not to. And many turned to empathetic older men who had gone through the same crises in their youth. Alex said his mother threatened to throw him out when she caught him with another boy when he was a teenager—but she needed the disability checks he receives because of nerve damage done at birth.

"I have three strikes against me: I'm black, I'm gay, and I'm in a wheelchair," he said. "All I wanted was love and comfort and being with someone in the world."

Sex with strangers was as close as he could get. His first time was in a stairwell of his housing project with a man he met on a black gay chat site.

"It was a hit and a bounce and leave," he said. "Unprotected oral and anal." When he was older, he sold himself on the Chelsea Piers.

Two scary events—getting syphilis and being raped by an older man he thought loved him—brought him to GMHC, which offers separate support groups for black and Hispanic men, teenagers and transsexuals. They offer advice, HIV tests and help on being openly gay.

For example, several men said they joined after being handed GMHC "I Love My Boo" fliers, which show young black male lovers holding hands and kissing in Central Park.

An Inadequate Response

But scattered local programs like these, and those offering housing, legal and medical help and other services, are not turning the tide of infections because the national response is fragmented and hesitant.

Few black political or religious leaders talk regularly about the problem—though there are exceptions, including Representative Barbara Lee, a Democrat from Oakland, Calif., and the Rev. Dr. Calvin O. Butts III of the Abyssinian Baptist Church in Harlem. Ms. Lee and several other congresswomen publicly take annual HIV tests. Few men in the Congressional Black Caucus agree to join them, one of her aides said.

Dr. Butts has endorsed home HIV tests from the pulpit and exhorted his congregants to accept gay relatives, but many black clergy members are far less accepting; some have fought same-sex marriage ballot measures.

Many programs have been proposed and tested, including financial incentives: paying parents who accept their gay sons to meet with parents who reject theirs; paying men who bring in friends for HIV tests; and paying older black men to give cooking lessons and safe-sex advice to younger ones. But none have been widely adopted.

At a recent GMHC forum on why its programs for young black men were being cut, Janet Weinberg, the agency's acting chief executive, said the epidemic was in some ways still where it was 30 years ago.

"We have the tools to end it," she said, "except for the government's indifference."

—*December 4, 2013*

The Mysterious Tree
of a Newborn's Life

By DENISE GRADY

Minutes after a baby girl was born on a recent morning at UCSF Medical Center here, her placenta—a pulpy blob of an organ that is usually thrown away—was packed up and carried off like treasure through a maze of corridors to the laboratory of Susan Fisher, a professor of obstetrics, gynecology and reproductive sciences.

There, scientists set upon the tissue with scalpels, forceps and an array of chemicals to extract its weirdly powerful cells, which storm the uterus like an invading army and commandeer a woman's body for nine months to keep her fetus alive. The placenta is the life-support system for the fetus. A disk of tissue attached to the uterine lining on one side and to the umbilical cord on the other, it grows from the embryo's cells, not the mother's. It is sometimes called the afterbirth: It comes out after the baby is born, usually weighing about a pound, or a sixth of the baby's weight.

It provides oxygen, nourishment and waste disposal, doing the job of the lungs, liver, kidneys and other organs until the fetal ones kick in. If something goes wrong with the placenta, devastating problems can result, including miscarriage, stillbirth, prematurity, low birth weight and pre-eclampsia, a condition that drives up the mother's blood pressure and can kill her and the fetus. A placenta much smaller or larger than average is often a sign of trouble. Increasingly, researchers think placental disorders can permanently alter the health of mother and child.

Given its vital role, shockingly little is known about the placenta. Only recently, for instance, did scientists start to suspect that the placenta may not be sterile, as once thought, but may have a microbiome of its own—a population of micro-organisms—that may help shape the immune system of the fetus and affect its health much later in life.

Dr. Fisher and other researchers have studied the placenta for decades, but she said: "Compared to what we should know, we know almost nothing. It's a place where I think we could make real medical breakthroughs that I think would be of enormous importance to women and children and families."

The National Institute of Child Health and Human Development calls the placenta "the least understood human organ and arguably one of the more

393

important, not only for the health of a woman and her fetus during pregnancy but also for the lifelong health of both."

In May, the institute gathered about 70 scientists at its first conference devoted to the placenta, in hopes of starting a Human Placenta Project, with the ultimate goal of finding ways to detect abnormalities in the organ earlier, and treat or prevent them.

Seen shortly after a birth, the placenta is bloody and formidable looking. Fathers in the delivery room sometimes faint at the sight of it, doctors say. It is bluish or dark red, eight or nine inches across and about an inch thick in the middle. The side that faced the fetus is covered by a network of branching blood vessels, the umbilical cord emerging like a fat stalk. The side that faced the mother, glommed onto the uterine wall, looks raw and meaty.

In some cultures, the organ has spiritual meaning and must be buried or dealt with according to rituals. In recent years in the United States, some women have become captivated by the idea of eating it—cooking it, blending it into smoothies, or having it dried and packed into capsules. Not much is known about whether this is a good idea.

When scientists describe the human placenta, one unsettling word comes up repeatedly: "invasive." The organ begins forming in the lining of the uterus as soon as a fertilized egg lands there, embedding itself deeply in the mother's tissue and tapping into her arteries so aggressively that researchers liken it to cancer. In most other mammals, the placental attachment is much more superficial.

"A parasite upon the mother" is how the placenta is described in the book *Life's Vital Link*, by Y. W. Loke, a reproductive immunologist. He goes on: "It has literally burrowed into the substance of her womb and is siphoning off nutrients from her blood to provide for the embryo."

An Invader's Intricacy

The placenta establishes a blood supply at 10 to 12 weeks of pregnancy. Ultimately, it invades 80 to 100 uterine vessels called spiral arteries and grows 32 miles of capillaries. The placental cells form minute fingerlike projections called villi, which contain fetal capillaries and come in contact with maternal blood, to pick up oxygen and nutrients and get rid of wastes.

Spread out, the tissue formed to exchange oxygen and nutrients would cover 120 to 150 square feet. Every minute, about 20 percent of the mother's blood supply flows through the placenta. The front line of the invasion is a cell called a trophoblast, from the outer layer of the embryo. Early in pregnancy, these cells multiply explosively and stream out like a column of soldiers.

"The trophoblast cells are so invasive from the get-go," Dr. Fisher said. "They just blast through the uterine lining to get themselves buried in there."

They shove other cells out of the way and destroy them with digestive enzymes or secrete substances that induce the cells to kill themselves. Michael McMaster, a professor of cell and tissue biology at the University of California, San Francisco, said that failures of this early process probably happened fairly often. People often assume that miscarriages and other problems arise from the fetus itself, but he said, "it's probably true that at this early stage, a lot of trophoblast malfunction can underlie pregnancy loss or future disease."

Trophoblasts are so invasive that they will form a placenta almost anywhere, even if they land on tissue other than the uterus. Occasionally, pregnancies begin outside the uterus, in fallopian tubes or elsewhere in the abdomen, and the rapid, penetrating growth of the placenta can rupture organs. Placentas that form over a scar on the uterus, where the lining is thin or absent—say, from a previous cesarean section—can invade so deeply that they cannot be safely removed at birth, and the only way to prevent the mother from bleeding to death is to take out the uterus.

Trophoblasts are a major focus of the research by Dr. Fisher's team, and her laboratory also acts as a bank, providing cell and tissue samples to other researchers around the country. One staff member is a recruiter, charged with the delicate task of asking women in labor to donate their placentas for research.

Dr. Fisher's lab discovered that as trophoblasts invade, they alter certain proteins on their surfaces, called adhesion molecules, to become more motile. Researchers later found that cancer cells do the same thing as they spread from a tumor to invade other parts of the body.

Trophoblasts change in other ways, mimicking cells of the blood vessels they invade. The spiral arteries, which feed the lining of the uterus, become paved with trophoblasts instead of the woman's own cells. This "remodeling" process dilates the arteries considerably to pour blood into the placenta and nourish the villi.

"When I first read this anatomy, I couldn't believe the whole world wasn't studying this," Dr. Fisher said.

Examining a micrograph of a remodeled artery, she said: "Look at the diameter of this vessel. It looks like some monster thing from the deep chasms of the sea."

What Can Go Wrong

Invasion and remodeling are essential: If they do not occur, the placenta cannot acquire enough of a blood supply to develop normally, and the results can be disastrous. One consequence can be pre-eclampsia, which affects 2 percent to 5 percent of pregnant women in the United States. Rates are higher in poor countries, particularly those in Africa. The condition brings high blood pressure and other abnormalities in the mother, and can be fatal.

Pre-eclampsia is considered a placental disease: Most women with the illness have abnormally small placentas, and when pathologists examine them after the delivery, they often find blood clots, discolorations and a poorly developed blood supply.

How and why the problem occurs is not entirely understood. For unknown reasons, the placenta does not form properly and cannot keep up with the demands of the growing fetus. The trophoblasts cannot fully change into artery cells and begin churning out an abnormal array of molecules that jack up the mother's blood pressure and may damage her blood vessels.

The rising blood pressure may be an attempt to compensate by forcing more circulation to the placenta. But it backfires. The only treatment is to deliver the baby, which probably works because it also removes the placenta.

At some hospitals, pathologists who specialize in the placenta examine the ones from troubled pregnancies or sickly newborns, looking for clues to what went wrong. Massachusetts General Hospital also keeps seemingly normal placentas in a refrigerator for about two weeks, until it is clear that the mother and the baby are healthy.

Dr. Drucilla J. Roberts, a placental pathologist there, said that relatively few hospitals had placental pathologists or the ability to train them. Nationwide, there are fewer than 100, she estimates. More are needed, she said. She and a colleague, Dr. Rosemary H. Tambouret, often examine specimens sent from other hospitals not equipped to do the work themselves.

"The placenta gives the answer in many term stillbirths," Dr. Roberts said. Half of those deaths are never explained, but many of them involve abnormalities in the placenta, including infections or unusual conditions in which the mother's immune system appears to have rejected the placenta.

"I can't tell you how important it is to the family just to have an answer," she said. Knowing can help ease the guilt that many parents feel when a child is stillborn. The information can also tell doctors what to watch for in future pregnancies.

In one case, Dr. Roberts said, examining the placenta helped diagnose an immune incompatibility between the parents that had caused multiple stillbirths and miscarriages. The mother was treated and went on to have a healthy child.

Another placental pathologist, Dr. Rebecca Baergen, the chief of perinatal and obstetric pathology at New York-Presbyterian Hospital, said that in some cases, particularly those involving fetal death or stillbirth, more could be learned from the placenta than from the fetus. She described a case in which a newborn was extremely small, had stunted limbs and did not survive. Doctors suspected a growth disorder, but bone samples revealed nothing.

The placenta was sent to Dr. Baergen. She found many problems with its blood supply and recommended a battery of tests for the mother. The tests found a hereditary blood disorder. The mother was treated and later gave birth to a healthy baby.

"The placenta has essentially been called the chronicle of intrauterine life," she said. "It really tells the story of what's been going on. It plays the role of many organs—liver, kidney, respiratory, endocrine. It can give you a lot of information about the baby's and the mom's health."

—July 15, 2014

Fast Way to Try Many Drugs on Many Cancers

By GINA KOLATA

Chemotherapy failed to thwart Erika Hurwitz's rare cancer of white blood cells. So her doctors offered her another option, a drug for melanoma. The result was astonishing.

Within four weeks, a red rash covering her body, so painful she had required a narcotic patch and the painkiller OxyContin, had vanished. Her cancer was undetectable.

"It has been a miracle drug," said Mrs. Hurwitz, 78, of Westchester County.

She is part of a new national effort to try to treat cancer based not on what organ it started in, but on what mutations drive its growth.

Cancers often tend to be fueled by changes in genes, or mutations, that make cells grow and spread to other parts of the body. There are now an increasing number of drugs that block mutations in cancer genes and can halt a tumor's growth.

While such an approach has worked in a few isolated cases, those cases cannot reveal whether other patients with the same mutation would have a similar experience.

Now, medical facilities like Memorial Sloan Kettering Cancer Center in New York, where Mrs. Hurwitz is a patient, are starting coordinated efforts to find answers. And this spring, a federally funded national program will start to screen tumors in thousands of patients to see which might be attacked by any of at least a dozen new drugs. Those whose tumors have mutations that can be attacked will be given the drugs.

The studies of this new method, called basket studies because they lump together different kinds of cancer, are revolutionary, much smaller than the usual studies, and without control groups of patients who for comparison's sake receive standard treatment.

Researchers and drug companies asked the Food and Drug Administration for its opinion, realizing that if the FDA did not accept the studies, no drugs would ever be approved on the basis of them. But the F DA said it sanctioned them and could approve drugs with basket study data alone.

Instead of insisting on traditional studies, said Dr. Richard Pazdur, who directs the FDA office that approves new cancer drugs, the agency will look at the data and ask, "Is the American population going to be better off with this drug than without it?"

These are the sorts of studies many seriously ill patients have been craving—a guarantee that if they enter a study they will get a promising new drug. And the studies move fast; it does not take years to see a big effect if there is one at all.

In Mrs. Hurwitz's case, the mutation in her rare cancer is in a gene, BRAF, found in about 50 percent of melanomas but rare in other cancers. She is among dozens of patients with the same mutation, but different cancers, in the new study that gives everyone the melanoma drug that attacks the mutation.

Basket studies became possible only recently, when gene sequencing became so good and its price so low that doctors could routinely look for 50, 60 or more known cancer-causing mutations in tumors. At the same time, more and more drugs were being developed to attack those mutations. So even if, as often happens, only a small percentage of patients with a particular tumor type have a particular mutation, it was possible to find a few dozen patients or more for a clinical trial by grouping everyone with that mutation together.

In a way, this is a leading edge of precision medicine that aims to target the drug to the patient. Unlike previous efforts that looked for small differences between a new treatment and an older one, with basket studies, researchers are gambling on finding huge effects.

"This is really a new breed of study," said Dr. David Hyman, a cancer specialist at Memorial Sloan Kettering who directs the study Mrs. Hurwitz is in and two similar ones.

And they are seeing some unprecedented responses, along with some failures. The responses, though, can be so striking that control groups might be unwarranted or infeasible, Dr. Pazdur said.

"Conventional therapy might give a response rate of 10 or 20 percent," Dr. Pazdur said. "The newer drug has a response rate of 50 or 60 percent. Does it make sense to do a randomized trial?" And even if a trial were planned, he said: "Who would go on that trial? Would you go on that trial?"

"When you are having a big effect, it is kind of jaw dropping," Dr. Pazdur added. "These are response rates we haven't seen before in diseases."

But these are still the early days, researchers caution. "It is a different world we are walking into," said Dr. Daniel Costa, a lung cancer researcher at Beth Israel Deaconess Medical Center in Boston. "And we are learning as we go along."

The new studies pose new problems. With no control groups, the effect has to be enormous and unmistakable to show it is not occurring by chance. When everyone gets a drug, it can be hard to know if a side effect is from the drug, a cancer or another disease. And gene mutations can be so rare that patients for a basket study are difficult to find.

The rarity of the mutations, in fact, is one reason for the new national effort, supported by the National Cancer Institute. Its study, called Match, is essentially a basket of basket studies. Doctors around the country will be sending tumor samples from at least 3,000 patients to central labs that will examine them for mutations. Those with any of a dozen or so mutations in their tumors can enroll in studies of drugs that target their tumor's mutation.

Dr. Keith Flaherty of Massachusetts General Hospital, principal investigator for the Match trial, said the number of baskets was uncertain—it would depend on the number of drugs. But he expects 12 to 15 baskets to start, expanding to perhaps 40 or more. There will be 31 patients per drug.

He anticipates mixed results. "We are exploring an unknown space here," Dr. Flaherty said. "But it is essentially impossible for this whole set of baskets to fail."

To show what is possible, Dr. José Baselga of Memorial Sloan Kettering points to preliminary results he presented in December for the basket study that includes Mrs. Hurwitz.

Among 70 patients, there are eight types of cancer. Eighteen patients had one of two very rare cancers, Erdheim-Chester disease or Langerhans disease, the cancer that struck Mrs. Hurwitz. Of them, 14 responded to the melanoma drug—their tumors vanished, shrank or stopped growing—and the remaining four have not been taking the drug long enough to say.

"Unbelievable," Dr. Baselga said.

"This is working in a way that is clear, that is unprecedented," he said. "I don't have enough patients to do a Phase 3 study," he added, referring to the large, randomized study traditionally used to test new drugs, "and I even question the morality of it."

But others in basket studies have not fared so well.

Eleni Vavas entered a basket study at Memorial Sloan Kettering hoping to stop the stomach cancer that was killing her. The study, said her husband, John Vavas, "was our last-ditch, Hail Mary effort." His wife, who was 36, entered it last spring, the only patient with stomach cancer. But, Mr. Vavas said, "she just didn't respond."

She died on July 1.

—*February 26, 2015*

CHAPTER 10

Neuroscience:
Secrets of the Brain

H. M., Whose Loss of Memory Made Him Unforgettable, Dies

By BENEDICT CAREY

He knew his name. That much he could remember.

He knew that his father's family came from Thibodaux, La., and his mother was from Ireland, and he knew about the 1929 stock market crash and World War II and life in the 1940s.

But he could remember almost nothing after that.

In 1953, he underwent an experimental brain operation in Hartford to correct a seizure disorder, only to emerge from it fundamentally and irreparably changed. He developed a syndrome neurologists call profound amnesia. He had lost the ability to form new memories.

For the next 55 years, each time he met a friend, each time he ate a meal, each time he walked in the woods, it was as if for the first time.

And for those five decades, he was recognized as the most important patient in the history of brain science. As a participant in hundreds of studies, he helped scientists understand the biology of learning, memory and physical dexterity, as well as the fragile nature of human identity.

On Tuesday evening at 5:05, Henry Gustav Molaison—known worldwide only as H. M., to protect his privacy—died of respiratory failure at a nursing home in Windsor Locks, Conn. His death was confirmed by Suzanne Corkin, a neuroscientist at the Massachusetts Institute of Technology, who had worked closely with him for decades. Henry Molaison was 82.

From the age of 27, when he embarked on a life as an object of intensive study, he lived with his parents, then with a relative and finally in an institution. His amnesia did not damage his intellect or radically change his personality. But he could not hold a job and lived, more so than any mystic, in the moment.

"Say it however you want," said Dr. Thomas Carew, a neuroscientist at the University of California, Irvine, and president of the Society for Neuroscience. "What H. M. lost, we now know, was a critical part of his identity."

At a time when neuroscience is growing exponentially, when students and money are pouring into laboratories around the world and researchers are mounting large-scale studies with powerful brain-imaging technology, it is easy to forget how rudimentary neuroscience was in the middle of the 20th century.

When Mr. Molaison, at nine years old, banged his head hard after being hit by a bicycle rider in his neighborhood near Hartford, scientists had no way to see inside his brain. They had no rigorous understanding of how complex functions like memory or learning functioned biologically. They could not explain why the boy had developed severe seizures after the accident, or even whether the blow to the head had anything do to with it.

Eighteen years after that bicycle accident, Mr. Molaison arrived at the office of Dr. William Beecher Scoville, a neurosurgeon at Hartford Hospital. Mr. Molaison was blacking out frequently, had devastating convulsions and could no longer repair motors to earn a living.

After exhausting other treatments, Dr. Scoville decided to surgically remove two finger-shaped slivers of tissue from Mr. Molaison's brain. The seizures abated, but the procedure—especially cutting into the hippocampus, an area deep in the brain, about level with the ears—left the patient radically changed.

Alarmed, Dr. Scoville consulted with a leading surgeon in Montreal, Dr. Wilder Penfield of McGill University, who with Dr. Brenda Milner, a psychologist, had reported on two other patients' memory deficits.

Soon Dr. Milner began taking the night train down from Canada to visit Mr. Molaison in Hartford, giving him a variety of memory tests. It was a collaboration that would forever alter scientists' understanding of learning and memory.

"He was a very gracious man, very patient, always willing to try these tasks I would give him," Dr. Milner, a professor of cognitive neuroscience at the Montreal Neurological Institute and McGill University, said in a recent interview. "And yet every time I walked in the room, it was like we'd never met."

At the time, many scientists believed that memory was widely distributed throughout the brain and not dependent on any one neural organ or region. Brain lesions, either from surgery or accidents, altered people's memory in ways that were not easily predictable. Even as Dr. Milner published her results, many researchers attributed H. M.'s deficits to other factors, like general trauma from his seizures or some unrecognized damage.

"It was hard for people to believe that it was all due" to the excisions from the surgery, Dr. Milner said.

That began to change in 1962, when Dr. Milner presented a landmark study in which she and H. M. demonstrated that a part of his memory was fully intact. In a series of trials, she had Mr. Molaison try to trace a line between two outlines of a five-point star, one inside the other, while watching his hand and the star in a mirror. The task is difficult for anyone to master at first.

Every time H. M. performed the task, it struck him as an entirely new experience. He had no memory of doing it before. Yet with practice he became proficient. "At one point he said to me, after many of these trials, 'Huh, this was easier than I thought it would be,'" Dr. Milner said.

The implications were enormous. Scientists saw that there were at least two systems in the brain for creating new memories. One, known as declarative memory, records names, faces and new experiences and stores them until they are consciously retrieved. This system depends on the function of medial temporal areas, particularly an organ called the hippocampus, now the object of intense study.

Another system, commonly known as motor learning, is subconscious and depends on other brain systems. This explains why people can jump on a bike after years away from one and take the thing for a ride, or why they can pick up a guitar that they have not played in years and still remember how to strum it.

Soon "everyone wanted an amnesic to study," Dr. Milner said, and researchers began to map out still other dimensions of memory. They saw that H. M.'s short-term memory was fine; he could hold thoughts in his head for about 20 seconds. It was holding onto them without the hippocampus that was impossible.

"The study of H. M. by Brenda Milner stands as one of the great milestones in the history of modern neuroscience," said Dr. Eric Kandel, a neuroscientist at Columbia University. "It opened the way for the study of the two memory systems in the brain, explicit and implicit, and provided the basis for everything that came later—the study of human memory and its disorders."

Living at his parents' house, and later with a relative through the 1970s, Mr. Molaison helped with the shopping, mowed the lawn, raked leaves and relaxed in front of the television. He could navigate through a day attending to mundane details—fixing a lunch, making his bed—by drawing on what he could remember from his first 27 years.

He also somehow sensed from all the scientists, students and researchers parading through his life that he was contributing to a larger endeavor, though he was uncertain about the details, said Dr. Corkin, who met Mr. Molaison while studying in Dr. Milner's laboratory and who continued to work with him until his death.

By the time he moved into a nursing home in 1980, at age 54, he had become known to Dr. Corkin's MIT team in the way that Polaroid snapshots in a photo album might sketch out a life but not reveal it whole.

H. M. could recount childhood scenes: Hiking the Mohawk Trail. A road trip with his parents. Target shooting in the woods near his house.

"Gist memories, we call them," Dr. Corkin said. "He had the memories, but he couldn't place them in time exactly; he couldn't give you a narrative."

He was nonetheless a self-conscious presence, as open to a good joke and as sensitive as anyone in the room. Once, a researcher visiting with Dr. Milner and H. M. turned to her and remarked how interesting a case this patient was.

"H. M. was standing right there," Dr. Milner said, "and he kind of colored— blushed, you know—and mumbled how he didn't think he was that interesting, and moved away."

In the last years of his life, Mr. Molaison was, as always, open to visits from researchers, and Dr. Corkin said she checked on his health weekly. She also arranged for one last research program. On Tuesday, hours after Mr. Molaison's death, scientists worked through the night taking exhaustive MRI scans of his brain, data that will help tease apart precisely which areas of his temporal lobes were still intact and which were damaged, and how this pattern related to his memory.

Dr. Corkin arranged, too, to have his brain preserved for future study, in the same spirit that Einstein's was, as an irreplaceable artifact of scientific history.

"He was like a family member," said Dr. Corkin, who is at work on a book on H. M., titled *A Lifetime Without Memory*. "You'd think it would be impossible to have a relationship with someone who didn't recognize you, but I did."

In his way, Mr. Molaison did know his frequent visitor, she added: "He thought he knew me from high school."

Henry Gustav Molaison, born on Feb. 26, 1926, left no survivors. He left a legacy in science that cannot be erased.

—December 5, 2008

Sizing Up Consciousness by Its Bits

By CARL ZIMMER

One day in 2007, Dr. Giulio Tononi lay on a hospital stretcher as an anesthesiologist prepared him for surgery. For Dr. Tononi, it was a moment of intellectual exhilaration. He is a distinguished chair in consciousness science at the University of Wisconsin, and for much of his life he has been developing a theory of consciousness. Lying in the hospital, Dr. Tononi finally had a chance to become his own experiment.

The anesthesiologist was preparing to give Dr. Tononi one drug to render him unconscious, and another one to block muscle movements. Dr. Tononi suggested the anesthesiologist first tie a band around his arm to keep out the muscle-blocking drug. The anesthesiologist could then ask Dr. Tononi to lift his finger from time to time, so they could mark the moment he lost awareness.

The anesthesiologist did not share Dr. Tononi's excitement. "He could not have been less interested," Dr. Tononi recalled. "He just said, 'Yes, yes, yes,' and put me to sleep. He was thinking, 'This guy must be out of his mind.'"

Dr. Tononi was not offended. Consciousness has long been the province of philosophers, and most doctors steer clear of their abstract speculations. After all, debating the finer points of what it is like to be a brain floating in a vat does not tell you how much anesthetic to give a patient.

But Dr. Tononi's theory is, potentially, very different. He and his colleagues are translating the poetry of our conscious experiences into the precise language of mathematics. To do so, they are adapting information theory, a branch of science originally applied to computers and telecommunications. If Dr. Tononi is right, he and his colleagues may be able to build a "consciousness meter" that doctors can use to measure consciousness as easily as they measure blood pressure and body temperature. Perhaps then his anesthesiologist will become interested.

"I love his ideas," said Christof Koch, an expert on consciousness at Caltech. "It's the only really promising fundamental theory of consciousness."

Dr. Tononi's obsession with consciousness started in his teens. He was initially interested in ethics, but he decided that questions of personal responsibility depended on our consciousness of our own actions. So he would have to figure out consciousness first. "I've been stuck with this thing for most of my life," he said.

Eventually he decided to study consciousness by becoming a psychiatrist. An early encounter with a patient in a vegetative state convinced Dr. Tononi that understanding consciousness was not just a matter of philosophy.

"There are very practical things involved," Dr. Tononi said. "Are these patients feeling pain or not? You look at science, and basically science is telling you nothing."

Dr. Tononi began developing models of the brain and became an expert on one form of altered consciousness we all experience: sleep. In 2000, he and his colleagues found that Drosophila flies go through cycles of sleeping and waking. By studying mutant flies, Dr. Tononi and other researchers have discovered genes that may be important in sleep disorders.

For Dr. Tononi, sleep is a daily reminder of how mysterious consciousness is. Each night we lose it, and each morning it comes back. In recent decades, neuroscientists have built models that describe how consciousness emerges from the brain. Some researchers have proposed that consciousness is caused by the synchronization of neurons across the brain. That harmony allows the brain to bring together different perceptions into a single conscious experience.

Dr. Tononi sees serious problems in these models. When people lose consciousness from epileptic seizures, for instance, their brain waves become more synchronized. If synchronization were the key to consciousness, you would expect the seizures to make people hyperconscious instead of unconscious, he said.

While in medical school, Dr. Tononi began to think of consciousness in a different way, as a particularly rich form of information. He took his inspiration from the American engineer Claude Shannon, who built a scientific theory of information in the mid-1900s. Mr. Shannon measured information in a signal by how much uncertainty it reduced. There is very little information in a photodiode that switches on when it detects light, because it reduces only a little uncertainty. It can distinguish between light and dark, but it cannot distinguish between different kinds of light. It cannot tell the differences between a television screen showing a Charlie Chaplin movie or an ad for potato chips. The question that the photodiode can answer, in other words, is about as simple as a question can get.

Our neurons are basically fancy photodiodes, producing electric bursts in response to incoming signals. But the conscious experiences they produce contain far more information than in a single diode. In other words, they reduce much more uncertainty. While a photodiode can be in one of two states, our brains can be in one of trillions of states. Not only can we tell the difference between a

Chaplin movie and a potato chip, but our brains can go into a different state from one frame of the movie to the next.

"One out of two isn't a lot of information, but if it's one out of trillions, then there's a lot," Dr. Tononi said.

Consciousness is not simply about quantity of information, he says. Simply combining a lot of photodiodes is not enough to create human consciousness. In our brains, neurons talk to one another, merging information into a unified whole. A grid made up of a million photodiodes in a camera can take a picture, but the information in each diode is independent from all the others. You could cut the grid into two pieces and they would still take the same picture.

Consciousness, Dr. Tononi says, is nothing more than integrated information. Information theorists measure the amount of information in a computer file or a cellphone call in bits, and Dr. Tononi argues that we could, in theory, measure consciousness in bits as well. When we are wide awake, our consciousness contains more bits than when we are asleep.

For the past decade, Dr. Tononi and his colleagues have been expanding traditional information theory in order to analyze integrated information. It is possible, they have shown, to calculate how much integrated information there is in a network. Dr. Tononi has dubbed this quantity "phi," and he has studied it in simple networks made up of just a few interconnected parts. How the parts of a network are wired together has a big effect on phi. If a network is made up of isolated parts, phi is low, because the parts cannot share information.

But simply linking all the parts in every possible way does not raise phi much. "It's either all on, or all off," Dr. Tononi said. In effect, the network becomes one giant photodiode.

Networks gain the highest phi possible if their parts are organized into separate clusters, which are then joined. "What you need are specialists who talk to each other, so they can behave as a whole," Dr. Tononi said. He does not think it is a coincidence that the brain's organization obeys this phi-raising principle.

Dr. Tononi argues that his Integrated Information Theory sidesteps a lot of the problems that previous models of consciousness have faced. It neatly explains, for example, why epileptic seizures cause unconsciousness. A seizure forces many neurons to turn on and off together. Their synchrony reduces the number of possible states the brain can be in, lowering its phi.

Dr. Koch considers Dr. Tononi's theory to be still in its infancy. It is impossible, for example, to calculate phi for the human brain because its billions of neurons and trillions of connections can be arranged in so many ways. Dr. Koch and Dr. Tononi recently started a collaboration to determine phi for a much more modest nervous system, that of a worm known as *Caenorhabditis elegans*. Despite the fact that it has only 302 neurons in its entire body, Dr. Koch and Dr. Tononi will be able make only a rough approximation of phi, rather than a precise calculation.

"The lifetime of the universe isn't long enough for that," Dr. Koch said. "There are immense practical problems with the theory, but that was also true for the theory of general relativity early on."

Dr. Tononi is also testing his theory in other ways. In a study published this year, he and his colleagues placed a small magnetic coil on the heads of volunteers. The coil delivered a pulse of magnetism lasting a tenth of a second. The burst causes neurons in a small patch of the brain to fire, and they in turn send signals to other neurons, making them fire as well.

To track these reverberations, Dr. Tononi and his colleagues recorded brain activity with a mesh of scalp electrodes. They found that the brain reverberated like a ringing bell, with neurons firing in a complex pattern across large areas of the brain for 295 milliseconds.

Then the scientists gave the subjects a sedative called midazolam and delivered another pulse. In the anesthetized brain, the reverberations produced a much simpler response in a much smaller region, lasting just 110 milliseconds. As the midazolam started to wear off, the pulses began to produce richer, longer echoes.

These are the kinds of results Dr. Tononi expected. According to his theory, a fragmented brain loses some of its integrated information and thus some of its consciousness. Dr. Tononi has gotten similar results when he has delivered pulses to sleeping people—or at least people in dream-free stages of sleep.

In this month's issue of the journal *Cognitive Neuroscience*, he and his colleagues reported that dreaming brains respond more like wakeful ones. Dr. Tononi is now collaborating with Dr. Steven Laureys of the University of Liège in Belgium to test his theory on people in persistent vegetative states. Although he and his colleagues have tested only a small group of subjects, the results are so far falling in line with previous experiments.

If Dr. Tononi and his colleagues can get reliable results from such experiments, it will mean more than just support for his theory. It could also lead to a new way to measure consciousness. "That would give us a consciousness index," Dr. Laureys said.

Traditionally, doctors have measured consciousness simply by getting responses from patients. In many cases, it comes down to questions like, "Can you hear me?" This approach fails with people who are conscious but unable to respond. In recent years scientists have been developing ways of detecting consciousness directly from the activity of the brain.

In one series of experiments, researchers put people in vegetative or minimally conscious states into fMRI scanners and asked them to think about playing tennis. In some patients, regions of the brain became active in a pattern that was a lot like that in healthy subjects.

Dr. Tononi thinks these experiments identify consciousness in some patients, but they have serious limitations. "It's complicated to put someone in a scanner," he said. He also notes that thinking about tennis for 30 seconds can demand a lot from people with brain injuries. "If you get a response I think it's proof that's someone's there, but if you don't get it, it's not proof of anything," Dr. Tononi said.

Measuring the integrated information in people's brains could potentially be both easier and more reliable. An anesthesiologist, for example, could apply magnetic pulses to a patient's brain every few seconds and instantly see whether it responded with the rich complexity of consciousness or the meager patterns of unconsciousness.

Other researchers view Dr. Tononi's theory with a respectful skepticism.

"It's the sort of proposal that I think people should be generating at this point: a simple and powerful hypothesis about the relationship between brain processing and conscious experience," said David Chalmers, a philosopher at Australian National University. "As with most simple and powerful hypotheses, reality will probably turn out to be more complicated, but we'll learn something from the attempt. I'd say that it doesn't solve the problem of consciousness, but it's a useful starting point."

Dr. Tononi acknowledged, "The theory has to be developed a bit more before I worry about what's the best consciousness meter you could develop." But once he has one, he would not limit himself to humans. As long as people have puzzled over consciousness, they have wondered whether animals are conscious as well. Dr. Tononi suspects that it is not a simple yes-or-no answer. Rather, animals will prove to have different levels of consciousness, depending on their integrated information. Even *C. elegans* might have a little consciousness.

"Unless one has a theory of what consciousness is, one will never be able to address these difficult cases and say anything meaningful," Dr. Tononi said.

—September 21, 2010

In Tiny Worm, Unlocking Secrets
of the Brain

By NICHOLAS WADE

In an eighth-floor laboratory overlooking the East River, Cornelia I. Bargmann watches two colleagues manipulate a microscopic roundworm. They have trapped it in a tiny groove on a clear plastic chip, with just its nose sticking into a channel. Pheromones—signaling chemicals produced by other worms—are being pumped through the channel, and the researchers have genetically engineered two neurons in the worm's head to glow bright green if a neuron responds.

These ingenious techniques for exploring a tiny animal's behavior are the fruit of many years' work by Dr. Bargmann's and other labs. Despite the roundworm's lowliness on the scale of intellectual achievement, the study of its nervous system offers one of the most promising approaches for understanding the human brain, since it uses much the same working parts but is around a million times less complex.

Caenorhabditis elegans, as the roundworm is properly known, is a tiny, transparent animal just a millimeter long. In nature, it feeds on the bacteria that thrive in rotting plants and animals. It is a favorite laboratory organism for several reasons, including the comparative simplicity of its brain, which has just 302 neurons and 8,000 synapses, or neuron-to-neuron connections. These connections are pretty much the same from one individual to another, meaning that in all worms the brain is wired up in essentially the same way. Such a system should be considerably easier to understand than the human brain, a structure with billions of neurons, 100,000 miles of biological wiring and 100 trillion synapses.

The biologist Sydney Brenner chose the roundworm as an experimental animal in 1974 with this goal in mind. He figured that once someone provided him with the wiring diagram of how 302 neurons were connected, he could then compute the worm's behavior.

The task of reconstructing the worm's wiring system fell on John G. White, now at the University of Wisconsin. After more than a decade's labor, which required examining 20,000 electron microscope cross sections of the worm's anatomy, Dr. White worked out exactly how the 302 neurons were interconnected.

But the wiring diagram of even the worm's brain proved too complex for Dr. Brenner's computational approach to work. Dr. Bargmann was one of the first biologists to take Dr. White's wiring diagram and see if it could be understood in other ways.

Cori Bargmann grew up in Athens, Ga., a small college town in the Deep South where her father taught statistics at the University of Georgia. Both her parents had been translators and met while Rolf Bargmann was working at the Nuremberg trials. Her mother, Ilse, would read to her in German the works of the Austrian animal behaviorists Konrad Lorenz and Karl von Frisch, planting the seeds of an interest in neuroscience.

"I went into science because I loved the labs," Dr. Bargmann says. She liked the machines and instruments, the fun of building things with one's own hands, of learning what no one else knew. An outstanding student, she chose for her PhD degree to work in the MIT lab of Robert A. Weinberg, a leading cancer biologist. The first mutated genes capable of causing cancer were being isolated. "It was an incredibly exciting time," she says.

Her task was to clone a rat gene called neu. When mutated, the gene causes a tumor, but one that the rat's immune system can attack and destroy. Several years later, the human version of neu, called HER-2, was found to be amplified in breast cancer, and its receptor protein product is the target of the artificial antibody known as Herceptin, a leading breast cancer drug.

For her postdoctoral work, Dr. Bargmann decided to work on animal behavior. The mouse is a standard organism for such studies, but she did not like hurting furry animals. "In Weinberg's lab I would start to cry every time I had to do anything with a mouse," she says. A nonfurry alternative was the fruit fly. She interviewed with a leading laboratory in California, but her husband at the time did not wish to move there.

That left the roundworm. There are now several hundred worm labs around the world, of which perhaps 30 or so, like Dr. Bargmann's, focus on the worm's nervous system. In 1987, "worms weren't entirely respectable," Dr. Bargmann says. But right there at MIT, H. Robert Horvitz had established one of the first serious worm labs in the United States. She joined his lab and read everything written on the worm, including all the back copies of the little field's informal journal, *The Worm Breeder's Gazette.*

She noticed that a particular behavior of *C. elegans* had been described but not well explored: it can taste waterborne chemicals and move toward those it finds attractive. Dr. White's wiring diagram had been published the year before, in 1986. With this in hand, she told Dr. Horvitz she planned to identify which of the worm's 302 neurons controlled its chemical-tracking behavior.

He thought the project was too ambitious, but said she could spend six months on the attempt. Each neuron in the worm's brain is known, and is assigned a three-letter name. Specific neurons can be identified under a microscope and zapped with a laser beam, allowing the neuron's role to be deduced from whatever function the worm may seem to have lost.

Dr. Bargmann slogged her way through the task of killing each neuron one by one. Telling one neuron from another under the microscope is not easy. "It's like knowing each grape in a bunch is different, but not quite being able to see it," Dr. Horvitz said. "The first thing she had to do was learn the worm's neuro-anatomy, and she did so in a way only one other person has ever done." (He was referring to John E. Sulston, who traced the lineage from the egg of all 959 cells in the adult worm's body).

She discovered, by accident, the neurons that control the worm's switch into hibernation, a survival strategy for when food is scarce or neighbors too many. Finally, she found the neurons that control taste, showing that without them the worm could not track chemicals, and that it retained this ability even if she killed all the other neurons in the worm's body.

She also discovered that the worms have a sense of smell—the ability to detect airborne chemicals—as well as a sense of taste. Since worms eat bacteria that feed on decaying plants and carcasses, she figured they should be able to detect and home in on the aromas of putrefaction. The redolent draft from these experiments caused a certain degree of complaint in Dr. Horvitz's lab. After she succeeded, she says, "Horvitz told me that my great strength as a scientist was that I could think like a worm."

"Cori is talented beyond thinking like a worm," Dr. Horvitz now says. "She can think like very few other people in a rigorous and creative way, and so has repeatedly developed new kinds of approaches."

Dr. Bargmann moved in 1991 to the University of California, San Francisco, to start her own lab. She began by following up her finding that worms have a sense of smell. In 1991, Richard Axel and Linda Buck discovered the molecular basis for the sense of smell: there are about a thousand genes, at least in rats, that make odorant receptors, proteins that stud the olfactory nerves' endings in the nose and respond to specific odors.

The *C. elegans* genome had just been decoded, and Dr. Bargmann was able to identify the worm's odorant receptor genes. In fact, they have 2,000 of them, twice as many as the rat.

"This is what they do," Dr. Bargmann says. The worm cannot see. Its world is one of smells, not sights. It needs to scent the soil bacteria that are its prey, while avoiding those that are poisonous to it. Ten percent of its genes are dedicated to making it a champion connoisseur of odors, mostly unpleasant.

With the odorant genes in hand, Dr. Bargmann could apply genetics to figuring out how the worm's sense of smell worked. By working with mutant worms, she showed that a specific odor receptor recognizes a specific odor, a finding that was implied by the Axel-Buck discovery but that no one had managed to nail down.

She found that worms with a mutation in a gene called odr-10 could not smell diacetyl, a chemical that gives butter its odor and is also produced by a bacterium that is a favorite worm food. The odr-10 gene, which makes the odor receptor protein that detects diacetyl, is active in neurons that guide the worm toward a scent.

Dr. Bargmann switched things around so that odr-10 was expressed only in a neuron that detected scents repulsive to the worm. These worms backed away from the buttery odor, showing that it is not the odor receptors but the wiring of the nervous system itself that determines whether the worm deems an odor delicious or detestable.

This was a surprising result because most people thought that sensory information was perceived as neutral, with the brain deciding later from the context whether it was good or bad. Some scientists said that only worms behave this way, but the same result was later obtained in mice.

Dr. Bargmann sees the arrangement in evolutionary terms. "The more reliable a piece of information is, the more it will be shifted into the genome," she says. That way, an organism does not have to risk learning what is good or bad; the genes will dictate the right behavior by wiring it into the nervous system. Worms are wired up to know that diacetyl means good eating.

Having studied the worm by mutating its genes, Dr. Bargmann then looked at natural variation in the genetic basis of worm behavior. Most worms in nature like to congregate in clumps, but the laboratory version of *C. elegans* has developed an unusual liking for being on its own. She linked this difference in behavior to the switch of a single amino acid unit in a protein called npr-1 (for neuropeptide Y receptor-1).

It took several more years to learn how the system worked. It turns out that social behavior in the worm is controlled by a pair of neurons called RMG. The two RMG neurons receive input from various sensory neurons that detect the several environmental cues that make worms aggregate. RMG integrates this information and sends signals to the worm's muscles.

The usual role of the RMG neurons is to promote social behavior, but when the npr-1 gene is active, the RMG neurons cannot receive input from their sensory neurons, and the worms switch to solitary behavior.

While working out the worm's sense of smell, Dr. Bargmann fell in love with another olfactory researcher, Richard Axel. Dr. Axel works at Columbia University, and she was able to join him in New York by finding a place at Rockefeller University. Dr. Axel was helping her clear out her apartment in San Francisco when he heard he had won the Nobel Prize.

Right after that pleasant news, he had to drive to the local Goodwill store to drop off the stuff to be given away. "People think that if you're married to a scientist you talk about science all the time," Dr. Bargmann says. They read each other's papers before publication, but they don't plan experiments together. Dr. Axel works on how olfactory information is handled in the cortex, the highest level of human and mouse brains.

"Probably once or twice a week we are sitting at dinner and Richard says, 'The cortex is hopeless,' and I say, 'That's why I work on the worm,'" Dr. Bargmann said.

After studying the little animal for 24 years, she believes she is closer to understanding how its nervous system works.

Why is the wiring diagram produced by Dr. White so hard to interpret? She pulls down from her shelves a dog-eared copy of the journal in which the wiring was first described. The diagram shows the electrical connections that each of the 302 neurons makes to others in the system. These are the same kind of connections as those made by human neurons. But worms have another kind of connection.

Besides the synapses that mediate electrical signals, there are also so-called gap junctions that allow direct chemical communication between neurons. The wiring diagram for the gap junctions is quite different from that of the synapses.

Not only does the worm's connectome, as Dr. Bargmann calls it, have two separate wiring diagrams superimposed on each other, but there is a third system that keeps rewiring the wiring diagrams. This is based on neuropeptides, hormonelike chemicals that are released by neurons to affect other neurons.

The neuropeptides probably help control the brain's general status, or mood. A strong hint of how they work comes from the npr-1 gene, which makes a protein that responds to neuropeptides. When the npr-1 gene is active, its neuron becomes unavailable to its local circuit.

That may be a reason why the worm's behavior cannot be computed from the wiring diagram: the pattern of connections is changing all the time under the influence of the worm's 250 neuropeptides.

The connectome shows the electrical connections, and hence the quickest paths for information to move through the worm's brain. "But if only a subset of neurons are available at any time, the connectome is ambiguous," she says.

The human brain, too, has neuropeptides that set mood and modify behavior. Neuropeptides are probably at work when the pain pathways are cut off in acute crises, allowing people to function despite serious wounds.

The human brain, though vastly more complex than the worm's, uses many of the same components, from neuropeptides to transmitters. So everything that can be learned about the worm's nervous system is likely to help with the human system.

Though the worm's nervous system is routinely described as simple, that is true only in comparison with the human brain. The worm has 22,000 genes, almost as many as a person, and its brain is a highly complex piece of biological machinery. The work of Dr. Bargmann's and other labs has deconstructed many of its operational mechanisms.

What would be required to say that the worm's nervous system was fully understood? "You would want to understand a behavior all the way through, and then how the behavior can change," Dr. Bargmann says.

"That goal is not unattainable," she adds.

—*June 20, 2011*

A Sense of Where You Are

By JAMES GORMAN

In 1988, two determined psychology students sat in the office of an internationally renowned neuroscientist in Oslo and explained to him why they had to study with him.

Unfortunately, the researcher, Per Oskar Andersen, was hesitant, May-Britt Moser said as she and her husband, Edvard I. Moser, now themselves internationally recognized neuroscientists, recalled the conversation recently. He was researching physiology and they were interested in the intersection of behavior and physiology. But, she said, they wouldn't take no for an answer.

"We sat there for hours. He really couldn't get us out of his office," Dr. May-Britt Moser said.

"Both of us come from nonacademic families and nonacademic places," Edvard said. "The places where we grew up, there was no one with any university education, no one to ask. There was no recipe on how to do these things."

"And how to act politely," May-Britt interjected.

"It was just a way to get to the point where we wanted to be. But seen now, when I know the way people normally do it," he said, smiling at the memory of his younger self, "I'm quite impressed."

So, apparently, was Dr. Andersen. In the end, he yielded to the Mosers' combination of furious curiosity and unwavering determination and took them on as graduate students.

They have impressed more than a few people since. In 2005, they and their colleagues reported the discovery of cells in rats' brains that function as a kind of built-in navigation system that is at the very heart of how animals know where they are, where they are going and where they have been. They called them grid cells.

"I admire their work tremendously," said Eric Kandel, the Nobel laureate neuroscientist who heads the Kavli Institute for Brain Science at Columbia and who has followed the Mosers' careers since they were graduate students.

John O'Keefe of University College London, whose discovery in the 1970s of so-called place cells in the brain that register specific places, like the corner deli or grandma's house, and who was one of the Mosers' mentors, said that the discovery of the grid cells was "incredibly significant."

The workings of the grid cells show that in the brain "you are constantly creating a map of the outside world," said Cori Bargmann, of Rockefeller University, who is one of the two leaders of a committee set up to plan the National Institutes of Health's contribution to President Obama's recently announced neuroscience initiative.

Often, the workings of billions of neurons that produce our thoughts are opaque. But electrical recordings of signals emitted by grid cells show a map "with a framework and coordinates that are completely intuitive," Dr. Bargmann said. And to find such a straightforward system is, in its own way, "just mind-boggling." What is the brain doing being so mysteriously unmysterious?

The implications of the discovery are both practical and profound. The cells have been proved to exist in primates, and scientists think they will be found in all mammals, including humans. The area in the brain that contains the grid cell navigation system is often damaged early in Alzheimer's disease, and one of the frequent early symptoms of Alzheimer's patients is that they get lost. The Mosers do not work on humans, but any clues to understanding how memory and cognitive ability are lost are important.

On the most profound level, Dr. O'Keefe, the Mosers and others speculate that the way the brain records and remembers movement in space may be the basis of all memory. This idea resonates with the memory palaces of the Renaissance, imagined buildings that used spatial cues as memory aids. The technique dates to the ancient Greeks. In this regard, neuroscience may be catching up with intuition.

A Welcome Ambush

Edvard, 51, and May-Britt Moser, 50, now direct the Kavli Institute for Systems Neuroscience and the Centre for the Biology of Memory at the Norwegian University of Science and Technology here in Trondheim. They have a steady stream of findings coming from their lab, and a slew of awards, the latest of which, the Perl-UNC Neuroscience Prize, they received April 16 at the University of North Carolina.

But they did not grow up in a center of academic ferment or intellectual competition. They were born and raised on islands off the coast of Norway a couple of hundred miles north of Bergen, part of an area known as Norway's Bible Belt. They went to the same high school, but didn't really get to know each other until they met again at the University of Oslo in the 1980s.

May-Britt, who grew up on a farm, remembers an environment in which drinking, card playing and dancing were all frowned upon. When she called home from Oslo announcing that she had been to a bar and had her first beer, her mother said, "And what's next?"

The Mosers married in 1985 while still undergraduates. By the time they had finished their doctorates, in 1995, they had two daughters, but they were ready to see the world, to train in laboratories outside Norway. And they did spend time in England, with Dr. O'Keefe, and in Scotland, with Richard Morris at the University of Edinburgh.

But the Mosers' travels were cut short when they were ambushed by a job offer too good to refuse, from the university in Trondheim, where they have been ever since.

"Without knowing it, we actually negotiated," May-Britt said, "because we were not interested if we only got one job, and we got two jobs. And we were not interested if we did not get the equipment we needed, and they gave us that." Suddenly, without having really planned it, they had their own lab.

Of course, nothing happens suddenly in research. They began in what Dr. May-Britt Moser described as a bomb shelter, and gradually, over time, built up their program. Similarly, they did not set out looking in the part of the brain where they ended up.

They began recording the activity of cells in the hippocampus, with electrodes implanted in the brains of rats as they roamed an enclosed area. This is still a main method, and the rats are intriguing to watch, pursuing little bits of chocolate cereal on the floor of an enclosure, seemingly oblivious to the implants attached to their skulls.

A Black Box

The Mosers wanted to find how information was flowing to the place cells, whether it was going from one area of the hippocampus to another. But even after they inactivated sections of this brain area, the place cells still functioned. So it seemed that information was flowing in from the nearby brain area, the entorhinal cortex.

They started looking there, and in their early work they were helped by Menno Witter, then in Amsterdam, now at Trondheim, in the delicate task of guiding the electrodes to the right spot.

"We didn't immediately find the grid cells," Dr. Edvard Moser said. At first they noticed cells that would emit a signal every time a rat went to a particular spot, and they thought that perhaps this was something like the place cells in the hippocampus that are tied to locations in the outside world. But gradually they learned that what they were seeing was a cell that tracked the rat's movement in the same way, no matter where the rat was. The cell was not responding to some external mark, it was keeping track of how the rat moved. And when they gave the rats enough room, a very regular pattern emerged.

"The first thing was that we thought there was something wrong with the equipment," Dr. Edvard Moser said.

"I thought, 'Is this a bug?'" Dr. May-Britt Moser said.

After a 2005 paper in *Nature*, in which they reported the discovery and named the cells, other labs confirmed the findings and more discoveries followed, in their lab and elsewhere.

It is now clear that the grid cells, in combination with cells that sense head direction and others that sense borders or boundaries—both originally identified in other parts of the brain by other labs—form a kind of dead-reckoning navigation system in the brain that maps movement.

Information flows from this part of the brain to the hippocampus, and then back. Exactly how the grid informs the place cells, and vice versa, is not known.

What scientists have now are two ends of a system with a black box in the middle that is not fully understood. At one end are place cells. At the other are grid cells. As to what exactly happens in between, and how the grid cells form in the first place, Dr. Edvard Moser said, "That's still a 10-, a 20-year research problem."

Or, as Dr. O'Keefe put it, "We are still in the pre-Newtonian phase of neuroscience."

The Mosers remain something of an anomaly. Not only are they off the beaten academic track, but they are a married couple who work together on the same scientific problems at the same institution at the highest levels of science, a true rarity.

They do have different spheres in their new, state-of-the-art lab. May-Britt is more hands-on with the experiments and the design, and Edvard is more involved in mathematical analysis and interpretation of the results.

"We have a common project and a common goal," he wrote in response to an e-mailed question, "and we both intensely burn for it. And we depend on each other for succeeding."

He continued, "Most couples manage to cooperate on child raising—for us, our brain project is our third child, so nothing different, really."

—*April 30, 2013*

CHAPTER 11

On Science
and Scientists:
People, Process and
Portrayals

A Severe Strain on Credulity

As a method of sending a missile to the higher, and even to the highest, part of the earth's atmospheric envelope, Professor Goddard's multiple-charge rocket is a practicable, and therefore promising, device. Such a rocket, too, might carry self-recording instruments, to be released at the limit of its flight, and conceivably parachutes would bring them safely to the ground. It is not obvious, however, that the instruments would return to the point of departure; indeed, it is obvious that they would not, for parachutes drift exactly as balloons do. And the rocket, or what was left of it after the last explosion, would have to be aimed with amazing skill, and in a dead calm, to fall on the spot whence it started.

But that is a slight inconvenience, at least from the scientific standpoint, though it might be serious enough from that of the always-innocent bystander a few hundred or thousand yards away from the firing line. It is when one considers the multiple-charge rocket as a traveler to the moon that one begins to doubt and looks again, to see if the dispatch announcing the professor's purposes and hopes says that he is working under the auspices of the Smithsonian Institution. It does say so, and therefore the impulse to do more than doubt the practicability of such a device for such a purpose must be—well, controlled. Still, to be filled with uneasy wonder and to express it will be safe enough, for after the rocket quits our air and really starts on its longer journey, its flight would be neither accelerated nor maintained by the explosion of the charges it then might have left. To claim that it would be is to deny a fundamental law of dynamics, and only Dr. Einstein and his chosen dozen, so few and fit, are licensed to do that.

His Plan Is Not Original

That Professor Goddard, with his "chair" in Clark College and the countenancing of the Smithsonian Institution, does not know the relation of action to reaction, and of the need to have something better than a vacuum against which to react— to say that would be absurd. Of course he only seems to lack the knowledge ladled out daily in high schools.

But there are such things as Intentional mistakes or oversights, and, as it happens, Jules Verne, who also knew a thing or two in assorted sciences—and had, besides, a surprising amount of prophetic power—deliberately seemed to make the same mistake that Professor Goddard seems to make. For the

Frenchman, having got his travelers to or toward the moon into the desperate fix of riding a tiny satellite of the satellite, saved them from circling it forever by means of an explosion, rocket fashion, where an explosion would not have had in the slightest degree the effect of releasing them from their dreadful slavery. That was one of Verne's few scientific slips, or else it was a deliberate step aside from scientific accuracy, pardonable enough in him as a romancer, but its like is not so easily explained when made by a savant who isn't writing a novel of adventure.

All the same, if Professor Goddard's rocket attains sufficient speed before it passes out of our atmosphere—which is a thinkable possibility—and if its aiming takes into account all of the many deflective forces that will affect its flight, it may reach the moon. That the rocket could carry enough explosive to make on impact a flash large and bright enough to be seen from the earth by the biggest of our telescopes—that will be believed when it is done.

—*January 13, 1920*

A Correction

On Jan. 13, 1920, "Topics of *The Times*," an editorial-page feature of *The New York Times*, dismissed the notion that a rocket could function in a vacuum and commented on the ideas of Robert H. Goddard, the rocket pioneer, as follows:

"That Professor Goddard, with his 'chair' in Clark College and the countenancing of the Smithsonian Institution, does not know the relation of action to reaction, and of the need to have something better than a vacuum against which to react—to say that would be absurd. Of course he only seems to lack the knowledge ladled out daily in high schools."

Further investigation and experimentation have confirmed the findings of Isaac Newton in the seventeenth century and it is now definitely established that a rocket can function in a vacuum as well as in an atmosphere. *The Times* regrets the error.

—*July 17, 1969*

5 Years After the Nobel:
Portrait of a Man Obsessed With Science

By HAROLD M. SCHMECK JR.

The picture on the screen, meaningless to a layman, looked like a row of vertical ink blotches on a dirty gray sheet. To the dozen or so young men and women clustered in the room, the blots were an unfinished mystery story.

The patterns showed cloned mouse cells that should have been identical. In fact they had grown into two genetically different populations. How? Why?

Orchestrating the discussion was a bearded man in his early 40s, smoking a pipe, wearing metal-rimmed glasses, a dark red sport shirt and suntan pants. His young colleagues addressed him as "David." He is known to the world at large as Dr. David Baltimore, American Cancer Society Professor of Microbiology at MIT, co-winner of a Nobel Prize and one of the most brilliant and inventive minds in modern American science.

As he and the young scientists tried to make sense of those mouse cells, Dr. Baltimore's comments were sparse but telling, terminating one line of investigation, encouraging another.

"I don't think it's going to be worthwhile to drive that into the ground," he said at one point. "Good, very good," he said almost inaudibly at another. One research worker said two specimens seemed to be identical. "No," Dr. Baltimore said quickly, "they are reversed." The meeting was informal, conducted at lunchtime with sandwiches from paper bags. Excitement and concentration seemed to wax and wane with the discussion.

Dr. Baltimore sat at the side, only occasionally asking a question or rising to go to the chalkboard. But the speakers seemed to be directing their words as much to him as to the rest of the group.

The young MIT scientists were seeking ways to harness this particular paradox—to make it reveal an underlying truth about the immune defense system of humans and animals.

The cells were precursors of those that produce the disease-fighting substances called antibodies. The experiments, still in progress, might explain something important about the development of these indispensable cells.

Dr. Baltimore's research style and the way he imparts it to a new generation reveal something of the creative genius of modern biology, something of the fire that has kindled a revolution in human understanding of the chemistry of life on earth.

"I think it's partly a habit of mind," he told a visitor. "It involves a kind of obsessiveness. Unless you are obsessed with scientific questions you are not going to get anywhere with them."

Also needed, he says, is a talent for thinking a logical train of experiments through to its long-term consequences. The layman's perception is often that of a scientist working doggedly, in lonely dedication, toward some distant goal. The reality, Dr. Baltimore says, is that a scientist must learn to find the path of least resistance through a maze of scientific unknowns, choosing the experiments that are ripe to be done, even backing off from a tough problem until some new insight or new technique softens it up.

He once backed off from an impasse in virus research for 17 years and then picked it up again when a new development made the question ripe for solving. The lapse had continued to bother him over the years until he ended it.

Brilliance in scientific research is not a simple talent, nor is it simply explained. The obsessive urge to find answers is clearly a part of it, as is the talent for choosing the right questions. Dr. Baltimore says he wakes up in the morning thinking—and assumes everyone else does, too. There were times in his early career, notably at Rockefeller University, when his waking hours were science and nothing but science; meals were the only breaks in the work, and even these functioned as opportunities to discuss science with colleagues.

A Compulsion to Protest

Once in the Vietnam War years he halted important research for a week in protest against the invasion of Cambodia, all the while feeling the compulsion to make the protest but also the agony over the delay in his experiments.

The ability to devise fruitful experiments is largely a learned talent, he says, that can be passed on to students. The meeting of his laboratory group showed that process at work.

Today Dr. Baltimore's creative role is mainly that of a catalyst, directing research rather than doing it himself. But his talents and style put an imprint on the work. He is a man of great energy and, some of his students say, a library of valuable information on many aspects of molecular biology.

His current obsessions embrace some of the key fields in molecular biology: study of how viruses reproduce and how some of them transform normal cells to a state like cancer; study of immunology, the internal defense system that tells friend from foe and fights back against invasion by germs and other intruders in the living body.

Central to much of the work is the sometimes controversial recombinant DNA technology. With it, scientists can make limitless copies of the pieces of deoxyribonucleic acid that serve as the genes of all forms of life; can snip and splice and rearrange genes from any species; can combine genetic material from man and mouse, and can grow human genes in bacteria.

Mathematics Came Easy

Dr. Baltimore was introduced to the real obsessions of science as a high school student in the middle 1950s. Mathematics and related fields came easy to him, he recalls, but fascination began to germinate at a summer session for high school students at the Jackson Laboratory in Bar Harbor, Me. The young visitors to the famous laboratory listened to lectures, did some research projects and discussed biology with experts.

Although Dr. Baltimore did not decide on a career in science until he was about halfway through his undergraduate studies at Swarthmore College, he now recognizes the Jackson laboratory experience as a key determinant.

The person he remembers as the "guru" of his group was Howard M. Temin, now of the McArdle Laboratory at the University of Wisconsin. The two never worked together after that summer. Neither had more than a passing knowledge of what the other was up to as their careers evolved in the next decade and a half. Their habits of mind, or at least their scientific styles, grew to be far different.

Yet, in one of the ironies of modern science, their research paths converged until, in 1970, they each, independently and unknown to the other, did almost the same experiments with viruses and thereby demolished what many considered a central dogma of modern molecular biology. Five years later they and Dr. Renato Dulbecco, one of the towering figures of modern biology, were awarded the Nobel Prize.

Looking Like Dr. Ehrlich

A photograph of Dr. Baltimore shaking hands with the King of Sweden at the Nobel Prize ceremonies that fall shows the young American scientist dressed in formal costume and looking a little like the actor who played Dr. Paul Ehrlich in the 1940 movie *Dr. Ehrlich's Magic Bullet.* The film was made approximately the year Dr. Baltimore was born.

In the research cited by the Nobel committee, Drs. Temin and Baltimore had identified a special class of enzymes through which certain viruses could subvert the genetic machinery of the cells they infected.

The two closely related master chemicals of heredity are DNA and ribonucleic acid (RNA). The arrangement of chemical subunits in the DNA serves as the genetic code spelling out each message of heredity. One of the key functions of RNA is to form a copy of the DNA and use this as the blueprint for the production of the specific protein coded for by the gene.

It had been almost universally assumed that the flow of information was always from a nucleic acid to protein. Some thought it always from DNA to RNA to protein. That concept left scientists puzzled over the ability of some of the viruses that contained RNA instead of DNA to transform the very nature of the cells they infected—to make them cancerous. Somehow, they reasoned, the virus RNA must be leaving its message permanently in the DNA of the infected cell.

The discovery of the enzymes now known as reverse transcriptases solved that puzzle. Such an enzyme was found by Dr. Temin in a virus that causes cancer in chickens. Dr. Baltimore found his in a virus that causes leukemia in mice. It soon became clear that RNA-containing viruses known to cause cancer in animals had these subversive enzymes while other viruses did not. Their existence, denied on theoretical grounds for years, proved to be a general phenomenon. Students since then have confirmed its profound importance.

On Predicting Creativity

Working closely with about 20 younger scientists today, Dr. Baltimore says he can see in some of them the possibility of creative greatness, although he adds quickly that creativity really defies prediction. Some scientists plug along for decades, then abruptly blossom into brilliant productivity. Others, seemingly racing ahead in the throes of genius, suddenly fade and lose the fire. Some concentrate narrowly. Others leap at creative opportunity where they perceive it.

Dr. Baltimore likes to be in the competitive forefront of a field and does not leave when the field becomes crowded with research workers following the current fashion. He has sometimes been criticized for his zeal in leaping into highly active fields.

The scientist credits association with many brilliant workers for helping shape his talents. Notable among these has been Dr. Alice Huang, a microbiologist at Harvard Medical School, an early collaborator on virus research and now his wife.

About six months after he had joined the laboratory of Richard Franklin at Rockefeller Institute (now University) in the early 1960s, he was doing significant work at the forefront of the science of that day. He says he does not know just how this happened, except that he was always allowed to follow his own creative path.

But there were older scientists at that time too who could see talent for creativity taking shape. "There are times in the development of a field of knowledge when the ground for the next major development is laid," said Dr. Igor Tamm of Rockefeller on an important occasion in 1964.

"David's teachers and associates have all been impressed with his broad grasp of concepts and the integrative quality of his mind," he continued. "I therefore think that David has ample qualifications not only for a productive life in research, but also for a rewarding life in teaching. I expect that his lively interests in science will fire enthusiasm in others; that his insights will illuminate many."

The occasion was the presentation of David Baltimore for the PhD degree, at the beginning of a creative career whose dimensions are still unfolding.

—August 26, 1980

Uncovering Science:
A Perpetual Student Charts a Course
Through a Universe of Discoveries

By MALCOLM W. BROWNE

After 22 years as a science writer I recently retired to cut firewood in Vermont and enjoy the memories of an exciting life, in which I covered a half-dozen wars before discovering the deeper satisfaction of observing and reporting the achievements of scientists.

I relished my 17 years as a foreign correspondent, but believe it or not, even the high drama of disaster, violence and political upheaval that dominates front pages can lose its luster for journalists seeking new experiences.

After a time, a news writer may begin to sense a kind of sameness in most of the events that pass as news. When that happens a lucky few of us discover that in science, almost alone among human endeavors, there is always something new under the sun.

In 1977, weary of the sameness of war and politics, I returned to the United States to become a science writer—a transition that almost overwhelmed me at first. Although I had earned my living in a chemical laboratory in the 1950s, I had almost forgotten how speedily science booms along.

As a trivial example, in the last quarter century alone the American Chemical Society has added more than 10 million chemical substances to its list of known molecules, most of them man-made. Stupendous strides in chemical synthesis have given the world a wealth of new materials and drugs, and have created entirely new classes of molecules, including hollow molecules shaped like cages that contain even smaller molecules. Perhaps most intriguing of all, chemists are turning up more and more hints of how life may have originated from the carbon spawned in the explosions of supernova stars.

The 1960s and '70s were a time of great ferment in all the sciences, and the momentum of those years has carried to the present. Among the major achievements in physics I was privileged to report were the discovery of the top quark at Fermilab—the last of six quarks predicted by theory—and the discovery in Japan that the elusive neutrino particle probably has some mass, a finding with profound implications for the fate of the universe.

Sometimes science writers watching the accelerating deluge of discoveries in physics, chemistry, molecular biology and astrophysics have actually outpaced the thinking of the scientists themselves, and science writing has bloomed as a major component of general journalism.

At *The New York Times*, for example, the science editor, Walter Sullivan, had been steeping himself in astrophysics for decades when in 1965 two scientists at Bell Labs, Dr. Arno A. Penzias and Dr. Robert W. Wilson, accidentally discovered a faint microwave radio signal coming from all directions in the sky. It was the first hard evidence that the universe had begun with a "Big Bang."

Dr. Penzias later paid Mr. Sullivan one of the warmest compliments ever given a science writer, after reading an article expanding on the implications of the discovery: "Only after reading Sullivan's story in *The New York Times*," Dr. Penzias said, "did we fully understand what we had done." In 1978 Dr. Penzias and Dr. Wilson, their discovery having radically changed man's view of the cosmos, were awarded a Nobel prize.

The raw material used in news coverage of important discoveries is often rather skimpy. . . . It is up to the science writer to judge the significance of the findings and place them in context.

This means that a science writer must be a perpetual student.

News stories about astronomy frequently have to do with black holes, for example, and for many astronomy fans it is enough to know that black holes suck in everything near them and won't let anything—even light—escape. But to understand black holes at a deeper level requires familiarity with Einstein's general theory of relativity, and some of the greatest minds in physics are still puzzling over some of relativity's implications.

The presumably lesser mind of the science writer has an even harder row to hoe than that of the scientist. But try, he or she must.

Curiously, as science floods the world with discoveries of variable quality— unhappily, the overwhelming majority of scientific papers fall into the category of junk science—the task of the writer seems to grow easier in some ways.

The science itself gets harder all the time, of course. The professional journals on which scientists (and writers) heavily depend are not easy to read, and they seem more difficult each year. To give some idea: several years ago the editor of a prestigious journal, *Physical Review Letters*, found it necessary to decree that at least the first couple of paragraphs of each published paper should be intelligible to an average PhD physicist not specializing in the subject of the paper.

But no discovery occurs in a vacuum, a fact that has helped many science writers find their way through the fog. Newton, the discoverer of much that is essential to modern physics and mathematics, wrote that he could not have seen so far without having stood on the shoulders of giants. In a small way, the science writer can also stand on the shoulders of giants. Most discoveries are incremental steps, and if a writer comes to terms with the earlier steps, new findings generally slide into an intelligible context.

The science writer also lives in dread of losing readers' interest, which happens all too often. For instance, it's hard to persuade a reader (or editor) to take seriously some gigantic experiment that produced only a null result. So the writer is obliged to point out that null results can have far-reaching scientific importance. The failure of an experiment by Albert Michelson and Edward Morley a century ago to detect the earth's passage through a hypothetical universal "ether" lent powerful support to relativity theory. Scientists will soon begin a quest for gravity waves in a strikingly similar experiment, and any writer unfamiliar with Dr. Michelson and Dr. Morley's work will have trouble reporting the story. . . .

One trick of the trade is the use of analogy to convey the flavor of an idea, discovery or equation. But effective though analogies may be, they are never exactly appropriate and sometimes are downright wrong. It's probably OK to call a proton a "beanbag" containing three quarks, but to call a proton accelerator an "atom smasher" makes physicists squirm.

Practice may not make perfect, but a science writer who stays in the game long enough is bound to get better. Unfortunately, it can happen that both the writer and reader may miss the significance of a scientific development; it is like being knocked down by a strong opponent.

Paul Gallico, a renowned sports writer in the 1930s, relished such encounters. Before writing about one of the boxing matches he covered, Mr. Gallico went into the ring with Jack Dempsey, who knocked him down. This, Mr. Gallico said, taught him all he needed to know about being hit, and that rich experience helped flavor his coverage.

Science writers and their readers sometimes get knocked down by hard ideas rather than by hard gloves. But the experience of grappling with such things as the fiendish mathematics of superstring theory or the complicated tactics of the AIDS virus is its own reward, even at the cost of some bumps.

—February 27, 2000

Colors Are Truly Brilliant in Trek Up Mount Metaphor

By GEORGE JOHNSON

Hovering above the ghoulish terrain, a visitor might feel transported to a distant planet. Rendered in black and white, the lay of the land would seem comfortably familiar: clusters of low, rounded foothills give way to higher, rougher ones, finally converging on majestic snowcapped peaks.

But the colors are all wrong. The alpine forests are a sickly chartreuse. The glaciers and snowfields are yellow at the bottom, orange in the middle, blood-red at the top. Elsewhere, a single peak, ascending through shades of bright yellow, fluorescent green and icy blue, juts above the crimson badlands like an obscenely protuberant Matterhorn.

Confronted with these images, which appeared in the journal *Science*, one might think they were digital photographs sent across space from a *Viking* or a *Voyager*, an exercise in extraterrestrial cartography.

But the territory exists only in the realm of abstraction, as arrangements of data in two experiments that have nothing to do with outer space. One involves genetics, the other quantum physics. In each, scientists are trying to get a better feel for their data by imagining it as a mathematical mountain range—one of the most dominant metaphors in science.

Explaining the strange in terms of the familiar—that is the essence of the scientific quest. In every field, from molecular biology to cosmology, data are sorted and analyzed mathematically. But in the end, the gray numbers are often translated into colorful three-dimensional pictures, the language human brains comprehend best. Using metaphor and analogy, the tools of artists and poets, abstract patterns take on substance and become lodged more firmly in the mind.

For many people, a "mountain of data" evokes a heaping pile of unorganized information. But in science, the phrase can mean data arranged with exquisite precision. Following a trend in the numbers becomes an ascent along a ridgeline leading to a rocky precipice and a stunning view over an expansive valley.

The first landscape, what scientists call a gene expression map, depicts the functioning of the genome of the worm *C. elegans*. Understanding how its DNA operates can lead to insights about the human genome, a biochemical structure commonly thought of as a map, a blueprint, an enciphered text or, more recently, as cellular software, the operating system for the cell.

Adopting instead the montane metaphor, scientists at the Stanford University School of Medicine distilled data from 553 experiments performed by 30 laboratories into an image they hoped would give an intuitive feel for how the worm's 19,000-plus genes interact. (The work, drawing on the computational talents of Stanford Medical Informatics and Sandia National Laboratories, in Albuquerque, was published in the Sept. 14 issue of *Science*.)

Each of the 44 mountains represents a group of genes that, though scattered throughout the worm's genome, become active under the same conditions, producing proteins that various cells need to conduct their affairs. (The significance of 14 of the peaks remains unknown, terra incognita.) The higher the mountain, the more genes it represents, ranging from the towering Mount Zero, a dizzying 2,703 genes high, down to Mount 43, a lowly hillock of five genes. As on a relief map, the arbitrary colors help the eye get a quick fix on the topography.

Though similar in contour, the second image, which appeared on the cover of *Science* in 1995, represents not genes but molecules of a substance called rubidium used in research that won this year's Nobel Prize in Physics. Here the altitude of the mountain indicates how fast the molecules are moving, with speed decreasing as the eye ascends into the chilly heights. The colors represent the number of atoms moving at each velocity, red being the fewest and white the most.

At the peak, most of the atoms are frozen near absolute zero, converging into a single superatom called a Bose-Einstein condensate. In this exotic substance, the rules of quantum mechanics allow thousands of atoms to crowd into the same place at the same time—resulting in a new state of matter.

Like pictures in *National Geographic*, the most arresting scientific images inspire feelings of wanderlust, a desire to lose oneself in a far-off land. A depiction of the data showing how high-speed laser pulses were used to manipulate the spins of electrons in a substance called zinc cadmium selenide becomes an eerily symmetrical iceberg, casting its lonely reflection in a frigid, impossibly still pond. The research by physicists at the University of California at Santa Barbara and Penn State University, earned the cover spot of the June 29 issue of *Science*.

High-speed computers and sophisticated "data mining" software are producing increasingly refined visualizations. But the practice of bringing substance to abstractions with pictures and analogies is as old as science itself.

An individual electron is an evanescent entity, acting something like a particle and something like a wave. Really it is neither, hovering in a metaphorical territory in between. But when electrons move en masse as electricity through a wire, they can be pictured as a liquid. Current, or amperage, becomes equivalent to the rate of flow, and voltage to the pressure in the "pipes."

The metaphor has its limits. Cut a wire and you won't get wet, any more than you'll freeze on top of Gene Mountain. But the analogy helps the mind get a more visceral grip.

The sophisticated procedure used to make thousands of atoms sit still long enough to merge into a Bose-Einstein condensate can be precisely described with mathematics. But it is much more evocative, with a bit of poetic license, to call it "optical cooling." Heat is defined, after all, as the random movement of atoms. So ambush each atom, hitting it from every direction with photon guns shooting tiny projectiles of light—striking it this way and that way until it is almost stationary. The resulting glob of slow-motion matter is called optical molasses.

Metaphors are always inexact; in the quantum realm they are stretched to the breaking point. Physicists talk about a subatomic particle's "spin." Like a top, a particle can rotate clockwise or counterclockwise. But take the analogy too far and it crumbles. An ordinary top can revolve faster or slower across a smooth range of speeds. Particles, being quantum in nature, can spin only at certain fixed velocities, preset by nature. And they can spin in various combinations—43 percent clockwise and 57 percent counterclockwise, for example—at the same time.

Less tangible still is a quality called isotopic spin. The nuclei of atoms are built from Janus-faced particles called nucleons. If a nucleon's "isospin" is counterclockwise, it acts as a positively charged proton; reverse the direction and it becomes a chargeless neutron. But these pirouettes take place in a purely mathematical realm, an artificial space whose dimensions have nothing to do with height, width or length.

Space itself has become a metaphor. Think of cyberspace, which can be explored but not measured, or the "desktop" of your personal computer, a simulated expanse across which you "drag" folders and icons. The motion is illusory. All your mouse strokes are really doing is rapidly switching pixels on and off.

You can construct your own "restaurant space" describing the dining in your neighborhood. Categorize them according to three parameters—price, quality and years in business—and plot the information on a three-dimensional graph. Each restaurant becomes a point in an abstract space in which nearness is a measure of similarity. Two adjacent establishments might be blocks apart in the physical world.

There is no need to stop with three dimensions. Imagine another axis measuring the number of tables and another measuring the variety of wines in the cellar. Now you have a five-dimensional "hyperspace," impossible to really picture but something that scientists use all the time.

It is not always clear whether a space is real or artificial. Super string theory holds that the particles making up matter and energy are secondary manifestations—epiphenomena—generated by tiny objects called strings and branes vibrating in a space of 10 dimensions. The theory is enormously successful on paper, but a question, perhaps unanswerable, lingers: are these extra dimensions physical or mental, like restaurant space?

Sometimes metaphors are outgrown. The biggest break in the Human Genome Project actually came half a century ago, when scientists realized that DNA could be thought of as a text, the chemical letters spelling out instructions for making and operating cells. But you can take a metaphor only so far. As experiments reveal how dynamic the genome is, with genes switching each other on and off, it begins to seem like a text that can read and edit itself—less like a book than a computer. But is DNA software or hardware? It is a little of each. As with wave/particle duality, neither metaphor exactly fits.

As the discoveries of science become part of popular culture, the metaphorical flow sometimes goes the other way. Novelists look to science for linguistic lenses to cast the familiar in a new light. The patterns of circuitry on a computer chip are commonly compared with the layout of a modern city. In *The Crying of Lot 49*, Thomas Pynchon turned the tables, comparing a sterile, overly planned Southern California community (called San Narciso) to a computer chip. In his best-known novel, the parabolic arc of a missile is memorably called "Gravity's Rainbow," a metaphor that seems especially apt if you remember a little college calculus.

In Jonathan Franzen's new novel, *The Corrections*, Arthur Lambert, a retired engineer, sits in his basement gloomily testing Christmas lights, only to discover in the depths of the tangle a blacked-out string of bulbs. A "substantia nigra," Mr. Franzen calls it.

The metaphor, if a little obscure, is pitch perfect. The substantia nigra ("black substance") is a region deep in the brain that produces the neurotransmitter dopamine. In a Pynchonian flip, electrical circuits are compared to neural circuits instead of the other way around. But the analogy cuts deeper. A burned-out substantia nigra is a symptom of Parkinsonism, the disease that afflicts Arthur. He is no more able to repair the Christmas lights than his doctors are able to fix his brain.

In another scene, visitors to a chic new restaurant, distinguished by its postindustrial design, sit inside a "glassed-in dining room, suspended in a blue Cherenkov glow." Cherenkov radiation, produced by rapidly moving charged particles, is responsible for the eerie luminescence in pools of water shielding nuclear reactors.

And here is how the novel describes the neurotic dependence Arthur's son Chip has developed on his sister, because of all the money he owes her: "He'd lived with the affliction of this debt until it had assumed the character of a neuroblastoma so intricately implicated in his cerebral architecture that he doubted he could survive its removal."

By daring to use such allusions, Mr. Franzen compliments his readers. Novels like his are a reminder that in literature, as well as science, illuminating the intangible with a good metaphor is a powerful art.

—December 25, 2001

The Birth of Science Times:
A Surprise, but No Accident

By JOHN NOBLE WILFORD

Twenty-five years ago, editors of *The New York Times* had a big problem: what to do about Tuesdays?

In a bold move to draw more readers and advertising revenue in a troubled economy, the newspaper was reinventing itself in format and content. The pages were redesigned to be six columns, instead of eight, giving the paper a more spacious look. But the most striking change was abandoning the two-section daily newspaper for one in four sections Monday through Friday.

A. M. Rosenthal, the managing editor and soon to be executive editor, asked Arthur Gelb, an assistant managing editor, to oversee the transformation, beginning in 1976. The first section continued to run foreign and national news, and the second, metropolitan news. The fourth section featured expanded coverage of business and financial news. The third section, it was decided, would be different each day of the week, though the specifics were left to fall into place over the next two years.

In his new memoir, *City Room*, Mr. Gelb writes that "virtually every executive by now viewed the forthcoming four-part paper as the lifeboat that would rescue *The Times* and secure its future."

The first of the new third sections was Weekend, devoted to the arts and entertainment events every Friday, starting in April 1976. This was followed in slow but steady progression by the introduction of Living on Wednesdays (dining, cooking and personal health), Home on Thursdays (furnishings, design and gardening) and Sports Monday, which had its debut in January 1978.

For months the third section on Tuesdays went without a theme, and no one seemed to agree on what it should be. Some on the business side of the newspaper argued for a style section that would emphasize fashion; they hoped it would attract more advertising. Abe Rosenthal resisted. He wanted something more serious.

"I felt if we put in a fashion section, it would tip the balance of the paper in its quality," Mr. Rosenthal recalled last week. "We had a lot of consumer stuff by then, and that was enough."

Colleagues said Mr. Rosenthal had become especially sensitive to outside criticism that with some of the new sections, *The Times* was going soft.

Talking over ideas with associates in the late summer of 1978, Mr. Rosenthal decided, as he now says, that a section of news and features about science and medicine would have "more strength and dignity." He made his case to the publisher, Arthur Ochs Sulzberger. "Punch was a very cooperative guy," he said, "and so we did it."

By this time, the newspaper was shut down by an 88-day strike of the pressmen. Mr. Gelb, working with Louis Silverstein, the assistant managing editor in charge of the art department, began shaping the concept and look of the new section. As editor of the science staff then, I joined in planning the types of articles and columns the section would carry. All the writers eagerly pitched in with ideas, sharing in the creation of Science Times.

In choosing science as the focus of the Tuesday section, *The Times* was dealing from strength. The newspaper had a long tradition of treating science as a dynamic part of modern culture. By this time, its staff of 10 science and medical reporters was the largest and most authoritative of any paper in the country. (Now the full-time science staff numbers five editors and 16 reporters, an art director, a graphics editor and a picture editor.)

For several years, the science staff had been moving beyond the daily fare of research news to write more comprehensive articles putting scientific advances in perspective and portraying scientists at their work. They were a taste of things to come.

The first issue of Science Times appeared on Nov. 14, 1978, shortly after the strike ended. In no time, the section was a hit. Teachers assigned it to classes, and doctors and scientists were impressed. Readership rose on Tuesdays, and to the surprise and delight of management, the section turned a profit with the eventual outpouring of advertising for personal computers.

Articles in Science Times won Pulitzer Prizes for two staff members. Among the other honors was a 2000 Lasker Foundation award to Science Times "for sustained, comprehensive and high-quality coverage about science, disease and human health."

Other newspapers responded with the sincerest form of praise: they started their own regular page or pages of science news.

A case can be made that Science Times was a significant step in communicating the work of science to the larger public, and to other scientists.

It was making a statement that science should be part of the well-informed person's regular reading diet. It was also saying to fellow journalists everywhere that science is news and that the responsible way to cover such a subject is to move beyond piecemeal reporting to more comprehensive and reflective articles that place new research in its broader context.

—November 11, 2003

Gray Matter and Sexes:
A Gray Area Scientifically

By NATALIE ANGIER and KENNETH CHANG

When Lawrence H. Summers, the president of Harvard, suggested this month that one factor in women's lagging progress in science and mathematics might be innate differences between the sexes, he slapped a bit of brimstone into a debate that has simmered for decades. And though his comments elicited so many fierce reactions that he quickly apologized, many were left to wonder: Did he have a point?

Has science found compelling evidence of inherent sex disparities in the relevant skills, or perhaps in the drive to succeed at all costs, that could help account for the persistent paucity of women in science generally, and at the upper tiers of the profession in particular?

Researchers who have explored the subject of sex differences from every conceivable angle and organ say that yes, there are a host of discrepancies between men and women—in their average scores on tests of quantitative skills, in their attitudes toward math and science, in the architecture of their brains, in the way they metabolize medications, including those that affect the brain.

Yet despite the desire for tidy and definitive answers to complex questions, researchers warn that the mere finding of a difference in form does not mean a difference in function or output inevitably follows.

"We can't get anywhere denying that there are neurological and hormonal differences between males and females, because there clearly are," said Virginia Valian, a psychology professor at Hunter College who wrote the 1998 book *Why So Slow? The Advancement of Women*. "The trouble we have as scientists is in assessing their significance to real-life performance."

For example, neuroscientists have shown that women's brains are about 10 percent smaller than men's, on average, even after accounting for women's comparatively smaller body size.

But throughout history, people have cited anatomical distinctions in support of overarching hypotheses that turn out merely to reflect the societal and cultural prejudices of the time.

A century ago, the French scientist Gustav Le Bon pointed to the smaller brains of women—closer in size to gorillas', he said—and said that explained the "fickleness, inconstancy, absence of thought and logic, and incapacity to reason" in women.

Overall size aside, some evidence suggests that female brains are relatively more endowed with gray matter—the prized neurons thought to do the bulk of the brain's thinking—while men's brains are packed with more white matter, the tissue between neurons.

To further complicate the portrait of cerebral diversity, new brain imaging studies from the University of California, Irvine, suggest that men and women with equal I.Q. scores use different proportions of their gray and white matter when solving problems like those on intelligence tests.

Men, they said, appear to devote 6.5 times as much of their gray matter to intelligence-related tasks as do women, while women rely far more heavily on white matter to pull them through a ponder.

What such discrepancies may or may not mean is anyone's conjecture.

"It is cognition that counts, not the physical matter that does the cognition," argued Nancy Kanwisher, a professor of neuroscience at the Massachusetts Institute of Technology.

When they do study sheer cognitive prowess, many researchers have been impressed with how similarly young boys and girls master new tasks.

"We adults may think very different things about boys and girls, and treat them accordingly, but when we measure their capacities, they're remarkably alike," said Elizabeth Spelke, a professor of psychology at Harvard. She and her colleagues study basic spatial, quantitative and numerical abilities in children ranging from five months through seven years.

"In that age span, you see a considerable number of the pieces of our mature capacities for spatial and numerical reasoning coming together," Dr. Spelke said. "But while we always test for gender differences in our studies, we never find them."

In adolescence, though, some differences in aptitude begin to emerge, especially when it comes to performance on standardized tests like the SAT. While average verbal scores are very similar, boys have outscored girls on the math half of the dreaded exam by about 30 to 35 points for the past three decades or so.

Nor is the masculine edge in math unique to the United States. In an international standardized test administered in 2003 by the international research group Organization for Economic Cooperation and Development to 250,000 15-year-olds in 41 countries, boys did moderately better on the math portion in just over half the nations. For nearly all the other countries, there were no significant sex differences.

But average scores varied wildly from place to place and from one subcategory of math to the next. Japanese girls, for example, were on par with Japanese boys on every math section save that of "uncertainty," which measures probabilistic skills, and Japanese girls scored higher overall than did the boys of many other nations, including the United States.

In Iceland, girls broke the mold completely and outshone Icelandic boys by a significant margin on all parts of the test, as they habitually do on their national math exams. "We have no idea why this should be so," said Almar Midvik Halldorsson, project manager for the Educational Testing Institute in Iceland.

Interestingly, in Iceland and everywhere else, girls participating in the survey expressed far more negative attitudes toward math.

The modest size and regional variability of the sex differences in math scores, as well as an attitudinal handicap that girls apparently pack into their No. 2 pencil case, convince many researchers that neither sex has a monopoly on basic math ability, and that culture rather than chromosomes explains findings like the gap in math SAT scores.

Yet Dr. Summers, who said he intended his remarks to be provocative, and other scientists have observed that while average math skillfulness may be remarkably analogous between the sexes, men tend to display comparatively greater range in aptitude. Males are much likelier than females to be found on the tail ends of the bell curve, among the superhigh scorers and the very bottom performers.

Among college-bound seniors who took the math SATs in 2001, for example, nearly twice as many boys as girls scored over 700, and the ratio skews ever more male the closer one gets to the top tally of 800. Boys are also likelier than girls to get nearly all the answers wrong.

For Dr. Summers and others, the overwhelmingly male tails of the bell curve may be telling. Such results, taken together with assorted other neuro-curiosities like the comparatively greater number of boys with learning disorders, autism and attention deficit disorder, suggest to them that the male brain is a delicate object, inherently prone to extremes, both of incompetence and of genius.

But few researchers who have analyzed the data believe that men's greater representation among the high-tail scores can explain more than a small fraction of the sex disparities in career success among scientists.

For one thing, said Kimberlee A. Shauman, a sociologist at the University of California, Davis, getting a high score on a math aptitude test turns out to be a poor predictor of who opts for a scientific career, but it is an especially poor gauge for girls. Catherine Weinberger, an economist at the University of California, Santa Barbara, has found that top-scoring girls are only about 60 percent as likely as top-scoring boys to pursue science or engineering careers, for reasons that remain unclear.

Moreover, men seem perfectly capable of becoming scientists without a math board score of 790. Surveying a representative population of working scientists and engineers, Dr. Weinberger has discovered that the women were likelier than the men to have very high test scores. "Women are more cautious about entering these professions unless they have very high scores to begin with," she said.

And this remains true even though a given score on standardized math tests is less significant for women than for men. Dr. Valian, of Hunter, observes that among women and men taking the same advanced math courses in college, women with somewhat lower SAT scores often do better than men with higher scores. "The SATs turn out to underpredict female and overpredict male performance," she said. Again, the reasons remain mysterious.

Dr. Summers also proposed that perhaps women did not go into science because they found it too abstract and cold-blooded, offering as anecdotal evidence the fact that his young daughter, when given toy trucks, had treated them as dolls, naming them "Daddy truck" and "baby truck."

But critics dryly observed that men had a longstanding tradition of naming their vehicles, and babying them as though they were humans.

Yu Xie, a sociologist at the University of Michigan and a co-author with Dr. Shauman of *Women in Science: Career Processes and Outcomes* (2003), said he wished there was less emphasis on biological explanations for success or failure, and more on effort and hard work.

Among Asians, he said, people rarely talk about having a gift or a knack or a gene for math or anything else. If a student comes home with a poor grade in math, he said, the parents push the child to work harder.

"There is good survey data showing that this disbelief in innate ability, and the conviction that math achievement can be improved through practice," Dr. Xie said, "is a tremendous cultural asset in Asian society and among Asian-Americans."

In many formerly male-dominated fields like medicine and law, women have already reached parity, at least at the entry levels. At the undergraduate level, women outnumber men in some sciences like biology.

Thus, many argue that it is unnecessary to invoke "innate differences" to explain the gap that persists in fields like physics, engineering, mathematics and chemistry. Might scientists just be slower in letting go of baseless sexism?

C. Megan Urry, a professor of physics and astronomy at Yale who led the American delegation to an international conference on women in physics in 2002, said there was clear evidence that societal and cultural factors still hindered women in science.

Dr. Urry cited a 1983 study in which 360 people—half men, half women—rated identical academic papers on a five-point scale. On average, the men rated them a full point higher when the author was "John T. McKay" than when the author was "Joan T. McKay." There was a similar, but smaller disparity in the scores the women gave.

Dr. Spelke, of Harvard, said, "It's hard for me to get excited about small differences in biology when the evidence shows that women in science are still discriminated against every stage of the way."

A recent experiment showed that when Princeton students were asked to evaluate two highly qualified candidates for an engineering job—one with more education, the other with more work experience—they picked the more educated candidate 75 percent of the time. But when the candidates were designated as male or female, and the educated candidate bore a female name, suddenly she was preferred only 48 percent of the time.

The debate is sure to go on.

Sandra F. Witelson, a professor of psychiatry and behavioral neurosciences at McMaster University in Hamilton, Ontario, said biology might yet be found to play some role in women's careers in the sciences.

"People have to have an open mind," Dr. Witelson said.

—January 24, 2005

Scientists Speak Up on Mix of God and Science

By CORNELIA DEAN

At a recent scientific conference at City College of New York, a student in the audience rose to ask the panelists an unexpected question: "Can you be a good scientist and believe in God?"

Reaction from one of the panelists, all Nobel laureates, was quick and sharp. "No!" declared Herbert A. Hauptman, who shared the chemistry prize in 1985 for his work on the structure of crystals.

Belief in the supernatural, especially belief in God, is not only incompatible with good science, Dr. Hauptman declared, "this kind of belief is damaging to the well-being of the human race."

But disdain for religion is far from universal among scientists. And today, as religious groups challenge scientists in arenas as various as evolution in the classroom, AIDS prevention and stem cell research, scientists who embrace religion are beginning to speak out about their faith.

"It should not be a taboo subject, but frankly it often is in scientific circles," said Francis S. Collins, who directs the National Human Genome Research Institute and who speaks freely about his Christian faith.

Although they embrace religious faith, these scientists also embrace science as it has been defined for centuries. That is, they look to the natural world for explanations of what happens in the natural world and they recognize that scientific ideas must be provisional—capable of being overturned by evidence from experimentation and observation. This belief in science sets them apart from those who endorse creationism or its doctrinal cousin, intelligent design, both of which depend on the existence of a supernatural force.

Their belief in God challenges scientists who regard religious belief as little more than magical thinking, as some do. Their faith also challenges believers who denounce science as a godless enterprise and scientists as secular elitists contemptuous of God-fearing people.

Some scientists say simply that science and religion are two separate realms, "nonoverlapping magisteria," as the late evolutionary biologist Stephen Jay Gould put it in his book *Rocks of Ages* (Ballantine, 1999). In Dr. Gould's view, science speaks with authority in the realm of "what the universe is made of (fact) and why does it work this way (theory)" and religion holds sway over "questions of ultimate meaning and moral value."

When the American Association for the Advancement of Science devoted a session to this idea of separation at its annual meeting this year, scores of scientists crowded into a room to hear it.

Some of them said they were unsatisfied with the idea, because they believe scientists' moral values must inevitably affect their work, others because so much of science has so many ethical implications in the real world.

One panelist, Dr. Noah Efron of Bar-Ilan University in Israel, said scientists, like other people, were guided by their own human purposes, meaning and values. The idea that fact can be separated from values and meaning "jibes poorly with what we know of the history of science," Dr. Efron said.

Dr. Collins, who is working on a book about his religious faith, also believes that people should not have to keep religious beliefs and scientific theories strictly separate. "I don't find it very satisfactory and I don't find it very necessary," he said in an interview. He noted that until relatively recently, most scientists were believers. "Isaac Newton wrote a lot more about the Bible than the laws of nature," he said.

But he acknowledged that as head of the American government's efforts to decipher the human genetic code, he had a leading role in work that many say definitively demonstrates the strength of evolutionary theory to explain the complexity and abundance of life.

As scientists compare human genes with those of other mammals, tiny worms, even bacteria, the similarities "are absolutely compelling," Dr. Collins said. "If Darwin had tried to imagine a way to prove his theory, he could not have come up with something better, except maybe a time machine. Asking somebody to reject all of that in order to prove that they really do love God—what a horrible choice."

Dr. Collins was a nonbeliever until he was 27—"more and more into the mode of being not only agnostic but being an atheist," as he put it. All that changed after he completed his doctorate in physics and was at work on his medical degree, when he was among those treating a woman dying of heart disease. "She was very clear about her faith and she looked me square in the eye and she said, 'what do you believe?'" he recalled. "I sort of stammered out, 'I am not sure.'"

He said he realized then that he had never considered the matter seriously, the way a scientist should. He began reading about various religious beliefs, which only confused him. Finally, a Methodist minister gave him a book, *Mere*

Christianity, by C. S. Lewis. In the book Lewis, an atheist until he was a grown man, argues that the idea of right and wrong is universal among people, a moral law they "did not make, and cannot quite forget even when they try." This universal feeling, he said, is evidence for the plausibility of God.

When he read the book, Dr. Collins said, "I thought, my gosh, this guy is me."

Today, Dr. Collins said, he does not embrace any particular denomination, but he is a Christian. Colleagues sometimes express surprise at his faith, he said. "They'll say, 'how can you believe that? Did you check your brain at the door?'" But he said he had discovered in talking to students and colleagues that "there is a great deal of interest in this topic."

Polling Scientists on Beliefs

According to a much-discussed survey reported in the journal *Nature* in 1997, 40 percent of biologists, physicists and mathematicians said they believed in God—and not just a nonspecific transcendental presence but, as the survey put it, a God to whom one may pray "in expectation of receiving an answer."

The survey, by Edward J. Larson of the University of Georgia, was intended to replicate one conducted in 1914, and the results were virtually unchanged. In both cases, participants were drawn from a directory of American scientists.

Others play down those results. They note that when Dr. Larson put part of the same survey to "leading scientists"—in this case, members of the National Academy of Sciences, perhaps the nation's most eminent scientific organization—fewer than 10 percent professed belief in a personal God or human immortality.

This response is not surprising to researchers like Steven Weinberg, a physicist at the University of Texas, a member of the academy and a winner of the Nobel Prize in 1979 for his work in particle physics. He said he could understand why religious people would believe that anything that eroded belief was destructive. But he added: "I think one of the great historical contributions of science is to weaken the hold of religion. That's a good thing."

No God, No Moral Compass?

He rejects the idea that scientists who reject religion are arrogant. "We know how many mistakes we've made," Dr. Weinberg said. And he is angered by assertions that people without religious faith are without a moral compass.

In any event, he added, "the experience of being a scientist makes religion seem fairly irrelevant," he said. "Most scientists I know simply don't think about it very much. They don't think about religion enough to qualify as practicing atheists."

Most scientists he knows who do believe in God, he added, believe in "a God who is behind the laws of nature but who is not intervening."

Kenneth R. Miller, a biology professor at Brown, said his students were often surprised to find that he was religious, especially when they realized that his faith was not some sort of vague theism but observant Roman Catholicism.

Dr. Miller, whose book, *Finding Darwin's God*, explains his reconciliation of the theory of evolution with his religious faith, said he was usually challenged in his biology classes by one or two students whose religions did not accept evolution, who asked how important the theory would be in the course.

"What they are really asking me is "do I have to believe in this stuff to get an A?,'" he said. He says he tells them that "belief is never an issue in science."

"I don't care if you believe in the Krebs cycle," he said, referring to the process by which energy is utilized in the cell. "I just want you to know what it is and how it works. My feeling about evolution is the same thing."

For Dr. Miller and other scientists, research is not about belief. "Faith is one thing, what you believe from the heart," said Joseph E. Murray, who won the Nobel Prize in medicine in 1990 for his work in organ transplantation. But in scientific research, he said, "it's the results that count."

Dr. Murray, who describes himself as "a cradle Catholic" who has rarely missed weekly Mass and who prays every morning, said that when he was preparing for the first ever human organ transplant, a kidney that a young man had donated to his identical twin, he and his colleagues consulted a number of religious leaders about whether they were doing the right thing. "It seemed natural," he said.

Using Every Tool

"When you are searching for truth you should use every possible avenue, including revelation," said Dr. Murray, who is a member of the Pontifical Academy, which advises the Vatican on scientific issues, and who described the influence of his faith on his work in his memoir, *Surgery of the Soul* (Science History Publications, 2002).

Since his appearance at the City College panel, when he was dismayed by the tepid reception received by his remarks on the incompatibility of good science and religious belief, Dr. Hauptman said he had been discussing the issue with colleagues in Buffalo, where he is president of the Hauptman-Woodward Medical Research Institute.

"I think almost without exception the people I have spoken to are scientists and they do believe in the existence of a supreme being," he said. "If you ask me to explain it—I cannot explain it at all."

But Richard Dawkins, an evolutionary theorist at Oxford, said that even scientists who were believers did not claim evidence for that belief. "The most they will claim is that there is no evidence against," Dr. Dawkins said, "which is pathetically weak. There is no evidence against all sorts of things, but we don't waste our time believing in them."

Dr. Collins said he believed that some scientists were unwilling to profess faith in public "because the assumption is if you are a scientist you don't have any need of action of the supernatural sort," or because of pride in the idea that science is the ultimate source of intellectual meaning.

But he said he believed that some scientists were simply unwilling to confront the big questions religion tried to answer. "You will never understand what it means to be a human being through naturalistic observation," he said. "You won't understand why you are here and what the meaning is. Science has no power to address these questions—and are they not the most important questions we ask ourselves?"

—*August 23, 2005*

A Sharp Rise in Retractions
Prompts Calls for Reform

By CARL ZIMMER

In the fall of 2010, Dr. Ferric C. Fang made an unsettling discovery. Dr. Fang, who is editor in chief of the journal *Infection and Immunity*, found that one of his authors had doctored several papers.

It was a new experience for him. "Prior to that time," he said in an interview, "*Infection and Immunity* had only retracted nine articles over a 40-year period."

The journal wound up retracting six of the papers from the author, Naoki Mori of the University of the Ryukyus in Japan. And it soon became clear that *Infection and Immunity* was hardly the only victim of Dr. Mori's misconduct. Since then, other scientific journals have retracted two dozen of his papers, according to the watchdog blog *Retraction Watch*.

"Nobody had noticed the whole thing was rotten," said Dr. Fang, who is a professor at the University of Washington School of Medicine.

Dr. Fang became curious how far the rot extended. To find out, he teamed up with a fellow editor at the journal, Dr. Arturo Casadevall of the Albert Einstein College of Medicine in New York. And before long they reached a troubling conclusion: not only that retractions were rising at an alarming rate, but that retractions were just a manifestation of a much more profound problem— "a symptom of a dysfunctional scientific climate," as Dr. Fang put it.

Dr. Casadevall, now editor in chief of the journal *mBio*, said he feared that science had turned into a winner-take-all game with perverse incentives that lead scientists to cut corners and, in some cases, commit acts of misconduct.

"This is a tremendous threat," he said.

Last month, in a pair of editorials in *Infection and Immunity*, the two editors issued a plea for fundamental reforms. They also presented their concerns at the March 27 meeting of the National Academies of Sciences committee on science, technology and the law.

Members of the committee agreed with their assessment. "I think this is really coming to a head," said Dr. Roberta B. Ness, dean of the University of Texas School of Public Health. And Dr. David Korn of Harvard Medical School agreed that "there are problems all through the system."

No one claims that science was ever free of misconduct or bad research. Indeed, the scientific method itself is intended to overcome mistakes and misdeeds. When scientists make a new discovery, others review the research skeptically before it is published. And once it is, the scientific community can try to replicate the results to see if they hold up.

But critics like Dr. Fang and Dr. Casadevall argue that science has changed in some worrying ways in recent decades—especially biomedical research, which consumes a larger and larger share of government science spending.

In October 2011, for example, the journal *Nature* reported that published retractions had increased tenfold over the past decade, while the number of published papers had increased by just 44 percent. In 2010 the *Journal of Medical Ethics* published a study finding the new raft of recent retractions was a mix of misconduct and honest scientific mistakes.

Several factors are at play here, scientists say. One may be that because journals are now online, bad papers are simply reaching a wider audience, making it more likely that errors will be spotted. "You can sit at your laptop and pull a lot of different papers together," Dr. Fang said.

But other forces are more pernicious. To survive professionally, scientists feel the need to publish as many papers as possible, and to get them into high-profile journals. And sometimes they cut corners or even commit misconduct to get there.

To measure this claim, Dr. Fang and Dr. Casadevall looked at the rate of retractions in 17 journals from 2001 to 2010 and compared it with the journals' "impact factor," a score based on how often their papers are cited by scientists. The higher a journal's impact factor, the two editors found, the higher its retraction rate.

The highest "retraction index" in the study went to one of the world's leading medical journals, the *New England Journal of Medicine.* In a statement for this article, it questioned the study's methodology, noting that it considered only papers with abstracts, which are included in a small fraction of studies published in each issue. "Because our denominator was low, the index was high," the statement said.

Monica M. Bradford, executive editor of the journal *Science,* suggested that the extra attention high-impact journals get might be part of the reason for their higher rate of retraction. "Papers making the most dramatic advances will be subject to the most scrutiny," she said.

Dr. Fang says that may well be true, but adds that it cuts both ways—that the scramble to publish in high-impact journals may be leading to more and more errors. Each year, every laboratory produces a new crop of PhD's, who must compete for a small number of jobs, and the competition is getting fiercer. In 1973, more than half of biologists had a tenure-track job within six years of getting a PhD. By 2006 the figure was down to 15 percent.

Yet labs continue to have an incentive to take on lots of graduate students to produce more research. "I refer to it as a pyramid scheme," said Paula Stephan, a Georgia State University economist and author of *How Economics Shapes Science,* published in January by Harvard University Press.

In such an environment, a high-profile paper can mean the difference between a career in science or leaving the field. "It's becoming the price of admission," Dr. Fang said.

The scramble isn't over once young scientists get a job. "Everyone feels nervous even when they're successful," he continued. "They ask, 'Will this be the beginning of the decline?'"

University laboratories count on a steady stream of grants from the government and other sources. The National Institutes of Health accepts a much lower percentage of grant applications today than in earlier decades. At the same time, many universities expect scientists to draw an increasing part of their salaries from grants, and these pressures have influenced how scientists are promoted.

"What people do is they count papers, and they look at the prestige of the journal in which the research is published, and they see how may grant dollars scientists have, and if they don't have funding, they don't get promoted," Dr. Fang said. "It's not about the quality of the research."

Dr. Ness likens scientists today to small-business owners, rather than people trying to satisfy their curiosity about how the world works. "You're marketing and selling to other scientists," she said. "To the degree you can market and sell your products better, you're creating the revenue stream to fund your enterprise."

Universities want to attract successful scientists, and so they have erected a glut of science buildings, Dr. Stephan said. Some universities have gone into debt, betting that the flow of grant money will eventually pay off the loans. "It's really going to bite them," she said.

With all this pressure on scientists, they may lack the extra time to check their own research—to figure out why some of their data doesn't fit their hypothesis, for example. Instead, they have to be concerned about publishing papers before someone else publishes the same results.

"You can't afford to fail, to have your hypothesis disproven," Dr. Fang said. "It's a small minority of scientists who engage in frank misconduct. It's a much more insidious thing that you feel compelled to put the best face on everything."

Adding to the pressure, thousands of new Ph.D. scientists are coming out of countries like China and India. Writing in the April 5 issue of *Nature*, Dr. Stephan points out that a number of countries—including China, South Korea and Turkey—now offer cash rewards to scientists who get papers into high-profile journals. She has found these incentives set off a flood of extra papers submitted to those journals, with few actually being published in them. "It clearly burdens the system," she said.

To change the system, Dr. Fang and Dr. Casadevall say, start by giving graduate students a better understanding of science's ground rules—what Dr. Casadevall calls "the science of how you know what you know."

They would also move away from the winner-take-all system, in which grants are concentrated among a small fraction of scientists. One way to do that may be to put a cap on the grants any one lab can receive.

Such a shift would require scientists to surrender some of their most cherished practices—the priority rule, for example, which gives all the credit for a scientific discovery to whoever publishes results first. (Three centuries ago, Isaac Newton and Gottfried Leibniz were bickering about who invented calculus.) Dr. Casadevall thinks it leads to rival research teams' obsessing over secrecy, and rushing out their papers to beat their competitors. "And that can't be good," he said.

To ease such cutthroat competition, the two editors would also change the rules for scientific prizes and would have universities take collaboration into account when they decide on promotions.

Even scientists who are sympathetic to the idea of fundamental change are skeptical that it will happen any time soon. "I don't think they have much chance of changing what they're talking about," said Dr. Korn, of Harvard.

But Dr. Fang worries that the situation could become much more dire if nothing happens soon. "When our generation goes away, where is the new generation going to be?" he asked. "All the scientists I know are so anxious about their funding that they don't make inspiring role models. I heard it from my own kids, who went into art and music respectively. They said, 'You know, we see you, and you don't look very happy.'"

—*April 17, 2012*

CHAPTER 12

Physics: Understanding the Inconceivable

THE ATOM AND ITS PARTS

Scientists Witness Smash-Up of Atoms

By ALVA JOHNSTON

Extraordinary photographs of the smash-up of atoms were displayed to chemists of the American Association for the Advancement of Science here [in Cambridge, Mass.] today in the course of a session at which many of the chief American investigators of atoms and the electrons that compose them were present.

New calculations as to the size of different atoms were laid before the chemical section by Dr. Theodore W. Richards of Harvard. A new periodic table of elements was explained by Dr. W. D. Harkins of the University of Chicago, which arranges the atoms and subvarieties of atoms called isotopes. Strictly speaking, there are no longer a mere group of 92 elements, according to the old periodic table, but a total of several hundred. For practical purposes, however, the old list of 92 holds good, because the new atoms, or isotopes, have no characteristics of their own except minute differences in mass. For instance, lead is never found as a single element, but is a combination of different varieties of lead atoms, each having apparently the same physical and chemical natures, except for the slight difference in mass.

Photographs of two new types of atomic collisions were showed by Dr. Richards. These were taken by the Wilson and Shimidzu methods of photographing an alpha particle, otherwise known as double-charged helium nucleus, during its flight through damp air.

One He Can't Explain

"I have not been able to explain this one," said Dr. Richards, showing a picture of the helium atom striking a hydrogen or nitrogen atom. This showed a helium atom travelling in a straight line after sideswiping a hydrogen atom, while the hydrogen atom which had been hit went backward instead of forward. Instantly on being struck the atom had started to travel backward in the direction from which the blow came. The atom which had delivered the blow in the meantime continued ahead on a straight course.

The atomic projectile made a speed of 20,000 miles an hour, and in other cases drove its targets in all directions except backward. Sometimes, in the case of a head-on collision, the projectile itself rebounded, but its target was knocked forward.

This unusual effect occurred only once in 10,000 photographs of crashing atoms, and thousands more are to be taken, Dr. Harkins said, to see whether repetitions of the effect throw any further light on the behavior of an atom.

The other type of photograph which is not understood resembles a follow shot in billiards. The projectile, after striking another atom, proceeds in almost the same path. The target atom also moves forward in almost the same path. There were several photographs of this effect.

Photographs Are Vivid

The atom itself is about a million times too small to be seen with the naked eye, but the high-speed alpha particle, or helium nucleus, which is used as a projectile, is able to produce effects which can be seen and photographed vividly. The particles are bombarded from radium or some other radioactive substances into an enclosure which is full of air supersaturated with water. All dust particles have been pumped out previously so that there is nothing for water to condense on and form drops.

When there is no dust for it to deposit itself on, water vapor will throw itself as drops on the trail of an alpha particle. Air molecules are electrically charged or transformed into ions by the speeding atom, and the water vapor will seize on an ion when dust particles are absent.

A fine mist is thus set up along the trail of the projectile and along the secondary trail made by the target after it has been knocked forward or to one side at high speed. Under a strong light these moisture lines can be beautifully photographed. The head-on collisions have been shown to knock some of the lighter atoms to pieces, thus producing the artificial disintegration of atoms first shown by Sir Ernest Rutherford. This rarely occurs, however, because of the smallness of the bodies involved. The alpha particle or double charged helium nucleus, occupies only about one million part of the space of the ordinary atom, which consists of one or more outer rings of atoms as well as the nucleus. The nucleus weighs 1,740 times as much as the outer shell, but is less than one-millionth as big.

The very small pellets have to penetrate the outer shell of another atom and hit the nucleus before the break-up of the atom can take place. Direct hits on this bull's-eye are seldom made. The helium nucleus is so small that on its flight through an inch of air it passes on the average 500,000 atoms before it hits one. Photographs and even motion pictures of these atomic collisions are being taken by the thousand in the hope of getting a good picture of a head-on collision resulting in a complete smash-up of the target and an artificial transmutation of the target atoms into hydrogen and helium.

The helium atom is apparently made of four hydrogen atoms, but the atom of helium has less mass than four atoms of hydrogen. This would mean that mass was lost, but that would conflict with the theory that mass is indestructible. But what has happened is that the lost mass is transformed into energy and energy is mass.

The new measurements which Dr. Richards gave for atoms were based on a study of their compressibility. Usually when atoms unite they pull toward each other and occupy a smaller space than they did when they existed separately. Dr. Richards sought to determine how many different elements could be compressed by their own force in these combinations.

Previous measurements of the atom had made the average about a hundred-millionth part of an inch. The measurement of Dr. Richards does not disturb these dimensions greatly, but is supposed to give them with great refinement.

Isotopes, or the variations in the atoms of the same element, were discussed at greater length by Dr. Robert S. Mulliker, of the Isotope Laboratory of the University of Chicago, who said that no laboratory method had yet been discovered of separating isotopes so completely that the isotopes could be individually studied. So far the quantities of isotopes which have been isolated have been too small for examination.

"While isotopes are very much like each other in their behavior," he said, "there are indications that this agreement is not quite complete and they will be studied for slight differences in their properties."

—December 28, 1922

Finds Two Particles Make Up Neutrons

By FERDINAND KUHN JR.

The belief that the neutron is a new ultimate particle like an electron or proton was challenged today by Dr. James Chadwick, young Cambridge scientist who discovered the elusive neutron last winter.

Delivering his first account of his discovery to the British Association for the Advancement of Science, Dr. Chadwick asserted the neutron was a new particle consisting of a proton and electron joined together, but he refused to make sweeping claims for it as some other scientists have done. He said he was not even sure after months of research how the proton and electron were joined, although he asserted positively that the existence of a joined particle had been confirmed.

His View Is Welcomed

Dr. Chadwick's modest claims for the neutron were welcomed by Professor O. W. Richardson, Nobel Prize winner in physics, who said:

"I am glad that Dr. Chadwick has stuck to the view that it is a combination of a proton and electron. Some people have said it was a new kind of ultimate particle. It was really too much to believe—that a new ultimate particle should exist with its mass so conveniently close to that of the proton and electron combined.

"It was nothing but a bad joke played on its creator and on the rest of us. Still, there is no doubt this neutron business is going to have many developments."

New facts about the neutron were revealed during today's discussion by Dr. Chadwick and scientists working under him at the famous Cavendish Laboratory. Dr. Chadwick told for the first time how the existence of the neutron had been confirmed.

In the original successful experiment, a piece of beryllium had been bombarded by alpha particles, thus creating strange radiation which could be explained only by neutrons. But the mass of beryllium is unknown and consequently it was impossible to establish with mathematical certainty that the neutron existed. Not until they had used a piece of boron, whose mass is known, were Dr. Chadwick and his assistants positive they had found the elusive particle, he said.

"The neutron can pass through very large thicknesses of matter without being deflected," Dr. Chadwick continued, "and, because it has no charge, it can enter right into the nucleus with very little disturbance. Sometimes the neutron comes out, and sometimes it is captured. We have not, however, observed a single case of collision between neutrons and electrons. Such collisions must occur, but they are extremely rare."

The Duc de Broglie, French Nobel Prize winner, said the existence of the neutron had been discovered from radium emanations as well as from bombardments of certain metals by alpha particles. He also revealed that neutrons emitted by boron had less energy than those from beryllium.

Neville Feather, who has photographed thousands of electron and proton tracks in the Cavendish Laboratory, asserted the neutron was the only particle which could disintegrate the oxygen nucleus. Earlier physicists struggled to solve the great problem of modern physics—the nucleus of the atom—especially to account for the queer behavior of certain nuclei, which apparently do not conform to general laws of conservation energy. Now, according to Professor Charles D. Ellis of Cambridge, new research during the past year has proved these offending nuclei behave much less queerly than has been supposed.

Links Energy to Problem

"We have clear evidence," said Professor Ellis, "that there is a vanishing point beyond which the energy of any electron cannot go. There is a definite limit—a fact of some significance, for it shows that energy still has some connection with the problem. There appears to be a close connection between the energies of alpha particles in the nucleus and the frequencies of gamma rays, which is entirely in accord with the validity of the energy principle and the principles of quantum mechanics."

Professor C. G. Darwin, descendant of the great exponent of evolution, refused to be downhearted over the apparent inconsistencies of nuclei.

"The misconduct of the nucleus, I think, is very much exaggerated," he said. "The nucleus is really very well behaved, and in time we will find all fitting in very nicely. If we dwell too much on their misconduct we are taking the attitude of the missionary in New Guinea, who deplores the cannibalism of a few natives and forgets all about their many virtues."

—*September 6, 1932*

Atom Bomb Based on Einstein Theory

By WILLIAM L. LAURENCE

A tomic energy, harnessed for the first time by our scientists for use in atomic bombs, is the practically inexhaustible source of power that enables our sun to supply us with heat, light and other forms of radiant energy without which life on earth would not be possible.

It also is the same energy, stored in the nuclei of the atoms of the material universe, that keeps the stars, bodies much larger than our sun, radiating their enormous quantities of light and heat for billions of years, instead of burning themselves out in periods measured only in thousands of years.

The existence of atomic energy first was discovered by Einstein about forty years ago on purely theoretical grounds, as a outgrowth of his famous relativity theory, according to which a body in motion has a greater mass than the same body at rest, this increase in motion bearing a direct relationship to the velocity of light. This meant that the energy of motion imparted an actual increase in mass.

From the formula for the relationship of this increase of mass to the velocity of light Einstein derived his famous mathematical equation that revealed for the first time an equivalence between mass and energy, one of the most revolutionary concepts in the intellectual history of mankind. The mass-energy equation showed that any given quantity of mass was the equivalent of a specific amount of energy, and vice versa.

Energy Highly Concentrated

Specifically this equation revealed the (at that time) incredible fact that very small amounts of matter contained tremendous amounts of energy. A piece of coal the size of a pea, the equation proved, contained enough energy to drive the largest ocean liner across the Atlantic and back. No one, however, least of all Einstein himself, believed at the time that any means ever could be found to tap this cosmic source of elemental energy.

In the mass-energy theorem Einstein showed the existence of a definite relationship between the Cosmic Trinity of matter, energy and the velocity of light. The relationship is so simple that, once arrived at, a grammar school student could work it out.

In this formula the letter "m" stands for mass in terms of grams; the letter "E" represents energy in terms of ergs (a small unit of energy or work); while the letter "c" stands for the velocity of light in terms of centimeters per second. The energy content of any given quantity of any substance, the formula states, is equal to the mass of the substance (in terms of grams), multiplied by the square of the velocity of light (in terms of centimeters per second). The velocity of light (in round numbers) is 300,000 kilometers, or 30,000,000,000 centimeters, per second.

Take one gram of any substance. According to the Einstein formula the amount of energy ("E") in ergs in this mass is equal to 1 (the mass of the substance in grams) multiplied by 30,000,000,000 squared. In other words, the energy content of one gram of matter equals 900 billion billion ergs. Translated in terms of pounds and kilowatt-hours this means that one pound of matter contains the energy equivalent of 10,000,000,000 kilowatt-hours.

Possible Accomplishments

If this energy could be fully utilized it would take only twenty-two pounds of matter to supply all the electrical power requirements of the United States for a year.

One-third of a gram of water would yield enough heat to turn 1,000 tons of water into steam.

A breath of air would operate a powerful airplane continuously for a year.

A handful of snow would heat a large apartment house for a year.

The pasteboard in a small railroad ticket would run a heavy passenger train several times around the world.

A cup of water would supply the power of a great generating station of 100,000-kilowatt capacity for a year.

One pound of any substance, if its atomic energy content could be utilized 100 percent, is equivalent in power-content to 3,000,000,000 pounds of coal, or 1,500,000 tons. The energy we now are able to utilize in the atomic bombs, at maximum efficiency, constitutes only one-tenth of 1 percent of the total energy present in the material. But even one-hundredth of 1 percent still would be the most destructive force by far on this earth.

Atomic energy, released through the splitting of atoms, differs radically from ordinary types of energy hitherto available to man in that it involves annihilation of matter. When an atom is split part of its matter is converted into energy.

This is materially different from obtaining power by the use of a water wheel, for example, or by the burning of coal or oil. In the case of the water wheel, the water molecules taking part remain entirely unchanged. They simply lose potential energy as they pass from the dam to the tailrace.

Atom's Identity Changes

In the case of burning coal or oil a more intense process takes place, as the atoms of carbon, hydrogen and oxygen (of which the coal and oil molecules are composed) are regrouped by combustion into new molecules forming new substances. The atoms themselves, however, still remain unchanged—they still are carbon, hydrogen and oxygen. None of them, as far as can be measured, loses any part of its mass.

In the case of atomic energy, however, the atom itself completely changes its identity, and in this process of change it loses part of its mass, which is converted into energy. The amount of energy liberated in this process is directly proportional to the amount of atomic mass destroyed. The sun, for example, obtains its energy through the partial destruction of its hydrogen, through a complex process in which the hydrogen is converted into helium.

In this process, four hydrogen atoms, each with an atomic mass of 1.008 (total, 4.032 atomic mass units) combine to form one helium atom, which has an atomic mass of 4.003. This represents a loss of mass on the part of the four hydrogen atoms (in addition to a loss of two positive electrons) of 0.029 atomic mass units, which is converted into pure energy. The amount of energy liberated in this process by the enormous quantities of hydrogen in the sun represents an actual loss of the sun's mass at the rate of 4,000,000 tons per second, a mere speck of dust in relation to the sun's total mass of two billion billion billion tons.

If the sun, however, were a mass of coal weighing the same amount, it would have to burn 3,000,000,000 times the mass it is burning now to produce the same amount of energy. If that were the case it would have used up the entire store of molecular energy contained in its body of coal in the course of 5,750 years. In other words, it would have burned out long before the earth was born.

Long Life for Earth

By the use of atomic energy, the sun has been able to give off its enormous amounts of radiation for a period estimated at 10,000,000,000 years, and its mass, at the present rate of burning, is enough to last 15,000 billion years more, although, of course, the amount of its radiation would be greatly reduced long before that in proportion to the decease of its mass. Radiations in amounts sufficient to support life on earth are estimated to continue for some 10,000,000,000 to 100,000,000,000 years longer.

Since the very existence of atomic energy was first discovered through the theory of relativity, the development of the atomic bomb constitutes the most dramatic proof so far offered for the correctness of the theory, and also marks the first time it has been put to practical use in mundane affairs.

It is one of the great ironies of history that the German war lords, who drove Einstein into exile, were forced to rely on the theory of relativity in their efforts to develop an atomic bomb to save them from defeat. America, of which Einstein now is an honored citizen, succeeded where the Nazis failed. When the bombs fell over Hiroshima and Nagasaki, they represented the fruition of what had been originally a pure mathematical concept.

Had that concept not come when it did the development of the atomic bomb also might have had to wait. This might have meant a prolongation of the war.

Thousands of young Americans thus may owe their lives to the theory of relativity. Which is another way of saying that pure science, no matter how impractical it may appear, pays high dividends in the end.

—*September 28, 1945*

The Forces That Shape
the Universe

Lights All Askew in the Heavens

Efforts made to put in words intelligible to the nonscientific public the Einstein theory of light proved by the eclipse expedition so far have not been very successful. The new theory was discussed at a recent meeting of the Royal Society and Royal Astronomical Society. Sir Joseph Thomson, president of the Royal Society, declares it is not possible to put Einstein's theory into really intelligible words, yet at the same time Thomson adds:

"The results of the eclipse expedition demonstrating that the rays of light from the stars are bent or deflected from their normal course by other aerial bodies acting upon them and consequently the inference that light has weight form a most important contribution to the laws of gravity given us since Newton laid down his principles."

Thomson states that the difference between theories of Newton and those of Einstein are infinitesimal in a popular sense, and as they are purely mathematical and can only be expressed in strictly scientific terms it is useless to endeavor to detail them for the man in the street.

"What is easily understandable," he continued, "is that Einstein predicted the deflection of the starlight when it passed the sun, and the recent eclipse has provided a demonstration of the correctness of the prediction."

"His second theory as to the anomalous motion of the planet Mercury has also been verified, but his third prediction, which dealt with certain sun lines, is still indefinite."

Asked if recent discoveries meant a reversal of the laws of gravity as defined by Newton, Sir Joseph said they held good for ordinary purposes, but in highly mathematical problems the new conceptions of Einstein, whereby space became warped or curled under certain circumstances, would have to be taken into account.

Vastly different conceptions which are involved in this discovery and the necessity for taking Einstein's theory more into account were voiced by a member of the expedition, who pointed out that it meant, among other things, that two lines normally known as parallel do meet eventually, that a circle is not really circular, that three angles of a triangle do not necessarily make the sum total of two right angles.

"Enough has been said to show the importance of Einstein's theory, even if it cannot be expressed clearly in words," laughed this astronomer.

Dr. W. J. S. Lockyer, another astronomer, said:

"The discoveries, while very important, did not, however, affect anything on this earth. They do not personally concern ordinary human beings; only astronomers are affected. It has hitherto been understood that light traveled in a straight line. Now we find it travels in a curve. It therefore follows that any object, such as a star, is not necessarily in the direction in which it appears to be astronomically.

"This is very important, of course. For one thing, a star may be a considerable distance further away than we have hitherto counted it. This will not affect navigation, but it means corrections will have to be made."

One of the speakers at the Royal Society's meeting suggested that Euclid was knocked out. Schoolboys should not rejoice prematurely, for it is pointed out that Euclid laid down the axiom that parallel straight lines, if produced ever so far, would not meet. He said nothing about light lines.

Some cynics suggest that the Einstein theory is only a scientific version of the well-known phenomenon that a coin in a basin of water is not on the spot where it seems to be and ask what is new in the refraction of light.

Albert Einstein is a Swiss citizen, about 50 years of age. After occupying a position as professor of mathematical physics at the Zurich Polytechnic School and afterward at Prague University, he was elected a member of Emperor William's Scientific Academy in Berlin at the outbreak of the war. Dr. Einstein protested against the German professors' manifesto approving of Germany's participation in the war, and at its conclusion he welcomed the revolution. He has been living in Berlin for about six years.

When he offered his last important work to the publishers he warned them there were not more than twelve persons in the whole world who would understand it, but the publishers took the risk.

—*November 10, 1919*

Einstein Expounds His New Theory

Now that the Royal Society, at its meeting in London on Nov. 6, has put the stamp of its official authority on Dr. Albert Einstein's much-debated new "theory of relativity," man's conception of the universe seems likely to undergo radical changes. Indeed, there are German savants who believe that since the promulgation of Newton's theory of gravitation, no discovery of such importance has been made in the world of sciences.

When *The New York Times* correspondent called at his home to gather from his own lips an interpretation of what to laymen must appear the book with the seven seals, Dr. Einstein himself modestly put aside the suggestion that his theory might have the same revolutionary effect on the human mind as Newton's theses. The doctor lives on the top floor of a fashionable apartment house on one of the few elevated spots in Berlin—so to say, close to the stars which he studies, not with a telescope, but rather with the mental eye, and so far only as they come with the range of his mathematical formulae; for he is not an astronomer but a physicist.

It was from his lofty library, in which this conversation took place, that he observed years ago a man dropping from a neighboring roof—luckily on a pile of soft rubbish—and escaping almost without injury. This man told Dr. Einstein that in falling he experienced no sensation commonly considered as the effect of gravity, which according to Newton's theory, would pull him down violently toward the earth. This incident, followed by further researches along the same line, started in his mind a complicated chain of thoughts leading finally, as he expressed it, "not to a disavowal of Newton's theory of gravitation, but to a sublimation or supplementation of it."

When he read in the message from *The Times* requesting the interview a reference to Dr. Einstein's statement to his publishers on the submission of his last book that not more than twelve persons in all the world could understand it, coupled with the editor's request that Dr. Einstein put his theory in terms comprehensible to a larger number than twelve, the doctor laughed good-naturedly, but still insisted on the difficulty of making himself understood by laymen.

"However," he said, "I am trying to talk as plainly as possible. To begin with the difference between my conception and Newton's law of gravitation: Please imagine the earth removed, and in its place suspended a box as big as a room or a whole house, and inside a man naturally floating in the center, there being no force whatever pulling him. Imagine, further, this box being, by a rope or other contrivance, suddenly jerked to one side, which is scientifically termed "difform motion, as opposed to "uniform motion." The person would then naturally reach bottom on the opposite side. The result would consequently be the same as if he obeyed Newton's law of gravitation, while, in fact, there is no gravitation exerted whatever, which proves that difform motion will in every case produce the same effects as gravitation.

"I have applied this new idea to every kind of difform motion and have thus developed mathematical formulas which I am convinced give more precise results than those based on Newton's theory. Newton's formulas, however, are such close approximations that it was difficult to find by observation any obvious disagreement with experience.

"One such case, however, was presented by the motion of the planet Mercury, which for a long time baffled astronomers. This is now completely cleared up by my formulas, as the Astronomer Royal, Sir Frank Dyson, stated at the meeting of the Royal Society.

"Another case was the deflection of rays of light when passing through the field of gravitation. No such deflections are explicable by Newton's theory of gravitation.

"According to my theory of difform motion, such deflections must take place when rays pass close to any gravitating mass difform motion then coming into activity.

"The crucial test was supplied by the last total solar eclipse, when observations proved that the rays of fixed stars, having to pass close to the sun to reach the earth, were deflected the exact amount demanded by my formulas, confirming my idea that what has so far has been regarded as the effect of gravitation is really the effect of difform motion. Elaborate apparatus and the closest and most indefatigable attention to the difficult task enabled that English expedition, composed of the most talented scientists, to reach those conclusions."

"Why is your idea termed 'the theory of relativity'?" asked the correspondent.

"The term 'relativity' refers to time and space," Dr. Einstein replied. "According to Galileo and Newton, time and space were absolute entities, and the moving systems of the universe were dependent on this absolute time and space. On this conception was built the science of mechanics. The resulting formulas sufficed for all motion of a slow nature; it was found, however, that they would not confirm to the rapid motions apparent in electrodynamics.

"This led the Dutch professor, Lorenz, and myself to develop the theory of special relativity. Briefly, it discards absolute time and space and makes them in every instance relative to moving systems. By this theory all phenomena in electrodynamics, as well as mechanics, hitherto irreducible by the old formulae—and there are multitudes—were satisfactorily explained.

"Till now it was believed that time and space existed by themselves, even if there was nothing else—no sun, no earth, no stars—while we now know that time and space are not the vessel for the universe, but could not exist at all if there were no contents, namely, no sun, earth, and other celestial bodies.

"This special relativity, forming the first part of my theory, relates to all systems moving with uniform motion; that is, moving in a straight line with equal velocity.

"Gradually I was led to the idea, seeming a very paradox in science, that it might apply equally to all moving systems, even of difform motion, and this I developed the conception of general relativity which forms the second part of my theory.

"It was during the development of the formulas for difform motions that the incident of the man falling from the roof gave me the idea that gravitation might be explained by difform motion."

"If there is no absolute time or space, supposedly forming the vessel of the universe," the correspondent asked, "what becomes of the ether?"

"There is no ether, as hitherto conceived by science, which is proved by the well-known experiment of the celebrated American savant, Michelson, showing that no influence by the motion of the earth on the ether is perceptible through change in velocity of light, such as ought to be produced if the old conception were true."

"Are you yourself absolutely convinced of the correctness of this revolutionary theory of relativity, or are there still reservations?"

"Yes, I am," Dr. Einstein answered. "My theory is confirmed by the two crucial cases mentioned before. But there is still one test outstanding, namely, the spectroscopic. According to my theory, the lines of the spectra of fixed stars must be slightly shifted through the influence of gravitation exerted by the very stars from which they emanate. So far, however, the results of the examinations have been contradictory; but I have no doubt of final confirmation, even through this test."

Just then an old grandfather's clock in the library chimed the midday hour, reminding Dr. Einstein of some appointment in another part of Berlin, and old-fashioned time and space enforced their wonted absolute tyranny over him who had spoken to contemptuously of their existence, thus terminating the interview.

—December 3, 1919

Researchers Slow Speed of Light to the Pace of a Sunday Driver

By MALCOLM W. BROWNE

When light travels through empty space, it zips along at a speed of 186,282 miles a second—the highest speed anything can attain, even in principle. A moonbeam takes only a little over one second to reach Earth.

But a Danish physicist and her team of collaborators have found a way to slow light down to about 38 miles an hour, a speed easily exceeded by a strong bicyclist.

The physics team, headed by Dr. Lene Vestergaard Hau, who works at the Rowland Institute for Science in Cambridge, Mass., and at Harvard University, expects soon to slow the pace of light still further, to a glacial 120 feet an hour— about the speed of a tortoise.

"We're getting the speed of light so low we can almost send a beam into the system, go for a cup of coffee and return in time to see the light come out," Dr. Hau said in an interview.

The achievement, by Dr. Hau, her two Harvard graduate students and Dr. Steve E. Harris of Stanford University, is being reported today in the journal *Nature*. Physicists said it had many potential uses, not only as a tool for studying a very peculiar state of matter but also in optical computers, high-speed switches, communications systems, television displays and night-vision devices.

One of the most desirable features of the apparatus that the researchers built for their work is that it does not transfer heat energy from the laser light it uses to the ultracold medium on which the light shines. This could have an important stabilizing effect on the functioning of optical computers, which operate using photons of light instead of conventional electrons. A switch using the system could be made so sensitive that it could be turned on or off by a single photon of light, Dr. Hau said.

The medium that Dr. Hau and her colleagues used in slowing light by a factor of 20 million was a cluster of atoms called a "Bose-Einstein condensate" chilled to a temperature of only 50 one-billionths of a degree above absolute zero. (Absolute zero is the temperature at which nothing can be colder. It is minus 273.15 on the Celsius scale, minus 459.67 on the Fahrenheit scale and zero on the Kelvin scale.)

Dr. Hau's group reached an ultralow temperature in stages, using lasers to slow the atoms in a confined gas and then evaporating away the warmest remaining atoms. The temperature they attained, one of the lowest ever reached in a laboratory, was far colder than anything in nature, including the depths of space.

Bose-Einstein condensates (named for the theorists who predicted their existence, Satyendra Nath Bose and Albert Einstein) were first prepared in a laboratory four years ago and became the objects of intense research in the United States and Europe. They owe their existence to some of the rules of quantum mechanics.

One of these is Werner Heisenberg's uncertainty principle, which states that the more accurately a particle's position can be known, the less accurately its momentum can be determined, and vice versa.

In the case of a Bose-Einstein condensate, atoms chilled to nearly absolute zero can barely move at all, and their momentum therefore approaches zero. But because zero is a very precise measure of momentum, the uncertainty principle makes the positions of these atoms very uncertain. In a condensate, as a result, such atoms are forced to overlap one another and merge into superatoms sharing the same quantum mechanical "wave function," or collection of properties.

It was such a superatom, made of a gas of superpositioned sodium atoms, that provided Dr. Hau and her associates the optical molasses they needed to slow down light.

Beginning their project last spring, the researchers tuned a "coupling" laser to the resonance of the atoms in their condensate, shot the laser into the cold cluster of atoms and thereby created a quantum mechanical system of which both the laser light and the condensate of atoms were components. At this stage, the system was no more transparent than a block of lead, Dr. Hau said.

The next step was to send a brief pulse of tuned laser light from a "probe" into the condensate, at right angles to the coupling laser, in such a way that the laser-condensate system interacted with the probe laser. Under these conditions, about 25 percent of the probe laser light passed through the "laser-dressed condensate," but at an astonishingly slow speed. The light that emerged from the apparatus was not visible to the naked eye, but detectors found that it had roughly the same color as the light that had entered.

The speed of light is reduced in any transparent medium, including water, plastic and diamond. Glass prisms and lenses, for example, slow light by differing amounts that depend on the thickness of the glass. The slowing of light causes the bending by which lenses focus images.

But the reduction of light speed in a laser-coupled Bose-Einstein condensate works in an entirely different, quantum-mechanical way. Not only is the speed brought to a crawl, but the refractive index of the condensate becomes gigantic.

Refractive index is a measure of the degree to which a medium bends light. The refractive index of the condensate created by Dr. Hau's group was about 100 trillion times greater than that of a glass optical fiber.

Although Dr. Hau said it might take 10 years before major applications were developed, the condensate's huge refractive index, which can be precisely controlled, may make it a basis for "up-shifting" devices that increase the frequencies of light beams from the infrared end of the spectrum up through visible light to ultraviolet. Possible applications include ultrasensitive night-vision glasses and laser light projectors that could create very bright projected images.

Laser-condensate combinations may also lead to ultrafast optical switching systems useful in computers that would operate using one light beam to control another light beam. Such a system could function as an optically switched logic gate, replacing the electronic logic gates computers now use.

Slow light could also be exploited in filtering noise from optical communications systems, Dr. Hau said.

Dr. Jene Golovchenko, a physics professor at Harvard familiar with Dr. Hau's work, said, "She has worked long and hard on this, and now she's really hit a home run."

—*February 18, 1999*

Trillions of Reasons to Be Excited

By DENNIS OVERBYE

It was late on an August evening when the proton wranglers at the Large Hadron Collider finally got five trillion high-energy particles under control, squeezed and tweaked them into tight bunches and started banging them together.

"Seven minutes too late," grumbled Darin Acosta, a physicist from the University of Florida, whose shift running a control room here [in Cessy, France], among sunflower fields and strip malls, had just ended. On the walls around him, computer screens were suddenly blooming with multicolored streaks and curling tracks depicting the primordial subatomic chaos of protons colliding 300 feet under his feet, in the bowels of the Compact Muon Solenoid, one of the four giant particle detectors buried around the collider ring.

A dozen or so physicists crowded around the screens, calling out the names of particles on the fly, trying to guess what others, as yet unknown to physics, were spraying from the mess in the middle, looking to see some sign from the universe. "This is way cool," one of them said.

"There's a muon," somebody else said as a spike darted out and into the void. "There's a jet."

"This is good," said Maria Spiropulu, a CERN and California Institute of Technology physicist. "This is very, very good."

It has been seven months and some six trillion collisions since physicists at CERN—as the European Organization for Nuclear Research is known—began running protons around their $10 billion, 18-mile electromagnetic race-track underneath the Swiss-French border outside Geneva and smashing them together in search of new particles and forces of nature. No new particles or forces have yet emerged, at least to the statistical satisfaction of the thousands of men and women now sifting through the debris from those collisions.

Nor, of course, has the world disappeared into a black hole.

The proton collisions are scheduled to end on Wednesday. The machine will collide lead ions later in November and then shut down for the holidays. The collider will resume banging protons in February and run until the end of 2011. But CERN physicists say that data has already been accumulating faster than they can analyze it, and that the collider has already begun to surpass its rival, Fermilab's Tevatron. "It's a really beautiful machine. It's performing far better than I expected," Lyn Evans, who oversaw the building of the collider, said recently.

In October, at a conference in Split, Croatia, Dr. Spiropulu showed fellow physicists a picture of a collision that could have produced one of the "dark matter" particles that astronomers say make up a quarter of the universe and are among the grand prizes of science these days. It is one of handful of "interesting events" popping out of the collider that could change the world—if in fact they are real.

But high-energy physics is a game of statistics, and one event is just a tantalizing hint, Dr. Spiropulu said. It will take trillions more collisions before physicists can know if events like these are the harbingers of an intellectual revolution in what the universe is made of.

Or if there is any new physics to be discovered in the collider at all.

"The stakes are violently high as we break new grounds," she said. "We must live up to the dream of 25 years with a lot of seriousness, even if we are like little kids in the candy store with all this data around."

But for all the euphoria in Geneva these days, the collider is still operating under the cloud of Sept. 19, 2008. That is when the electrical connection between two of the collider's powerful superconducting electromagnets exploded, turning one sector of the collider ring into a car wreck and shutting down the newly inaugurated machine for more than a year.

As a result, the machine is operating at only half power, at 3.5 trillion electron volts per proton instead of the 7 trillion electron volts for which it was designed, so as not to blow out the delicate splices. At the end of 2011, all the CERN accelerators will shut down for 15 months, so that the suspect splices—some 10,000 of them—can be strengthened and an unknown number of magnets that have mysteriously lost the ability to handle the high currents and produce the high fields needed to run the collider at close to full strength can be "retrained."

CERN has been under pressure lately to trim its budget, and stopping all the accelerators instead of just the collider will save $25 million, said Rolf Heuer, CERN's director general.

The collider will start up again in 2013 with proton energies of 6.5 trillion electron volts, but it is not likely to reach full power until 2014, if ever.

In interviews recently, scientists and managers said that they had been too eager to get the collider running at full power in 2008. "In perspective, we may say we started too ambitious," said Lucio Rossi, a superconductivity expert who joined CERN from the University of Milan in 2001. One reason, in his opinion, was arrogance.

In order to steer protons racing at more than 99 percent the speed of light around the underground track, the collider's electromagnets have to carry currents of some 12,000 amperes, which they can do only by being cooled by superfluid liquid helium to less than 2 degrees Celsius above absolute zero. At that point, their niobium-titanium wires conduct electricity without resistance. The engineers had been able to cool each of the 10,000 or so superconducting magnets and test them before putting them in the collider, but they could not test the connections between them, Dr. Rossi said.

Those connectors are sandwiches of superconducting wire and copper glued together with solder. The copper is there to take over carrying the enormous current if the superconductor heats up in a so-called quench and loses its superconductivity—but only for the 100 seconds or so that it takes for the magnets to dump their enormous and dangerous energy.

Dr. Rossi said that several reviews had established that the connector design was very good, but those reviews had missed the point that it was not robust against faulty construction, like missing solder, which apparently is what caused the 2008 disaster.

The current had no place to go. Sparks punctured a surrounding vessel of supercold helium, which flooded out, pushing the 30-ton magnets around like toys. Soot spread for two miles along the pipe that carries the proton beams.

"Superconductivity calls for total quality; one mistake will undo the whole system," Dr. Rossi said. "It becomes a fuse."

The fix, as described by Dr. Rossi, is fairly simple: an extra set of copper shunts on the outsides of the splices, providing an extra bridge across any divisions or junctions that might lack solder, as well as an improved system to spot trouble and bigger valves to release helium less explosively.

The result should be "supersafe for life," said Steve Myers, who is in charge of running the collider. "And we can dispense with talking about them, because I'm fed up with talking about connectors."

Indeed, undeterred by past disasters, CERN recently laid out plans for the next 20 years of running and upgrading the collider and its detectors, including an idea to swap out all its magnets in 2030 to increase the total proton energies to 33 trillion electron volts—almost as much as the ill-fated American superconducting super collider, a project canceled by Congress in 1993, would have had.

The latter suggestion raised eyebrows among physicists in and out of CERN, who wondered, among other things, what it would mean for the International Linear Collider, which has long been presumed to be the next big physics machine.

"To speak of 33 trillion electron volts is premature," Dr. Evans said. The long hiatus has had a dramatic effect on the hunt for the collider's main quarry, a particle known as the Higgs boson, which theory says is responsible for imbuing other elementary particles with mass. The Higgs supposedly has a mass somewhere between 114 billion electron volts and 185 billion electron volts—in the units of mass-energy favored by physicists.

By the time it shuts down in 2011, the CERN collider should have amassed about 20 times as much data as it now has, enough to make a dent in the Higgs hunt. The lead in that quest currently belongs to the Tevatron, until last year the world's largest accelerator, which has been colliding protons and antiprotons with energies of a trillion electron volts for the last two decades at the Fermi National Accelerator Laboratory in Batavia, Ill., piling up data. Last summer, Fermilab physicists announced that they had eliminated the region between 158 billion and 175 billion electron volts.

The Tevatron was scheduled to shut down in 2011, but Pier Oddone, Fermilab's director, recently said he would seek financing to keep the Tevatron running until 2014, by which time it could gather enough data to examine the whole energy range over which the Higgs is or is not hiding. But he said he needed at least $35 million a year to avoid hurting other Fermilab projects. CERN and Fermilab both deny they are in a race to find the Higgs or for predominance in physics.

"Of course we feel a healthy competition with the Tevatron, let's put it this way," said Dr. Heuer.

In September Dr. Heuer and Dr. Oddone issued a joint statement deploring what they said was a news media emphasis on competition between the labs, pointing out that Europeans and Americans have worked at one another's labs and that Fermilab played a major role in constructing the Large Hadron Collider.

"Both CERN and Fermilab directors are committed to supporting each other and the global particle physics community in addressing the most important fundamental questions of our era," they said.

John Ellis, a CERN theorist, said the future looked bright.

"The vise is closing in inexorably," he said of the Higgs. As for dark matter, he said the CERN collider would soon exceed the Tevatron in exploring for new particles: "I can hardly contain my enthusiasm."

Those sentiments were echoed by Fabiola Gianotti, a CERN physicist and leader of a collaboration of some 3,000 physicists, whose Atlas detector is the prime rival to the Compact Muon Solenoid experiment. The Atlas building is across the street from the entrance to CERN, in Meyrin, Switzerland, and on a late August day the artist Josef Kristofoletti was finishing a giant mural of the Atlas detector that is visible from the surrounding countryside.

Showing off the Atlas control room, Dr. Gianotti said that from the moment the collisions began last spring, she noticed that they were richer, with more particles coming out. That richness is only now beginning to be plumbed.

"We have been waiting so long," she said. "Only good and beautiful things are coming."

—November 2, 2010

CHAPTER 13

Technology:
Invention and Revolution

THE SCIENCE OF INVENTION

Sound and Electricity

Lecture By PROF. ALEXANDER GRAHAM BELL—
An Exhibition of the Speaking Telephone

At Chickering Hall last evening, before about 300 persons, Prof. Alexander Graham Bell lectured on "Sound and Electricity," and gave an exhibition of the speaking telephone. The address was mainly devoted to an explanation of the different qualities of sound produced, as the speaker expressed it, by different shades of vibration. He then went into an exhaustive discussion of the transmission of sound by electricity and a history of the telephone, illustrated by a number of complex and not very intelligible figures cast upon a prepared background by means of a stereopticon. At the end of the lecture Prof. Bell stated that he had intended to give what he believed would have been interesting and instructive illustrations of the power of the telephone, but, unfortunately, his improved instruments had not arrived from Boston, and he would have to content himself with displaying a telephone of inferior power. This, he explained, was connected by an ordinary telegraph wire with an improved telephone in New-Brunswick, N.J., 32 miles away, and this in turn was attached to an ordinary organ, upon which a number of simple tunes would be played, and the sounds transmitted to New York by means of his instruments. He then telegraphed to New Brunswick, and shortly after, from a little box on the stage, and from other instruments in different parts of the hall, came the music of the song known as "The Sweet By and By." This was followed by "Home, Sweet Home," and afterward by "Hold the Fort," sung by a strong baritone voice, and plainly audible. After this, Prof. Bell and Mr. C. W. Field asked a number of questions through the telephone. They were all answered satisfactorily by those at the end of the wire in New Brunswick. The performance was concluded by the transmission of a song, which many of those present were enabled to hear with great distinctness by passing by the instrument on the stage and placing their ears in close proximity to its mouth.

—May 18, 1877

The Röntgen Discovery

Some of the photographs of objects invisible to the naked eye, made by Prof. Röntgen of Wurzburg, have reached England, and are causing comment by men skilled in photography in that country.

The name "cathode rays" seems already the accepted term for the form of radiant energy which has the peculiarity of being stopped through transparent glass, and of passing through ground glass, wood, metal, and human flesh. The use of this term comes, of course, from the fact that the rays are excited at the cathode, or negative, pole of a Crooke's tube in action.

Crooke's tube is simply a modification of Geissler's tube. It consists of an egg-shaped bulb of glass, from which the air has been almost exhausted. At one end the positive current is brought in to the tube by means of a fused platinum wire and a small disk-shaped piece of aluminum is placed at the end of the wire. On the lower side of the tube is the spot where a similar disk of aluminum receives the current which has been transmitted through the vacuum. Where the current enters is called the anode, and where it leaves is called the cathode. These are otherwise known as the positive and negative poles, and are often indicated by a plus and minus sign, respectively.

Prof. Crookes passed an electric current through this tube and it gave out brilliant phosphorescent effects, as usual in such cases. Near the cathode, however, was a small dark spot, all the more remarkable on account of the brilliant light around it. Prof. Crookes, investigating certain effects on bodies in the tube, such as a small Maltese cross of aluminum, found that some remarkable energy was manifesting itself there. This was the beginning which led up to Prof. Röntgen's discovery.

Arthur Bowes of Salford, England, who has seen photographs made by Prof. Röntgen, has written a letter to *The London Photographic News*, part of which is here reproduced. Mr. Bowes is a civil engineer, employed by the city of Salford, and is an amateur photographer of some distinction. He says in his letter:

"By the kindness of Prof. Schuster of the Owens College, Manchester, I have been able to examine the photographs sent to this country, and obtain some particulars of the marvelous discoveries they illustrate, made by Prof. Röntgen of Wurzburg. As some of your readers may be aware, these photographs demonstrate the possibility of photographing through substances which are opaque to ordinary light.

"It has long been known that in experimenting with electric discharges through glass tubes, which have been exhausted more or less completely of air, there is produced a strong phosphorescence of fluorescence in the glass walls of the tube. Some years ago Heriz showed that these phosphorescent emanations would permeate thin metal, and in 1893 Dr. Lenard described before the Royal Prussian Academy of Sciences at Berlin an arrangement by which the rays were made to pass through a plate of aluminum .003 millimeters thick. This plate, while quite opaque to ordinary light, permitted the rays from the Geissler tube to permeate it, rendering the air faintly luminous and creating a strong odor of ozone.

"Prof. Röntgen's discovery is that, in addition to this phosphorescence, another radiation of a hitherto unknown nature is produced, which is capable of penetrating through all bodies, though not to the same extent. Wood and flesh are more easily penetrated by it than glass, and although its effects do not make themselves visible to the human eye, they can be recorded photographically.

"Thin plates of metal offer little opposition to its passage, and can be photographed through almost as readily as can a pane of glass with ordinary daylight. The first photograph which Prof. Schuster showed me was a half-plate print on albumen paper, exhibiting the effect of interposing a human hand between the source of light and the sensitive plate.

"The outlines of the flesh were only faintly defined because the flesh was comparatively transparent to the radiations; the bones of the fingers were very plainly shown, with the knuckles clearly defined; a signet ring on one finger was the most distinct feature in the picture. The whole effect was that of a badly defined skeleton hand with a ring on one finger.

"No camera had been used, because as Prof. Schuster explained, the new light—or radiation, as the professor preferred to call it—differed from ordinary light in some of its most essential features. So far as is yet known it can neither be reflected or refracted; the lens has no power to concentrate it and form an image in the usual way. All the photographs taken with it are in the nature of shadows, formed by interposing various substances in the path of the rays.

"Another photograph showed a mariner's compass or similar dial which had been photographed while enclosed in a metal case. The dial was about two inches in diameter and well defined. In another experiment the source of light had been placed at one side of a wooden door, while the plate had been exposed on the other side, and in the resulting picture the internal markings of the wood were revealed, as well as the outlines of a metal hasp or fastener.

"A photograph of a piece of zinc, composed of several strips of zinc rolled into one apparently homogeneous mass, revealed distinct striations in the interior of the metal. One of the most instructive of the photographs was produced by passing radiations through thin slabs of various materials, such as glass, Iceland spar, iron and aluminum. The relative amount of obstructions offered to the rays was made manifest by the varying depths of tint in the photograph, and Prof. Schuster pointed out that the opacity appeared to vary in the same relation as the density.

"The heavy substances, such as iron, lead, glass, and Iceland spar, offered more obstruction than did lighter substances, such as wood, paper, aluminum, or flesh, and this holds good, quite irrespective of the behavior of these substances with ordinary light. As thin layers of wood are comparatively transparent to the new emanations, it will be seen that a photograph can be taken on a plate enclosed in a dark slide without drawing the shutter on the slide."

Mr. Bowes makes the remarkable suggestion that the spiritualistic mediums must have known all along of the new light. He says:

"Surely the spiritualistic mediums who produce ghostly photographs for the edification of Mr. W. T. Stead and others will rejoice that science has begun to confirm their assertions at last. I asked the professor if the photographing of an object enclosed in a metal box did not look like clairvoyance or something more incredible, but he smilingly refused to commit himself to any opinion.

"As to the genuineness of the photographs there is no room for doubt. Prof. Röntgen holds a high position in the scientific world, and while himself entirely beyond suspicion, he is sufficiently astute not to have been made the victim of others. There is no doubt that the discovery he has made is one of the utmost importance, and its ultimate development in the fields of physical and medical science may lead to the extension of our knowledge over vast fields hitherto undreamed of."

Another correspondent of the same journal, Leslie Miller, gives his view of the matter as follows:

"To conclude, cathode rays are disturbances set up in the luminiferous ether by an electric current, which are neither light rays nor electric rays, but have some of the properties of both. They can be deflected by a magnet, and are obstructed by the molecules of air and all gases, glass, metals, and other substances to varying degrees, but quite differently to light, as the substances that offer the greatest obstruction to light may offer little to them. They are also able, unlike light rays, to pass round an obstruction.

"Finally, cathode rays are not all alike. They possibly differ in the same way that light does from violet, red, or green, &c., and it is probable that Prof. Röntgen has hit on a particular number of vibrations or other quality having strong chemical effects, and therefore, of importance photographically.

As regards the necessary apparatus, from the point of view of a maker of such things, it offers no particular difficulty, especially now that alternating current from the street mains is at hand, so that transformers, to transform up to the necessary tension, can be employed. M. Lenard used a coil which gave about an inch spark in air between spheres; but the length of the spark modifies the nature of the rays, and from the increased effects obtained by Prof. Röntgen he must have made some modifications."

In a more technical view of the photographic importance of the new form of radiant energy, E. J. Wall, in the *British Journal of Photography*, says:

"It is, of course, well known that the haloid salts of silver are not only sensitive to light, but that the invisible images capable of being developed are produced by pressure, by electricity, and by various chemicals, such, for instance, as the hypophosphites.

"The following extract from Guillemin's *Electricity and Magnetism*, page 371, edited by Prof. Sylvanus Thompson, is not without interest or bearing on this subject, as at least, leading to considerable speculation as to whether these experiments of Prof. Röntgen do not support the theory that the action of light on the silver haloid is not rather photo-electrical than photo-chemical. In speaking of Faraday's discovery of the magnetic power of magnets on light it is stated that Clerk Maxwell deduced the theory 'that light itself is simply an electro-magnetic phenomenon; that which we call waves of light are not mechanical waves at all, but are immensely rapid electric displacements taking place in the all-pervading ether of space. . . . It is believed by the editor of this work that this theory, which is now in the main accepted by all the younger generation of physicists, will be able, with proper modification and development, to explain the curious relation existing between electricity and the phosphorescent, fluorescent, and chemical properties of light.'

"Crookes, in his presidential address to the British Association in 1886, drew attention to an extremely graphic illustration of the grouping of the elements according to the periodic law, and roughly speaking, this might be likened to a pendulum with ever-decreasing amplitude, and at various points the elements fall into line, being respectively electro-negative and electro-positive.

"This being the case, then, might it not happen that, accepting the electromagnetic theory of light as enunciated above, a film of silver haloid may be so affected by the electric displacement called light as to be disassociated? Minchin and Waterhouse have proved the passages of a current of electricity when light falls on a film of silver salts, and Mr. Bolas, in his affiliation lecture, suggested that the developer might merely close an electric circuit."

A New York physician, who is an ardent and distinguished amateur photographer, said yesterday that he thought one of the great values of Prof. Röntgen's discovery would be the impetus which it would give to work on color photography. In his opinion it was the invisible rays at the violet end of the spectrum which kept color photography from being now in everyday use. What he hoped from Prof. Röntgen's discovery was that photographic inventors would be able to use it so as to exclude the invisible rays from the camera.

—*January 29, 1896*

Television Is Brought Nearer the Home

By ORRIN E. DUNLAP JR.

Television is moving toward the American home by a series of important steps. A notable advance was made during the last week when Dr. E. F. W. Alexanderson, in the laboratory of the General Electric Company, demonstrated that the magic images can now be made to perform on a theater screen. Various television demonstrations have been conducted, but the significance of this latest one is found in the fact that the pictures can now be shown on a much larger screen than heretofore and that television is ready to be booked as a vaudeville act.

Each advance in the science of seeing by radio arises from what the engineers call an elaborate and highly coordinated series of researches. Each year the process is simplified. Those close to the development of this new science declare that, despite the remarkable advances up to the present time, the broadcasting of sight needs still further refinement before the elusive images can be made to dance into the living rooms of America to entertain the family on a silver screen. With each simplification in mechanical and electrical parts the experts discover that the faces that travel through space become larger and clearer.

These scientists have three goals ahead of them. First, the pictures must be clear. They must be large enough so that the entire family or a theater audience can watch the images act on a screen just as the motion-picture actors do. And third, the television receiver that is eventually designed for home use must be as simple and foolproof as an ordinary broadcast receiver.

Two Goals in Sight

Dr. Alexanderson proved in his latest demonstration that the engineers are within sight of two of the goals. They have succeeded in clarifying the pictures. They have nurtured them until they have grown from miniature dimensions to the size of a theater screen. Now they assert that a few more wrinkles remain to be ironed out of the ethereal faces. Ahead of them, however, is the ultimate goal—the home.

"The possibilities for new inventions in this art of television are inspiring," said Dr. Alexanderson. "Just think of it when you can put an electric eye wherever you wish and you can see through this eye just as if you were there. An airplane with a news reporter will fly to see whatever is of interest and the whole theater audience will be with him, seeing what he does, and yet the audience will be perfectly safe and comfortable.

"Or what will this mean in the wars the future when a staff officer can see the enemy through the television eyes of his scouting planes or when they can send a bombing plane without a man on board which can see the target and be steered by radio up to the moment when it hits? Or what will it mean for peaceful aviation when the ships of the air approach a harbor in fog, take on a local pilot, not from a little craft that comes to meet the ship, but by television, whereby the trained eyes of the pilots functioning by television will guide the ship to the airport in safety?"

Three Years' Progress

Each year sees the television images, the children of science, grow. It was in April 1927, that Herbert Hoover, then secretary of commerce, was televised in Washington and seen at the end of a wire line in New York. Some believed that television was about to enter the home. The engineers warned against immediate hopes. They saw numerous obstacles confronting them despite the fact that the "fruition of years of study on the problem of seeing at a distance as though face to face" was an achievement.

Public imagination grasps each television show without seeing the hazards strewn in the path of the research men whose ambition it is to build television for commercial use. The Bell Laboratories' engineers have, since 1927, developed a direct-scanning system or portable radio camera which photographs outdoor scenes so that they can be reproduced on a distant screen. This instrument dispensed with the glaring lights under which a person had to sit in order to be televised. This step forward was credited to the men who discovered how to make the photoelectric cells or "eyes" more sensitive. They made it possible to illuminate the subject broadly by daylight. A tennis player was seen in action on the screen.

It was in June 1929, that Dr. Herbert E. Ives of the Bell Telephone Laboratories demonstrated television in colors. The picture was about the size of a postage stamp. And the looker-in had to peer through a peek-hole in a dark booth. Those who witnessed this demonstration went away feeling that radio vision performances in the home would not be practical until the images could be flashed on a screen large enough for the entire family to see the show at once.

Last Autumn Dr. Alexanderson brought his vision machine to the New York radio show. Thousands stood in line to be televised, while others filed through a darkened room to see the faces of friends smile at them from the pinkish-colored screen. The images were only fourteen inches square. They were not large enough to convince the visitors at the show that television was practical for their living rooms. Some were heard to remark that they thought seeing by radio would remain more or less of a scientific dream. Dr. Alexanderson and his corps of experts went to their research laboratory in the Mohawk Valley confident that the day was not far distant when the television images would grow up.

Two-Way Television

And in the meantime Dr. Ives also continued his work. It was only a few weeks ago that this electro-optical research director of the Bell Laboratories demonstrated that two-way television is practical. The speakers go into soundproof booths at the ends of the telephone line. They see and hear each other as they speak. Dr. Ives explained that the system can be used on wire or radio and over long distances. But distance adds to the expense and so does radio at the present time. The two-mile wire demonstration was conducted chiefly to reveal the progress up to this time.

Dr. Ives does not believe that a stage of perfection has been reached where television images can be entrusted to radio, therefore he is keeping them on the wire. He sees limitations in radio carrying the images. He points to static and fading as the worst enemies that streak the ethereal faces and shoot them full of holes, unless the atmospheric conditions are almost perfect. And they seldom are. Dr. Ives asserts that to send television images through space today is like expecting a brook running through a populous area to flow clear and uncontaminated. Nevertheless, he is hopeful that man's inventive genius will conquer the elements and save television from its unseen enemies that lurk in the vastness of space.

Over in England John L. Baird, a Scotsman, is in the television race. He is broadcasting sound and sight for the London audience, featuring songs, monologues, cartoons and news. Those who have witnessed the Baird performances report that head and shoulder images are surprisingly clear and steady.

Dr. Alfred N. Goldsmith, chief engineer of the Radio Corporation of America, has a television receiver in his home in West End Avenue. He has entertained friends by tuning in on a face in a television studio there miles away. The pictures were very clear. Dr. Goldsmith said that the images were remarkable considering the fact that they traveled through New York's skyscrapers, which absorb the strength of the waves. Outside of New York, away from the influence of the street structures, he said the images would probably travel clearly for at least forty-five miles.

C. Francis Jenkins of Washington, D.C., a pioneer in radio vision, has erected a station for image transmission experiments at Jersey City. Vladimir Zworykin at the Westinghouse Electric and Manufacturing Company at Pittsburgh has developed a television receiver based upon a new type of cathode ray tube. He predicted not long ago that a television bulb for attachment to an ordinary broadcast receiver could be built in quantity for $100. Baird's sets sell for $125 complete in London. The Zworykin tube eliminates the scanning disk and other moving parts.

Further Research Ahead

Philo T. Farnsworth, a research worker in San Francisco, is also reported to be developing a cone-shaped vacuum tube device that dispenses with the revolving disk. The image is seen on a fluorescent screen. In Boston the Short Wave and Television Laboratories, cooperating with station WEEI, are working on color television. Films are scanned instead of an actual person or object.

Dr. Alexanderson, or course, has a television receiver in his home, which is three miles from the television station. However, he is an expert and it is a foolproof instrument under his careful handling. But it is not ready for all homes where mother, father, sister and brother will all tune it. That is why Dr. Alexanderson has turned to the theater to disclose this 1930 step forward in television. He would rather not talk about seeing by radio in the home. It is an entirely different proposition that must be looked at from a different angle.

It is apparent, however, that Dr. Alexanderson is very hopeful of television, far more than he was a year or two ago. He predicts that a wave of amateur television will sweep the country and fire the imagination in much the same way that broadcasting did in 1920 and 1921.

It will be recalled that the music in the air ten years ago inspired thousands to build broadcast receivers. And now Dr. Alexanderson foresees a multitude of radio amateurs and technicians dusting off their workbenches in the attics and cellars of America to build machines that will put them in tune with the elusive images that jump about in the emptiness of space. These magic images have life. They speak, sing, play and act just like motion picture artists. Television is being introduced to the public as a talking picture, not a silent drama.

Television's screen is wheeled out on the stage just as easily as a piano for a novelty act. When the curtains parted last Thursday the radio men in the audience realized that Dr. Alexanderson had triumphed in creating a large sized picture that makes radio vision practical for the theater at least. No longer is the image restricted to miniature dimensions. No longer must the looker-in squint through a tiny peek-hole to catch a glimpse of the fleeting images. And behind or at the side of the screen is television's voice, a giant loudspeaker, which reproduces the voice of the speaker several miles away, at the same time that his actions and facial expressions are flashed on the screen.

Unlike the movies, no beam of light streaks across the auditorium above the heads of the audience. Television's projector is located backstage. The pictures are thrown on the screen from behind. This new art of entertainment seems destined to introduce innovations in theatrical entertainment.

Audiences at the afternoon and evening show at the Schenectady theater last Thursday walked to their seats unsuspecting that they were to witness a historic performance. They saw the orchestra come into the pit as usual. But the director was missing. He was several miles away, listening to his men tune up over a telephone wire. Then the theater darkened. The curtains parted. A flood of light swept over the silver screen on the stage. An image appeared. It was the conductor with his baton raised ready to direct his orchestra by television. Then performers flashed on the screen.

The entertainers were televised in an improvised studio which is now a part of Dr. Alexanderson's laboratory. The light impulses reflected from their faces were converted into electrical impulses and then into radio waves broadcast by a laboratory transmitter tuned to release the images on the 140-meter wave length. Microphones close to the actors picked up the speech, music and songs, and converted the sound into electricity, which was carried by wire to the short-wave transmitter at South Schenectady for broadcasting on the 92-meter wavelength. The sound and images were scattered through space at the speed of sunlight. Over at the theater, an assistant to Dr. Alexanderson acted as the control operator. It was his duty to manipulate the apparatus that intercepted the moving pictures that were somewhere in the air. A small device called a monitor telopticon transferred the impulses to a light valve, at which point the light was broken up to produce the image that corresponds in every detail to the person or object being televised several miles away. Only the head and shoulders were seen.

The Delicate Light Valve

This light valve is based upon the invention of Dr. August Karolus of Leipzig, Germany. It is the heart of an intricate system of lenses, which is in front of a high intensity arc lamp similar to those used for the projection of motion pictures. The light valve is a delicate device. It must function with the utmost accuracy to permit the passage of light that corresponds perfectly with the impulses received from the television transmitter. These light emissions are passed on through lenses to a disk corresponding in size, design and rate of rotation to a disk at the radio "camera" or originating point. Other lenses pass the light forward to the screen, on which the light impulses, at the rate of 40,000 per second, wash or paint an active, life-like motion picture.

The arc lamp, with the associated lenses and light valve, which all comprise the television projector, is placed seventeen feet back from the screen. A heavy black cloth from the projector to the screen forms an effective light tunnel which eliminates any stray light beams from acting like static to blot or blur the pictures. The entire apparatus at the theater is mounted on wheels to facilitate assembly and disassembly when used as part of the vaudeville show.

A second radio receiver is on duty at the theater to detect the words or music, which is fed into the large loudspeaker—television's voice.

The life-like image is not a silhouette, nor is it merely a black-and-white picture. All the gray shades between black and white are reproduced, registering every shadow and shade of the original scene, giving both depth and detail to the image.

The engineers point out that in radio broadcasting the frequencies of speech and music modulate or shape the current sent out from the aerial wires. But in television the aerial radiation is modulated or formed to correspond to the image by a succession of light impulses. The person to be televised stands in front of an incandescent lamp. Between the person and the light is a metal disk about the size of a bicycle wheel and drilled with 48 holes. The disk is made to revolve so that it covers the person's face 20 times in a second. That creates twenty complete pictures made up of light and shade. A large square frame contains four photo-electric cells or radio "eyes," sensitive to light just as the microphone is to sound. These "eyes" respond 40,000 times in a second to the light impulses reflected from the person being televised.

Future of Television

"Looking back over the development of the electrical industry," said Dr. Alexanderson, "we can clearly trace the forces which have enabled the science of electricity to give birth to the electrical industry. We see how later the electrical industry took hold of another branch of science and created the radio industry. We are able to some extent to project into the future the working of the forces that give birth to new epochs, but as to the destiny and significance of these new movements, after they have been launched, the engineer is peculiarly blind. Owen D. Young has repeatedly said that he has the great advantage of not being handicapped by scientific knowledge. His predictions of the future also have been much more far-flung and correct than those of the engineers associated with him.

"For fifteen years radio was simply an auxiliary to navigation. In 1915 and 1916 we held daily communication by radio telephone from Schenectady to New York. We found that many amateurs adopted the habit of listening in, and our noon hour of radio became the first regular broadcasting. But we had no idea what it would lead to. Our idea was to telephone across the ocean, and so we did at the close of the war, but we failed to see the great social significance of broadcasting.

"Television is today in the same state as radio telephony was in 1915. We may derive some comfort from this experience of the past, but, on the other hand, we are not sure that the analogy is justifiable and that television will repeat the history of radio telephony. We must then fall back upon our conviction that the development of television is inevitable on account of the forces working in the scientific world today, and that it is a satisfaction to make one's contribution to this evolution even if, in this case, the results should prove to be only a stepping-stone to something else." . . .

A 20,000-Mile Journey

"In this interest the amateurs and the professional experimenters are on common ground. We got a real thrill out of sending a television wave to Australia and have it come back and tell its tale, even though it was a simple one. We observed that after traveling 20,000 miles a rectangle still had four corners, which was more than we had expected. As a matter of fact, it was broken up into pieces most of the time. But there were glimpses of encouragement and a fertile field for the imagination. These are the incentives of the explorer, whether he is an amateur or a professional.

"Whether the general public will be enough interested or get enough satisfaction out of television to make it possible to commercialize home sets for television is still to be seen. A new technique of entertainment will be required. As a supplement of broadcasting it can make a reality of the radio drama. Political and educational speakers may use it as a medium, and entertaining personalities like Will Rogers will tell the latest wisecracks and comment on the news of the day. You have seen our test at Proctor's Theatre in this city. It is likely that every moving-picture theater in the large cities will have to be equipped to give a short television act.

One of Many Steps

"What we have demonstrated is just one of the many steps that must be taken in our efforts to conquer distance by television. The improvement of light control which makes it possible for us now to show a picture of theater size is due to an invention by Dr. Karolus, whom I visited in Leipzig some years ago and whose inventions we have been endeavoring to perfect. In our past exhibits the improvements of light control have been due to Dr. D. McFarlan Moore and his neon lamps, and I should not be surprised if next time Dr. Moore would go one better than Dr. Karolus. Invention is a delightful and friendly sport, and if we did not have competition we should not have inventions, just as you could not have a race unless you have somebody to race with."

Dr. Alexanderson said that perhaps some might think that he was letting his imagination run riot. Those acquainted with him know differently. He is very modest. He dislikes the role of prophet in radio or television. But to show the seriousness of his conviction, he is now on his way to Panama on the airplane carrier *Saratoga*, at the invitation of the United States Navy, in order to study some of the practical conditions for the realization of some of his ideas, among which is the control of a pilotless plane by radio. The plane will "see" by a television eye, and those who control its actions in the sky will keep sight of it, although far below the horizon.

—May 25, 1930

Contact With Moon Achieved by Radar in Test by the Army

By JACK GOULD

The first man-made contact with the moon was achieved on Jan. 10 when the Army Signal Corps beamed a radar signal on it and 2.4 seconds later received an echo reflected by the celestial body, it was announced yesterday. The signal, covering a round-trip distance of an estimated 450,000 miles, was sent out from the Evans Signal Laboratories at Belmar, N.J.

Applications almost beyond immediate comprehension were foreseen as a result of the electronic achievement. New and far more accurate study of the universe, perhaps ultimately resulting in the detailed topographical mapping of distant planets, was anticipated. Detection of enemy missiles flying through cosmic space also was expected to be possible from the new definitive proof that radio waves could penetrate beyond the earth's ionosphere.

The sound that the moon sent back to the earth took the form of a 180-cycle note, or somewhat higher in pitch than the hum to be heard on a home radio receiver when a station is not tuned in. It lasted half a second. The Army also recorded the echo visually on an oscilloscope. There the epic-making peep appeared as a series of jagged, saw-tooth lines.

Army Announces Feat

The official announcement that a radio signal had been bounced off the moon was made by Maj. Gen. George L. Van Deusen, Chief of the Engineering and Technical Service, Office of the Chief Signal Officer, at the annual dinner of the Institute of Radio Engineers at the Hotel Astor.

The first word to reach the public, however, came several hours earlier under circumstances anything but formal. A group of reporters crowded into a small upstairs reception room in the hotel and a quiet 39-year-old officer, Lieut. Col. J. H. DeWitt, who supervised the experiment, announced what had been done.

As the men who had finally "reached the moon," Colonel DeWitt and his four chief associates in the venture were modest in the extreme. Only upon the reporters' insistence was there any revelation of biographical material on the quintet.

Colonel DeWitt, a former broadcast engineer in Nashville, Tenn., and a "ham" (amateur) radio operator, acknowledged that the results were the climax of his peacetime hobby to put a signal up to the moon. He said he failed in an attempt in 1940.

Jacob Mofsenson, 32, a graduate of City College, who entered the Signal Corps in April, 1942, was even more hesitant, but finally consented to tell his peacetime occupation.

"I was a diamond dealer," he said, with a laugh.

The other principal participants were Dr. E. K. Stodola, 31, a graduate of Cooper Union, who was in charge of research; Dr. Harold Webb, 36, a former teacher of physics and mathematics at West Liberty College, West Liberty, Va., and Herbert Kauffman, 31, who had worked in radio in New Orleans.

Two Conducted First Test

Dr. Webb and Mr. Kauffman were actually the only two at the radar receiving equipment when the first echo came back from the moon, but said they had betrayed no particular emotion at the time over the event.

"We looked for it and got the results," Dr. Webb said.

Work on reaching the moon by radar was started soon after V-J Day, according to Colonel DeWitt, and on Jan. 10 preparations had been completed for a test. On that day, he said, the moon rose at 11:48 a.m and a few minutes later the initial radar impulse was beamed heavenward on a frequency of 111.6 megacycles. It was at 11:58 a.m., as the scientists remembered it, that the first flick of light appeared on the oscilloscope denoting success.

The tests were continued for five days, three tests being made as the moon rose and one as it set. Tests of the moon's "receptiveness" when it was higher in its arc were not possible because of lack of suitable antenna equipment and some days no signal came back, apparently because of propagation characteristics within the earth's atmospheric region.

The peak power of the transmitter was three kilowatts but through use of a special antenna giving a gain of 200 its radiation effectiveness thereby being vastly increased. The strength of the signal received back from the moon was calculated at about three watts.

Colonel DeWitt emphasized the "real trick" in making contact with the moon was not so much in the transmission but in the construction of a receiver of exceptional sensitivity to pick up the feeble echo from the planet. He estimated the sensitivity at .01 microvolts.

The radar waves traveled at the speed of light—186,000 miles a second. The mean distance between the moon and the earth is calculated at 238,857 miles, but the greatest problem for the scientists at Belmar was to allow also for the distance variation involved in the relationship between the speed of the moon and the earth's movement. The moon's speed, it was explained, varies from 750 miles faster than the earth's rotation to 750 miles slower.

Having demonstrated that a signal can reach the moon and return, Colonel DeWitt said the only problem left was the calculation of the time interval. When the echo came back in 2.4 seconds, the scientists were convinced they had achieved their aim "because there was nothing else there but the moon."

To make sure that there might be no error, a small group of scientists, not identified, visited Belmar and verified the conclusions.

"Hello" From Moon Expected

Colonel DeWitt professed a dislike for speculation of a "Buck Rogers or Jules Verne" character, but acknowledged that the Army scientists hoped to increase their transmitter's power so that it could be modulated by voice.

"We should be able to say "hello" and hear the moon say "hello" back," he said.

He quickly added: "I hope the moon doesn't answer, 'Good-bye.'"

In connection with the announcement in New York, the War Department in Washington issued a statement on the implications of the feat. Maj. Gen. Harry C. Ingles, Chief Signal Officer of the Army, noted that it could have "valuable peacetime as well as wartime applications, although it is impossible at this stage to predict with certainty what these will be."

"One obvious possibility is the radio control of long-range jet or rocket-controlled missiles, circling the earth above the stratosphere," the War Department continued. "The German V-2 missiles already are believed to have reached an altitude of sixty miles.

"The primary significance of the Signal Corps achievement is that this is the first time scientists have known with certainty that a very high frequency radio wave sent out from the earth can penetrate the electrically charged ionosphere which encircles the earth and stratosphere. The several layers of the ionosphere start about thirty-six miles above the surface of the earth and extend to approximately 250 miles.

"On this basis, the V-2 projectiles already have risen above the lower ionosphere levels, and it is now known that radio waves can completely penetrate the ionosphere."

"The new technique will also be valuable for studying the effects of the ionosphere upon radio waves. Scientists already have learned that low and medium frequency waves are reflected by the ionosphere, and these reflections form the 'skywaves' used for long-distance broadcasting. The ionized layers also sometimes distort and bend radio waves, much as a prism distorts light waves."

"Another valuable application may be the provision of new astronomical information. Not only may it be possible to construct detailed topographical maps of distant planets with the aid of radar data, but scientists may be able to determine the composition and atmospheric characteristics of other celestial bodies by this means.

"A less likely application of the new technique will be the possibility of radio control from the earth's surface of 'space ships' venturing thousands of miles from the earth, and the radio reporting of astronomical data electronically computed aboard such vessels."

For the project, which Army officers informally labeled "Diane," the chief deviation from conventional radar application was in the use of a much longer pulse-repetition rate, somewhere between three and five seconds, compared with the usual pulse rate of thousands of times a second. The length of time each pulse of energy existed varied from one-tenth to one-half a second, an "enormously long interval" compared with wartime standards.

It was the factor of the rate and duration of the pulse, plus companion antenna problems, that led Colonel DeWitt and his associates to a relatively cautious approach in speculating on making contacts with Mars and Venus, as different distance ranges present new sets of problems.

Termed "Interesting Tool"

Dr. Harlow Shapley, director of the Harvard College Observatory, described the Army's radar contact with the moon tonight as "an interesting tool in exploring the solar system" and predicted that more startling war-born developments would be revealed within the next few years.

"I believe radar contact with the moon is extremely helpful," he said, "because it will aid us in the study of meteoric material in the vicinity of the earth. I don't think it will ever help us to fine another planet."

Indicating that the radar experiments had been known to scientists for at least two years, Dr. Shapley said the importance of this advance was not at all comparable to the atomic bomb. However, he said astronomers were working on other discoveries made during the war, and he predicted they would be far more startling than the radar contact when they were announced.

—January 25, 1946

Tiny Tube Excites Electronics Field

A tiny crystal device called the transistor, about the size of a kernel of corn, has electronics engineers and military men excited over its vast potentials. Some rate it as one of the most important developments in electronics since the vacuum tube, which made possible the modern telephone, radio, television, radar and countless other electronics marvels, according to the Associated Press.

The transistor can do many of the things a vacuum tube can do; it can do some better. Besides being a fraction the size of a vacuum tube, it's many times as rugged, takes only a minute amount of electricity, requires no warm-up before starting to operate, and—having no filament to burn out—will last indefinitely.

The transistor, first developed by Bell Telephone laboratories in 1948 and greatly improved since then, will go into first practical use in about a year in long distance telephone equipment. Other early applications may be in tubeless radios, smaller, more efficient hearing aids that will work for a year without a change of batteries, and in your telephone set so that you can hear and be heard clearly and loudly even on the remotest rural line.

Seen Promoting Giant Industry

But even greater things are expected of the transistor in making possible the application of electronics for many uses now impossible with the vacuum tube. Robert M. Burns, chemical director of Bell Laboratories, foresees a new electronics industry based on the transistor that "will become the only rival in size of the chemical industry, that growing industrial group that seems destined to take over all other manufacturing industry."

"The principal operating industries, electric power, transportation and communications, will continue to depend equally upon the electronic and chemical industries, he declares. "So we may look ahead to the time when the electron will do man's work, including thinking, while chemistry supplies him with food, clothing and everything else."

The transistor was invented by Dr. William Shockley, of Bell Laboratories. The first model consisted of two tiny wires pressed against a speck of germanium, a substance belonging to a family of materials known as semi-conductors. They are enclosed in a metal cylinder about the size of a 22-caliber rifle bullet.

Last July, the laboratories announced a radically new type called a junction transistor.

It consists of a tiny piece of germanium treated so that it provides a thin electrically positive layer sandwiched between two electrically negative layers, with wires connected to each layer. It is inside a tiny plastic case and takes up only one-fiftieth the space of a typical sub-miniature vacuum tube. With the wires protruding, the device resembles a spider.

One Bell System publication, in assaying the potentials of the transistor, points out its application in mathematical computing machines—the fabulous "electronic brains." Despite the massive amounts of calculations they can accomplish, the publication said, "present machines boast about one-millionth of man's mental power; but, according to at least one scientist, transistors can increase this by a hundred thousand fold."

—January 13, 1952

Light Amplifier

A new electronic device that makes possible true amplification of light was demonstrated here last week by Dr. Theodore H. Maiman of the Hughes Aircraft Company's research laboratories at Culver City, Calif. It is named a "laser," an acronym for "light amplification by stimulated emission of radiation."

In ordinary light sources the atoms emitting the light radiate individually at random, thus producing light waves that are "incoherent," like a jumble of radio waves that are out of phase. But in the laser, an ordinary light source stimulates a ruby crystal to emit its radiations in phase, resulting for the first time in a beam of "coherent light," a goal sought for many years.

The device, it was stated, is a new scientific tool for investigating matter. When perfected, it will make possible the focusing of light into powerful high-intensity beams for space communication, and will lead to vast increases in the number of available communications channels, using light waves instead of radio waves. Because it generates "coherent" light a million times purer than the purest frequencies now available, the laser could generate a beam that would spread only 100 feet if sent from Los Angeles to San Francisco, while a search light beam would spread fifty miles. If sent to the moon, a beam would illuminate a lunar area less than ten miles wide, while an ordinary searchlight, if it could reach the moon, would spread over of area 25,000 miles.

—July 10, 1960

The Computer Revolution

Electronic Computer Flashes Answers, May Speed Engineering

By T. R. KENNEDY JR.

One of the war's top secrets, an amazing machine which applies electronic speeds for the first time to mathematical tasks hitherto too difficult and cumbersome for solution, was announced here tonight by the War Department. Leaders who saw the device in action for the first time heralded it as a tool with which to begin to rebuild scientific affairs on new foundations.

Such instruments, it was said, could revolutionize modem engineering, bring on a new epoch of industrial design, and eventually eliminate much slow and costly trial-and-error development work now deemed necessary in the fashioning of intricate machines. Heretofore, sheer mathematical difficulties have often forced designers to accept inferior solutions of their problems, with higher costs and slower progress.

The "Eniac," as the new electronic speed marvel is known, virtually eliminates time in doing such jobs. Its inventors say it computes a mathematical problem 1,000 times faster than it has ever been done before.

The machine is being used on a problem in nuclear physics.

The Eniac, known more formally as "the electronic numerical integrator and computer," has not a single moving mechanical part. Nothing inside its 18,000 vacuum tubes and several miles of wiring moves except the tiniest elements of matter—electrons. There are, however, mechanical devices associated with it which translate or "interpret" the mathematical language of man to terms understood by the Eniac, and vice versa.

Ceremonies dedicating the machine will be held tomorrow night at a dinner given a group of Government and scientific men at the University of

Pennsylvania, after which they will witness the Eniac in action at the Moore School of Electrical Engineering, where it was built with the assistance of the Army Ordnance Department.

The Eniac was invented and perfected by two young scientists of the school, Dr. John William Mauchly, 38, a physicist and amateur meteorologist, and his associate, J. Presper Eckert Jr., 26, chief engineer of the project. Assistance also was given by many others at the school.

Army ordnance men had been on the lookout for a machine with which to prepare a large volume of ballistic data, which in turn was needed to break a threatened bottleneck in the production of firing and bombing tables for new offensive weapons going overseas. Without the tables the guns could not be used effectively.

Project Took Thirty Months

Capt. H. H. Goldstine, Army ordnance mathematician, then at the school, heard of Dr. Mauchly's ideas, told Col. Paul N. Gillon of the Aberdeen (Md.) Proving Ground, enlisted his enthusiastic support and the project went forward with Government aid. Thirty months to the day later it was finished and operating, doing easily what had been done laboriously by many trained men. The Eniac soon will be permanently installed at Aberdeen.

"A very difficult wartime problem" was sent through its intricate circuits soon after it was completed. The Eniac completed the task in two hours. Had it not been available the job would have kept busy 100 trained men for a whole year. So clever is the device that its creators have given up trying to find problems so long that they cannot be solved.

This resolver of difficult problems is what computing experts call a "digital" counter. Basically, it does nothing more than add, subtract, multiply and divide. It does this by generating very accurately timed electrical impulses at a speed of 100,000 per second, and can do one operation every twentieth pulse, thereby adding, for instance, at the rate of 5,000 per second.

Since all mathematical tasks, no matter how abstruse or involved, can be resolved to basic arithmetic if enough time is available, the Eniac can reverse the process, eliminate time, and arrive at an answer to virtually any problem. So say its inventors.

Machine Has Memory, Too

The machine, however, can do much more. It has the human faculty of "memory," four kinds of it, to perform certain tasks in the proper sequence. It also has "control" elements, and can, up to a point, dictate its own action. It can, for instance, compare two numbers and, depending on which one is larger, choose one of two possible courses.

First, it gets its original numbers from a series of cards in which holes are punched to indicate the "initial and boundary conditions" of the problem. One of the Eniac "minds" performs this job.

When the problem is punched on the cards they are dropped into a slot in a "reader." The man who wants the answers may then sit down and await results. He seldom has to wait long; the Eniac does most of its tasks in seconds.

A unit called "a master programmer" oversees the whole computation and makes sure it is carried out.

The Eniac has some 40 panels nine feet high, which bristle with control and indicating material. Pink neon lights blink on several panels as buttons are pressed. Numbers are printed beside the lights.

Those who witnessed the demonstration entered a 30-by 60-foot room. The computer took up most of the space.

Dr. Arthur W. Burks of the Moore School explained that the basic arithmetical operations, if made to take place rapidly enough, might in time solve almost any problem.

"Before You Can Say . . ."

"Watch closely, you may miss it," he asked, as a button was pressed to multiply 97,367 by itself 5,000 times. Most of the onlookers missed it—the operation took place in less than the wink of an eye.

To demonstrate the Eniac's extreme speed, Dr. Burks next slowed down the action by a factor of 1,000 and did the same problem. Had the visitors been content to wait 16⅔ minutes they could have observed the answer in neon light. The next was multiplication—13,975 by 13,975. In a flash the quotient appeared—195,300,625. A table of squares and cubes of numbers was generated in one-tenth of a second. Next, a similar one of sines and cosines. The job was finished and printed on a large sheet before most of the visitors could go from one room to another.

The Eniac was then told to solve a difficult problem that would have required several weeks' work by a trained man. The Eniac did it in exactly fifteen seconds.

All problems must first be re-solved to their essentials, punched on cards and run through an International Business Machines unit called a "reader." The reader translates the mathematical language to that of the Eniac, and vice versa. When this is done the machine is ready to operate. Numerical values covering a wide range of scientific "constants" are interjected as and when they are needed. There are four kinds of "memory" in the Eniac to accomplish this. Constant adjustments are made in advance for each type of problem.

Normally the Eniac handles ten-digit numbers—a billion, for instance—but it can handle twenty-digit numbers just as easily, resulting in numbers running to astronomical size.

Machine Cost $400,000

More than 200,000 man-hours went into the building of the machine. It contains more than half a million soldered joints, and cost about $400,000. Three times as much electricity is required to operate it as for one of our largest broadcasters—150 kilowatts.

Little more than three years ago the Eniac was only an idea; today it is perhaps the greatest marvel of electronic ingenuity. Dr. Mauchly joined the Moore School staff in 1941, hoping he might be able to realize his ambition, to revolutionize the art of dealing with huge numbers in complex form. He believed, for instance, that something could be done about long-range weather predicting.

In the field of peacetime activities Dr. Mauchly foresees not only better weather-predicting—months ahead—but also better airplanes, gas turbines, micro-wave radio tubes, television, prime movers, projectiles operating at supersonic speeds carrying cargoes in peace and even more and better accuracies in studying the movements of the planets.

According to Capt. Goldstine, "mountainous" computational burdens have been carried by scientists in the past, which will be largely removed by electronic computers. He pointed out that the solution of equations of motion has been a hindrance in the past and that studies of shell flight, high-speed planes, rockets and bombs are "a few of the fields that will benefit hugely through electronic computing."

Mr. Eckert predicted an era which, with electronic speeds available, problems that have been thought impossible because they might require a lifetime will be readily resolved for man's use.

"The old era is going, the new one of electronic speed is on the way, when we can begin all over again to tackle scientific problems with new understanding," he told reporters.

Mr. Eckert briefly described the Harvard and Massachusetts Institute of Technology mechanical and electro-mechanical computing machines, the most recent of which was announced only a few months ago.

—February 15, 1946

The "Chip" Revolutionizes Electronics

By CHARLES LEEDHAM

When Lemuel Gulliver awoke on a November morning in 1699 to find himself confronted by six-inch humans walking about on his chest, he accepted without question the size of his miniature captors. With equal readiness we in the 1960s have accepted a world of Lilliputian miniaturization—the pocket radio, the battery TV portable, the invisible hearing aid and other offspring of the ubiquitous transistor—which would have been marvels beyond words a very few years ago. And now we are on the threshold of not just miniature electronics, but microelectronics, of devices built of circuits so unbelievably tiny that they can literally be passed through the eye of a needle.

Wristwatch radios? This dream of a decade ago is now an antiquated concept. A complete radio can today be comfortably fitted into a signet ring. Pocket-sized television? Given the advances in electronics, it is now nearly possible to put the entire circuitry of a color television set inside a man's wristwatch case. A computer to fit a desk top? It is already here, in General Electric's 18-pound A-212, an electronic digital computer capable of 166,000 operations per second, and in a package the size of a standard office typewriter. There is almost nothing electronic which can any longer be called a dream.

Behind it all is the "chip," or integrated circuit, a fantastic development in technology which has been under way for several years with surprisingly little fanfare in the laboratories of such major electronics manufacturers as Motorola, Texas Instruments, Westinghouse and Sylvania. At the heart of the integrated circuit is the transistor, already responsible for its own electronic revolution.

Electronics began with the vacuum tube, developed immediately after the turn of the century to serve the infant science of radio. The major function of the tube is to amplify; without amplification useful radio could not exist.

Radio waves leaving a transmitting station are enormously powerful, but they may arrive at the receiving set reduced over the distance to fractions of a millionth of a volt. Fed to the input of a vacuum tube, however, these feebly fluctuating wavelets of radio energy are used by the tube to control larger voltages and to produce at the output of the tube an amplified version of the input. In this way the original faint radio signal has enough force to activate a loudspeaker and produce the sound we hear.

The early vacuum tubes were large, clumsy affairs, but improvement began almost immediately. Tubes got smaller, and the giant, room-dominating Atwater-Kent of fond memory shrank over the years to the table-model radio. Tubes were as small as seemed possible, about the size of the first joint of a little finger. Then came the transistor.

It was in 1948 that Bell Telephone Laboratories announced the first practical transistor, and miniaturization prepared to take a major jump forward. Within a space no bigger than a pinhead of germanium or silicon crystal, the first transistors managed the marvel of amplification like tubes; their later, more sophisticated descendants have taken over almost every other vacuum-tube function in electronic circuits. Transistors are much smaller than tubes, and their other advantages allow even further size reductions. They need far less voltage—6 to 12 volts on the average compared to more than 100 volts for a tube—and thus smaller batteries and power supplies.

Because they do not break or wear out like tubes, they can be permanently mounted in a circuit without the need for tube-type sockets or the space that is required for the removal, testing and replacement of tubes. Theoretically a transistor, given reasonably good treatment, will perform perfectly forever. Other things go wrong with transistor radios, not the transistors themselves.

Printed circuits made the next small contribution to miniaturization by eliminating most of the wiring between components. In place of wires, small pathways of conducting material are laid down—printed in complex patterns directly on a thin board of nonconducting material. The transistors and other components are then mounted on these boards cheek by jowl, their connections soldered to the conducting pathways, and another step downward in size is possible.

The final step—so far at least—is the integrated circuit, or chip, so called because of its minuscule size. In a crude comparison, the integrated circuit is an ultraminiature printed circuit, in which not only the connections but the components themselves are imprints. But even this sort of comparison cannot give a true impression of the incredible minuteness of the chips, or the startling advance in electronic technology they represent.

In one process used by Sylvania, slices of pure silicon (a major transistor material) the size of a quarter are the raw stock. By a process of microphotoengraving amazingly sophisticated in itself, not one but several hundred tiny, identical circuits are "printed" on this single piece of silicon. Microscopically small channels are cut in the material, and other materials are added in nearly invisible amounts, layer by layer and pattern by pattern, until complete, minuscule transistors and other necessary components are created, each in place and connected together properly within the space of a few hundredths of an inch.

The individual minute chip circuits cut out of the basic silicon slice are so small that the necessary input and output connections of ultrafine metal strips or wires cannot be made by human fingers. Instead, miniature machinery is used to mount each chip in the center of a more manageable "flatpack" or "package," a wafer of nonconducting material about the size of a large postage stamp. In this process the chip connections are attached to narrow metal pathways which radiate from the center of the pack and are large enough to be handled by humans. In building a computer, for example, where several chips must be linked in a multi-part circuit, human operators solder the large connectors together in operating sequence, or slip the connectors into spring clips which make the necessary hookups.

Even in these early stages of the development of integrated circuits, a chip five one-hundredths of an inch square can contain as many as 22 transistors plus associated components. A thimbleful of chips can provide enough circuitry for a dozen computers, or thousands of radios. Credibility falters, but there the chips are, in production and already in limited use.

Chips have other assets besides size. Their reliability is vastly increased because there are no longer any soldered connections between the many parts within the chips to break or otherwise fail. They require fractions of a volt for power and present no heat problems.

The space-age implications of microelectronics are immediately apparent, where ounces and cubic inches saved are far more precious than pounds and cubic feet of gold, and where assured reliability is worth a scientist's ransom. Guidance computers aboard rockets and satellites already make partial use of chips, permitting dramatic size and weight reductions. One radio-guidance system has been reduced from 250 pounds to under 50, and the extensive use of chips promises to reduce this still further to 20 pounds. When man goes to the moon and beyond, he will be riding on chips almost as much as rockets.

In earthbound military applications of the future, microelectronics will even transform the basic fighting infantryman. Microelectronic receivers and transmitters weighing less than an ounce and fitted into helmet liners will give instant communication between units, between commanders and men, and between individual men on patrols. Microradar sets no larger than a flashlight will probe enemy terrain, and subminiature heat detectors will locate even the most carefully hidden guerrilla fighter. Sophisticated combinations of detectors and guidance systems may even aim machine guns with deadly accuracy at distant, invisible enemies, even as larger radar systems now control the fire of anti-aircraft batteries.

Civilian uses of chips will be slower in coming, but already one chip is being used in the building of ever-tinier hearing aids—which were also the first consumer adaptation of the transistor. Late in 1963, Zenith put on the market a quarter-ounce, behind-the-ear unit built around an audioamplifier chip containing the equivalent of 6 transistors and 16 resistors in a piece of silicon .05 by .15 inches. And earlier this year, a smaller version was released, a complete hearing aid so tiny that it fits entirely into the ear, in a custom case molded to the individual ear canal.

Neither medical nor electronic researchers like to speculate in public on the further application of microelectronics to medical problems for fear that false hopes might be raised by such authoritative predictions no matter how carefully they were qualified. But within the foreseeable future, it is likely that new techniques will make it possible to embed an audio-amplifier chip surgically in the ear structure itself, with the tiny amount of power needed to operate the chip drawn from body heat, pressure changes or any of several other power sources within the body. Since their connectionless structure reduces their failure rate to near zero, chips are specially suited to permanent surgical implantation.

There is already in existence a tiny radio transmitter in a capsule which reports on acid balance, pressure and other factors to a consulting-room receiver as the capsule travels through a patient's digestive system. Microelectronic TV cameras in pills could show doctors the actual state of the stomach or intestines: In routine hospital care, cheap and tiny monitors could give floor nurses continuous reports on the heartbeat, pulse, respiration and temperature of their patients while miniature TV cameras watched each room.

Further micromedical applications are in the works or at least are now possible with microelectronics. Russian research teams are reported to be experimenting with miniature computers and control devices connected to tiny motors within artificial arms. If the minute electrical impulses from remaining nerve ends can be properly amplified and controlled, it should be entirely possible to make an artificial arm with full movement, activated by nothing more than the nerve impulses of normal motion, and eventually with near-normal hand and finger movement.

It is in such computer circuitry that chips hold the greatest promise, for computers require hundreds and thousands of reduplications of similar and identical simple circuits. Basically, a computer is made up of a large number of small, rather stupid circuits capable of nothing more than elementary counting. But put enough of them together in the right sequences and, unlike humans in committee, the whole turns out to be astronomically greater than the sum of the parts—a machine capable of making thousands, millions and even billions of calculations and operations per second.

The first vacuum-tube computer ever built weighed 30 tons, a balky, cranky, overheated monster which filled a 30-by-50-foot floor space and was, by present standards, not particularly bright. Today, using chips, much more capable computers can be built into desk-sized units, most of the space being taken up by controls and accessories necessary for macroscopic humans to work with it.

The current computers are not only smarter, but almost infinitely faster as their size is reduced. Because of its small size—as well as its design factors—a standard Sylvania computer chip unit, called the "J-K flip-flop," can switch a signal in one 10-billionths of a second, infinitely faster than a bulky tube, and appreciably faster than transistor-plus components circuits.

Major business procedures are already being revolutionized by computers. Giant retail chains like J. C. Penney keep fingertip control of their sales, stock and ordering by feeding the daily sales records from each store into data-processing computers. The day is not far distant when microelectronics will have brought down the size and price of computers to a point where even a small retailer will be able to rent an elementary model for his store. It will sit on the retailer's counter in the place of the old-fashioned cash register and will be wired into the memory and processing banks of a large central computer. The latter will take over from the merchant all the routine decisions of business, acting as his buyer, stock clerk, accountant and paymaster and performing half a dozen other functions.

For example, when the owner or clerk sells one pair of black shoelaces, the code meaning "shoelaces, black, one pair" will be punched into the computer on the counter along with the cash amount. Late that night, when the master computer scans the retailer's unit, it will note this sale along with all others. Instantly it will scan the merchant's stock of shoelaces and, taking into account the season, the item's sales rate over the past years, the weather and every other known factor affecting black shoelace sales, it will decide whether or not to put in a reorder.

With receipts and payments punched in, the computer will also give the merchant a daily report on his bank balance, his cash on hand, his accounts payable, his receivables. It will figure his sales taxes and other taxes, fill in the tax forms and make out the payment checks. It will compute his payroll, deduct all applicable taxes and print those checks, too.

Or, in a different and even more complicated system of computerization—one including a computer at the shoelace factory—the retailer's unit will immediately relay the sale information to the central master computer. Then, if a reorder is indicated, the central computer will so inform the computer at the factory, and that unit will have the invoice, bill and shipping label printed even before the retailer's customer walks out the door with the pair of laces that triggered the reorder process.

In large corporations, this kind of operation is already almost a reality. Orders coming into the Owens-Illinois Glass Company are popped immediately into a computer which checks stock and either issues withdrawal orders or initiates automatic production runs. If a customer's computer has printed the original order, it is only one step removed from direct computer-to-computer ordering and order-filling.

Finally, when bank computers are hooked into the central memory and processing units, not even checks will change hands. The retailer's account will be debited and the supplier's credited in the same fraction of a second that the order is placed and filled. All the owner will have left to do is wander about the aisles, smiling at his customers.

In the more distant future, much the same computer setup may become available to the housewife, who will then be able to leave it to her kitchen computer to keep the refrigerator stocked, her checkbook balanced, her menus planned and her calories unerringly counted.

More immediately, there are countess applications of the miniaturization principle for the home. An oven with devices highly sensitive to temperature will cook a roast or bake a cake perfectly every time. In washing machines, light-sensitive scanners will keep the washing cycle going until the clothes are indisputably clean; a drier with humidity sensors will turn itself off at the exact moment its load is completely dry. Fine-control air-conditioning and heating systems will analyze the temperature and humidity at several points in each room and keep them at a constant, desired level by producing minute adjustments in burners, compressors, blowers and duct openings.

Another computer application might be the completely automatic vacuum cleaner housed in a small, self-powered, wheeled unit. At 3 a.m. such a unit could turn itself on, wheel out of a small "garage" built into a wall and then methodically move back and forth across the floor of the various rooms, silently vacuuming, automatically avoiding obstacles as it went. Its task completed, it would return to the "garage" to empty its load of dust into an incinerator and turn itself off until the next night.

Unlikely? In 1948, British neurologist Dr. Grey Walker built and demonstrated a similar small machine as an experiment in electronically duplicated animal-response patterns. Dr. Walker used vacuum tubes—but reduce the steering circuits to chips, fill the remainder of the space with suction equipment, and the fully automatic vacuum cleaner would be a reality.

Integrated circuits in computer and communications circuitry will soon create a revolution in personal communications as great as the change from horseback messenger to the telephone. Through the use of chips, wristwatch radios will become commonplace, each unit capable of being dialed directly from every other unit, with the signals relayed, if necessary, by satellites like *Early Bird.*

With microelectronic computer circuitry it will also be possible to set every communicator so as to allow calls only from certain other numbers and to shut off all calls during sleeping hours unless a caller equipped with a special personal code dials a level of urgency high enough to override the receiver's shutoff.

Well before the advent of such personal radio communication, though, miniature computer circuitry will have made today's telephone systems seem incredibly antiquated. There is already in existence a circuit that will take over once you have dialed a number and got a busy signal—the circuit keeps trying the number until it is free and answers; then it rings you back.

It will be possible, whenever you are at any place equipped with a telephone, to punch into that phone your own personal code. Then, when anyone calls your home number and gets no answer, computer circuitry will in seconds scan every telephone in the city until it finds the one where your code is temporarily registered and proceed to ring that number.

The circuitry could, at the caller's command, continue its search through the phone systems of the entire world, until it had either reached you or reported that your code was not punched into any phone on earth at the moment.

Leslie Solomon, an editor of *Electronics World,* once said of these future personal communication and search systems, "We will soon be at the point where, if your party doesn't answer, he is dead."

The list of applications of miniaturization runs on seemingly without end. And around the electronics industry there are rumors that somewhere, in some laboratory, really miniature work is being done which will soon make integrated chips seem big and old-fashioned. Exactly what that means, or will mean, is still a matter for guessing. One thing is certain: Miniaturization will unalterably change our lives and the lives of our children probably far beyond recognition. From tiny chips, and perhaps tiny chips of chips, the world of the future will grow.

—*September 18, 1965*

A Free and Simple Computer Link

By JOHN MARKOFF

Think of it as a map to the buried treasures of the Information Age.

A new software program available free to companies and individuals is helping even novice computer users find their way around the global Internet, the network of networks that is rich in information but can be baffling to navigate.

Since its introduction earlier this year, the program, called Mosaic, has grown so popular that its use is causing data traffic jams on the Internet. That worries some computer scientists. But Mosaic's many passionate proponents hail it as the first "killer app" of network computing—an applications program so different and so obviously useful that it can create a new industry from scratch.

"Mosaic has given me a sense of limitless opportunity, which is the reason that I went into computer science in the first place," said Brian Reid, a computer researcher who is the director of the Digital Equipment Corporation's Network System's Laboratory in Palo Alto, Calif.

Digital, a leading computer maker, is exploring ways of using Mosaic as the basis for a whole new system of electronic commerce, letting customers easily browse through on-line product catalogues. Other companies—including Xerox, the software company Novell Inc. and the publisher R. R. Donnelley—are also exploring business opportunities they see springing from Mosaic. And in California, a government and private industry consortium called Smart Valley Inc. is using Mosaic to create an electronic marketplace for Silicon Valley high-tech companies.

Before Mosaic, finding information on computer data bases scattered around the world required knowing—and accurately typing—arcane addresses and commands like "Telnet 192.100.81.100." Mosaic lets computer users simply click a mouse on words or images on their computer screens to summon text, sound and images from many of the hundreds of data bases on the Internet that have been configured to work with Mosaic.

Click the mouse: there's a NASA weather movie taken from a satellite high over the Pacific Ocean. A few more clicks, and one is reading a speech by President Clinton, as digitally stored at the University of Missouri. Click-click: a sampler of digital music recordings as compiled by MTV. Click again, et voila: a small digital snapshot reveals whether a certain coffee pot in a computer science laboratory at Cambridge University in England is empty or full.

Other databases searchable with Mosaic include the card catalogues of the Library of Congress and hundreds of American and foreign university libraries, Federal Government archives, various NASA computers and the University of California at Berkeley paleontology museum. The French Government is also considering using Mosaic to display digitized versions of paintings and other art exhibits from its national galleries.

"Mosaic is the first window into cyberspace," said Larry Smarr, the director of the National Center for Supercomputing Applications in Champaign, Ill., where Mosaic was developed. The center in Champaign, which is one of the nation's four federally financed supercomputer research centers, is receiving more than 600,000 electronic information queries each week from Mosaic users.

Available free to Internet users willing to download it to their computers, Mosaic has been acquired by several hundred thousand computer networkers in less than a year, according to several industry estimates. The users include computer scientists, librarians, software developers, magazine publishers, record companies and catalogue distributors, all of whom see it as the first general-purpose navigational tool for the emerging data highway.

Helping Shape a Debate

So sudden and dramatic has been Mosaic's success in attracting commercial software developers that the program may play a decisive role in determining the shape of the national "information infrastructure" now being debated by Government officials and telecommunications and computer executives.

One evangelist for Mosaic is Mitchell D. Kapor, founder of the Lotus Development Corporation, the company whose Lotus 1-2-3 spreadsheet helped ignite the personal computer revolution in the early 1980s. Mr. Kapor, who heads the Electronic Frontier Foundation, a public interest group on computer issues, visited a cable television industry show last week in Anaheim, Calif., and demonstrated Mosaic. He sees it as a tool in his crusade to cajole the telephone, cable television and computer industries to establish an open and accessible national data highway rather than a private toll road that many of the private companies seem to prefer.

"For me Mosaic was a turning point," Mr. Kapor said. "It's like C-Span for everyone."

A Better Way to Browse

Mosaic was created by a small group of software developers and students at the supercomputer center in Champaign, who set out 18 months ago to create a system for browsing through the World-Wide Web. The Web is an international string of computer databases that uses an information-retrieval architecture developed in 1989 by Tim Berners-Lee, a British computer specialist at the CERN physics laboratory in Geneva.

Mr. Berners-Lee's system originally permitted many of the world's high-energy physics researchers to exchange information represented in a form known as hypertext. With hypertext, highlighted key words and images are employed to point a user to related sources of information.

"I realized that if everyone had the same information as me, my life would be easier," Mr. Berners-Lee said.

From a small electronic community of physicists, the World-Wide Web has grown into an international system of data base "server" computers offering diverse information. The Web has also fundamentally changed the way information is obtained over the Internet. In the past it has been largely necessary to connect to a remote host computer using a complicated software program that fooled the computer into thinking the far-away visitor was using one of the host machine's own local terminals. But the Web simplifies things by using a networking model called client-server computing, which allows remote requests for information from any personal computer or workstation.

"It's like the difference between the brain and the mind," Mr. Berners-Lee said. "Explore the Internet and you find cables and computers. Explore the Web and you find information."

Much of the Web data now available to Mosaic users is being made available by university researchers who are constructing demonstration projects to explore the technology.

But commercial applications are being worked on by a wide range of corporate developers. Novell, the world's largest developer of computer network software, has organized its technical reference literature so that any computer network user can retrieve it. Such browsing represents a form of advertising that is noninvasive and that seems to fit well with the culture of the Internet, whose users tend to be easily infuriated by electronic junk mail.

"I'm convinced that very quickly we'll have a new Madison Avenue kind of industry devoted to this style of advertising," said Tony Rutkowski, an executive at the Sprint Corporation, one of the companies whose long-distance lines provide the backbone for the Internet.

Already, O'Reilly & Associates, a technical book publisher based in Sebastapol, Calif., has used Mosaic and the Web to create an on-line magazine that includes advertising.

There remain, however, significant barriers to using Mosaic. It requires that the user have a computer that is directly connected to the global Internet. Many businesses and almost all universities now have such connections, but the majority of personal computer users currently connect to the Internet only indirectly through on-line information services like Delphi or America Online.

Many companies are now working on software that will make connecting to the Internet no more difficult than dialing into these on-line services. Earlier this week, for example, O'Reilly & Associates, in partnership with a Seattle software developer called Spry Inc., announced that next year it will begin distributing a software package for less than $100 called Internet-in-a-Box. The product, which will permit computer users to connect directly to the Internet, will also probably include a version of Mosaic to permit network novices to browse data stored in the Web.

"The problem with Mosaic is that it's currently for the haves of the Internet," said Tim O'Reilly, president of the publishing company. "This is an attempt to bring this exciting end of the Internet to the average user. We think this is the future of on-line publishing."

—*December 8, 1993*

Look Officer, No Hands: Google Car Drives Itself

By JOHN MARKOFF

Anyone driving the twists of Highway 1 between San Francisco and Los Angeles recently may have glimpsed a Toyota Prius with a curious funnel-like cylinder on the roof. Harder to notice was that the person at the wheel was not actually driving.

The car is a project of Google, which has been working in secret but in plain view on vehicles that can drive themselves, using artificial-intelligence software that can sense anything near the car and mimic the decisions made by a human driver.

With someone behind the wheel to take control if something goes awry and a technician in the passenger seat to monitor the navigation system, seven test cars have driven 1,000 miles without human intervention and more than 140,000 miles with only occasional human control. One even drove itself down Lombard Street in San Francisco, one of the steepest and curviest streets in the nation. The only accident, engineers said, was when one Google car was rear-ended while stopped at a traffic light.

Autonomous cars are years from mass production, but technologists who have long dreamed of them believe that they can transform society as profoundly as the Internet has.

Robot drivers react faster than humans, have 360-degree perception and do not get distracted, sleepy or intoxicated, the engineers argue. They speak in terms of lives saved and injuries avoided—more than 37,000 people died in car accidents in the United States in 2008. The engineers say the technology could double the capacity of roads by allowing cars to drive more safely while closer together. Because the robot cars would eventually be less likely to crash, they could be built lighter, reducing fuel consumption. But of course, to be truly safer, the cars must be far more reliable than, say, today's personal computers, which crash on occasion and are frequently infected.

The Google research program using artificial intelligence to revolutionize the automobile is proof that the company's ambitions reach beyond the search engine business. The program is also a departure from the mainstream of innovation in Silicon Valley, which has veered toward social networks and Hollywood-style digital media.

During a half-hour drive beginning on Google's campus 35 miles south of San Francisco last Wednesday, a Prius equipped with a variety of sensors and following a route programmed into the GPS navigation system nimbly accelerated in the entrance lane and merged into fast-moving traffic on Highway 101, the freeway through Silicon Valley.

It drove at the speed limit, which it knew because the limit for every road is included in its database, and left the freeway several exits later. The device atop the car produced a detailed map of the environment.

The car then drove in city traffic through Mountain View, stopping for lights and stop signs, as well as making announcements like "approaching a crosswalk" (to warn the human at the wheel) or "turn ahead" in a pleasant female voice. This same pleasant voice would, engineers said, alert the driver if a master control system detected anything amiss with the various sensors.

The car can be programmed for different driving personalities—from cautious, in which it is more likely to yield to another car, to aggressive, where it is more likely to go first.

Christopher Urmson, a Carnegie Mellon University robotics scientist, was behind the wheel but not using it. To gain control of the car he has to do one of three things: hit a red button near his right hand, touch the brake or turn the steering wheel. He did so twice, once when a bicyclist ran a red light and again when a car in front stopped and began to back into a parking space. But the car seemed likely to have prevented an accident itself.

When he returned to automated "cruise" mode, the car gave a little "whir" meant to evoke going into warp drive on *Star Trek*, and Dr. Urmson was able to rest his hands by his sides or gesticulate when talking to a passenger in the back seat. He said the cars did attract attention, but people seem to think they are just the next generation of the Street View cars that Google uses to take photographs and collect data for its maps.

The project is the brainchild of Sebastian Thrun, the 43-year-old director of the Stanford Artificial Intelligence Laboratory, a Google engineer and the co-inventor of the Street View mapping service.

In 2005, he led a team of Stanford students and faculty members in designing the Stanley robot car, winning the second Grand Challenge of the

Defense Advanced Research Projects Agency, a $2 million Pentagon prize for driving autonomously over 132 miles in the desert.

Besides the team of 15 engineers working on the current project, Google hired more than a dozen people, each with a spotless driving record, to sit in the driver's seat, paying $15 an hour or more. Google is using six Priuses and an Audi TT in the project.

The Google researchers said the company did not yet have a clear plan to create a business from the experiments. Dr. Thrun is known as a passionate promoter of the potential to use robotic vehicles to make highways safer and lower the nation's energy costs. It is a commitment shared by Larry Page, Google's co-founder, according to several people familiar with the project.

The self-driving car initiative is an example of Google's willingness to gamble on technology that may not pay off for years, Dr. Thrun said. Even the most optimistic predictions put the deployment of the technology more than eight years away.

One way Google might be able to profit is to provide information and navigation services for makers of autonomous vehicles. Or, it might sell or give away the navigation technology itself, much as it offers its Android smart phone system to cellphone companies.

But the advent of autonomous vehicles poses thorny legal issues, the Google researchers acknowledged. Under current law, a human must be in control of a car at all times, but what does that mean if the human is not really paying attention as the car crosses through, say, a school zone, figuring that the robot is driving more safely than he would?

And in the event of an accident, who would be liable—the person behind the wheel or the maker of the software?

"The technology is ahead of the law in many areas," said Bernard Lu, senior staff counsel for the California Department of Motor Vehicles. "If you look at the vehicle code, there are dozens of laws pertaining to the driver of a vehicle, and they all presume to have a human being operating the vehicle."

The Google researchers said they had carefully examined California's motor vehicle regulations and determined that because a human driver can override any error, the experimental cars are legal. Mr. Lu agreed.

Scientists and engineers have been designing autonomous vehicles since the mid-1960s, but crucial innovation happened in 2004 when the Pentagon's research arm began its Grand Challenge.

The first contest ended in failure, but in 2005, Dr. Thrun's Stanford team built the car that won a race with a rival vehicle built by a team from Carnegie Mellon University. Less than two years later, another event proved that autonomous vehicles could drive safely in urban settings.

Advances have been so encouraging that Dr. Thrun sounds like an evangelist when he speaks of robot cars. There is their potential to reduce fuel consumption by eliminating heavy-footed stop-and-go drivers and, given the reduced possibility of accidents, to ultimately build more lightweight vehicles.

There is even the farther-off prospect of cars that do not need anyone behind the wheel. That would allow the cars to be summoned electronically, so that people could share them. Fewer cars would then be needed, reducing the need for parking spaces, which consume valuable land.

And, of course, the cars could save humans from themselves. "Can we text twice as much while driving, without the guilt?" Dr. Thrun said in a recent talk. "Yes, we can, if only cars will drive themselves."

—*October 10, 2010*

ACKNOWLEDGMENTS

The New York Times Book of Science is the work of many hands: the reporters who wrote the articles (largely anonymous before the 1920s and '30s); the men and women who assigned and edited them (usually anonymous even today); the photographers and graphics editors who illustrated them; and the journalistic archaeologists who unearthed them, especially—and invaluably—Susan Campbell Beachy of *The Times* research department. Three *Times* science editors deserve special thanks for their support: Cornelia Dean, Laura Chang, and the late Barbara Strauch.

Alex Ward, editorial director, book development, at *The Times*, and Barbara Berger, executive editor at Sterling Publishing, oversaw the project with patience and expertise. Thanks also to the following people at Sterling: Marilyn Kretzer, editorial director; Kimberly Broderick, production editor; Christine Heun, interior design; David Ter-Avanesyan, cover design; Elizabeth Lindy, cover art director; and Terence Campo, production.

Contributors' Biographies

Lawrence K. Altman has covered medicine for *The New York Times* since 1969. A physician specializing in internal medicine and epidemiology, he served in U.S. immunization programs against smallpox and measles in Africa in the 1960s. He writes the "Doctor's World" column for Science Times.

Natalie Angier joined *The New York Times* in 1990 and won a Pulitzer Prize ten months later for science reporting. She writes the "Basics" column for Science Times. Her books include the best-selling *Woman: An Intimate Geography*.

William Beebe (1877–1962) was a naturalist, explorer and author famed for his bathysphere dives to the ocean floor in the 1930s.

Sandra Blakeslee, a longtime science writer for *The New York Times* and other publications, is the author or co-author of many books, including *The Good Marriage: How and Why Love Lasts* and *The Body Has a Mind of Its Own*.

William J. Broad is a science writer for *The New York Times* who has won two Pulitzer Prizes as well as an Emmy Award. He is the author of *The Science of Yoga: The Risks and the Rewards*.

Jane E. Brody, who joined *The New York Times* in 1965, is the longtime "Personal Health" columnist for Science Times. Her books include the best sellers *Jane Brody's Nutrition Book* and *Jane Brody's Good Food Book*.

Malcolm W. Browne (1931–2012), a longtime science writer for *The New York Times*, won a Pulitzer Prize in 1964 for his coverage of the Vietnam War for the Associated Press. His 1963 photograph of the self-immolation of a Buddhist monk became one of the war's most memorable images.

John F. Burns was a foreign correspondent for *The New York Times* for forty years, retiring in 2015. He won Pulitzer Prizes for international reporting in Bosnia and Afghanistan.

Benedict Carey covers behavior and psychology for *The New York Times*. His books include *How We Learn: The Surprising Truth About When, Where, and Why It Happens*.

Osgood Caruthers (1915–85) was a foreign correspondent for *The New York Times* from 1955 to 1961.

Kenneth Chang has been a science reporter for *The New York Times* since 2000. He covers a wide variety of subjects, notably spaceflight and mathematics.

Cornelia Dean, science editor of *The New York Times* from 1997 to 2003, has written widely about the environment. She is the author of *Against the Tide: The Battle for America's Beaches* and editor of a previous book in this series, *The New York Times Book of Physics and Astronomy*.

Maureen Dowd joined *The New York Times* as a reporter in 1983 and has been an Op-Ed columnist since 1995. She won a Pulitzer Prize for commentary in 1999.

R. L. Duffus (1888–1972) was an editorial writer and reporter for *The New York Times* from 1930 to 1962.

Orrin E. Dunlap Jr. (1896–1970) was radio editor of *The New York Times* from 1922 to 1940.

Leonard Engel (1916–64) was a contributor to many national magazines and the author of several books on science.

Tom Ferrell was a longtime editor at *The New York Times Book Review* and the *Week in Review*.

John W. Finney (1923–2014) was a senior correspondent and editor in the Washington bureau of *The New York Times* for thirty years.

Henry Fountain, an environmental reporter for *The New York Times*, originated the popular "Observatory" column in Science Times. He is at work on a book about the 1964 Alaska earthquake.

James Gleick, a former reporter and editor for *The New York Times*, is an author and biographer whose books include *Chaos: Making a New Science* and *Genius: The Life and Science of Richard Feynman*.

James Gorman, a science reporter for *The New York Times*, writes about evolutionary biology, brain science, conservation, and other subjects, and narrates the Science Take video series. His books include *How to Build a Dinosaur*, with Jack Horner.

Jack Gould (1914–93) was *The New York Times*'s chief television reporter and critic from 1948 to 1972.

Denise Grady is a science reporter for *The New York Times* and the author of *Deadly Invaders*, a book about emerging viruses.

Amy Harmon is a domestic correspondent for *The New York Times* whose series "The DNA Age" won a Pulitzer Prize for explanatory reporting in 2008.

Gladwin Hill (1914–92) was a *The New York Times* reporter for forty-four years and its first full-time environmental correspondent.

Lindsey Hoshaw, an environmental journalist, is an online producer for KQED Science in San Francisco.

James H. Jeans (1877–1946) was a British astronomer, mathematician, and physicist.

George Johnson, a longtime science writer for *The New York Times* and many other publications, writes the monthly "Raw Data" column in Science Times. His books include *The Cancer Chronicles* and *The Ten Most Beautiful Experiments*.

Alva Johnston (1888–1950) was a reporter for *The New York Times* from 1912 to 1928. He won a Pulitzer Prize in 1923 for his coverage of science.

William J. Jorden (1923–2009) was a diplomatic correspondent of *The New York Times* who became a diplomat himself, serving as the U.S. ambassador to Panama in the 1970s.

Waldemar Kaempffert (1877–1956) was a science writer for *The New York Times* for three decades. He was also a managing editor of *Scientific American* and the first director of Chicago's Museum of Science and Industry.

T. R. Kennedy Jr. (1891–1980) was a radio editor and technology writer for *The New York Times* from 1927 to 1955.

Gina Kolata has covered science and medicine for *The New York Times* since 1987 and has twice been a Pulitzer Prize finalist. She is the author of *Rethinking Thin: The New Science of Weight Loss and the Myths and Realities of Dieting*, and the editor of two previous books in this series, *The New York Times Book of Mathematics* and *The New York Times Book of Medicine*.

Ferdinand Kuhn Jr. (1905–78) was *The New York Times*'s chief London correspondent before World War II.

William L. Laurence (1888–1977) joined *The New York Times* in 1930 as the first reporter assigned to daily science coverage. He received two Pulitzer Prizes and, as the official historian of the Manhattan Project, was the only journalist to witness the 1945 test of the atom bomb in New Mexico and the nuclear bombing of Japan.

Charles Leedham (1926–99) wrote about science and technology for *The New York Times Magazine* in the 1960s. He later became New York City's marine and aviation commissioner.

John Markoff is a longtime technology correspondent for *The New York Times* and the author of the new book *Machines of Loving Grace: The Quest for Common Ground Between Humans and Robots*. He was part of the team that won the 2013 Pulitzer Prize for explanatory reporting.

Donald G. McNeil Jr., a medical reporter for *The New York Times*, covers infectious disease, particularly in the developing world.

Lorus Milne (1912–87) and Margery Milne (1912–2006) were educators and biologists. Their books included *The Senses of Animals and Men*.

Jon Nordheimer joined *The New York Times* in 1969 and was a reporter and national correspondent for three decades.

Dennis Overbye, a science reporter for *The New York Times*, was a finalist for the 2014 Pulitzer Prize for "Chasing the Higgs," a Science Times special report on the search for the Higgs boson. His books include *Einstein in Love*.

Robert K. Plumb (1922–72) was a reporter for *The New York Times* for eighteen years.

Boyce Rensberger, a science reporter at *The New York Times* from 1971 to 1979, later became director of the Knight Science Journalism Fellowship program at MIT.

A. M. Rosenthal (1922–2006) was executive editor of *The New York Times* from 1977 to 1988 and an Op-Ed columnist until 1999. He won a Pulitzer Prize in 1960 for reporting from Poland.

Harold M. Schmeck Jr. (1923–2013) was a science reporter for *The New York Times* who specialized in medical research.

Daniel Schwarz (1908–2005) worked at *The New York Times* for forty-four years, rising from clerk to top Sunday editor.

Robert B. Semple Jr., associate editor of *The New York Times*'s editorial page, has been with the paper since 1963. He won a Pulitzer Prize in 1996 for his editorials on environmental issues.

Walter Sullivan (1918–96), who joined *The New York Times* as a copy boy in 1940, covered many of the twentieth century's leading science stories over five decades and was considered the nation's dean of science writers. His books included *We Are Not Alone* (1964), on the search for extraterrestrial intelligence.

William K. Stevens, a reporter for *The New York Times* from 1968 to 2000, was the paper's lead climate writer in the 1990s.

Manil Suri, a mathematics professor at the University of Maryland, Baltimore County, is also a novelist and a contributing opinion writer for *The New York Times*.

Sabrina Tavernise is a science reporter in the Washington bureau of *The New York Times*. She has been a foreign correspondent in Iraq, Lebanon and Russia.

John Tierney, a former reporter and columnist for *The New York Times*, is co-author, with Roy Baumeister, of *Willpower: Rediscovering the Greatest Human Strength*.

Nicholas Wade, a science reporter and editorial writer for *The New York Times*, was its science editor from 1990 to 1996. His books include *A Troublesome Inheritance: Genes, Race, and Human History* and *Before the Dawn: Recovering the Lost History of Our Ancestors*.

John Noble Wilford has been a science reporter at *The New York Times* for fifty years, covering the space program and, later, archaeology and paleontology. He won Pulitzer Prizes in 1984 and 1987. His books include *The Mapmakers*, on the history of cartography.

Kurt Wegener (1878–1964) was a German polar explorer and meteorologist.

Richard Witkin (1918–2013), a longtime aviation reporter for *The New York Times*, shared a Pulitzer Prize in 1987 for coverage of the *Challenger* space shuttle disaster.

Carol Kaesuk Yoon writes about biology for *Science Times*. She is the author of *Naming Nature*, a book about taxonomy.

Carl Zimmer, who writes the weekly "Matter" column for Science Times, is the author of twelve science books, including *Soul Made Flesh: The Discovery of the Brain—and How It Changed the World*.

INDEX